Food Science, Nutrition and Health

Brian A Fox

Allan G Cameron

Fifth Edition

Edward Arnold
A division of Hodder & Stoughton
LONDON MELBOURNE AUCKLAND

D0257072

© 1989 Brian A. Fox and Allan G. Cameron

First published in Great Britain 1961

British Library Cataloguing in Publication Data

Fox, Brian A.
    Food science: a chemical approach.—4th ed.
    1. Food—Composition
    I. Title   II. Cameron, Allan G.

    ISBN 0 340 49675 4

First published as *A Chemical Approach to Food and Nutrition* 1961
Second edition published as *Food Science: a chemical approach* 1970
Fifth impression 1976
Third edition 1977
Fourth impression 1981
Fourth edition 1982
Fifth impression 1987
First edition published as *Food Science, Nutrition and Health* 1989
Fifth impression 1993

Typeset in Great Britain by Mathematical Composition Setters Ltd.,
Salisbury, UK.
Printed and bound in Great Britain
for Edward Arnold, a division of Hodder & Stoughton Limited,
Mill Road, Dunton Green, Sevenoaks, Kent TN13 2YA by Clays
Ltd., St Ives plc.

Preface to fifth edition

This fifth edition is very largely new. As implied by the change in its title – from *Food Science: a chemical approach* to *Food Science, Nutrition and Health* – the book has been restructured to take into account an approach to food and nutrition that is more concerned with discovering what constitutes healthy eating than with analysing food to discover what nutrients it contains. Since the book was first published in 1961 changes have occurred in nearly every area discussed in the book and this has required not only the updating of every section but a radical change of emphasis. Thus many of the chapters are new to allow the relation between food and health and disease to be fully explored.

While so much of the book has been rewritten, including many new tables and diagrams, it is to be hoped that its clear and concise style has been retained so that, like previous editions, it 'can be commended as a useful introduction of an important subject to all who are interested in the effect of diet on human welfare' (*British Medical Journal*).

The broad intention of this new edition remains the same as that of earlier editions, namely to provide an elementary but comprehensive and up-to-date account of food science. To this end it begins with three chapters which present a simple summary of the nature and functions of food and its digestion together with an explanation of how our understanding of the diet–health and diet–disease links has developed.

The main body of the book is given over to a description of the nature of nutrients and the foods which contain them and to a description of what happens to food when it is grown, stored, processed, preserved, cooked and eaten. Complete chapters are devoted to the important topics of cooking, food spoilage and preservation and to food poisoning and food hygiene. A chapter is also devoted to diet and health which explores the ways in which diets can be changed to promote health and combat modern 'diseases of affluence'. In this chapter the importance of the nutritional guidelines recommended by the National Advisory Committee on Nutrition Education (NACNE) is fully explained.

The last chapter deals with the important and controversial topic of food contaminants and food additives and seeks to present a balanced view and give a comprehensive account of what has become an emotive subject. An attempt is made to evaluate both the risks and the advantages of using additives and to explain how Britain's membership of the European Economic Community has led to a reappraisal of additives and the need for much new legislation, including the adoption of the distrusted E numbers to signify permitted additives.

Finally, it is hoped that this book will prove of interest and value to all who are concerned with elementary food science. In particular, it is intended for students of Food Science and Technology, Home Economics, Catering and Nutrition as well as for GCE 'A' level studies in schools and for BTEC courses in Colleges of Further Education. Students of Medicine and Nursing should find it helpful as background reading.

Brian A Fox
Allan G Cameron

Acknowledgement

The authors wish to thank the Controller of Her Majesty's Stationery Office for permission to reproduce Appendix I and to use data from *The Composition of Foods* and the COMA *Report on the Prevention of Coronary Heart Disease*.

Contents

Some people have a foolish way of not minding, or of pretending not to mind, what they eat. For my part, I mind my belly very studiously and very carefully; for I look upon it that he who does not mind his belly will hardly mind anything else.

Samuel Johnson

CHAPTER 1

Food and its functions

The basic function of food is to keep us alive and healthy, and in this book we shall consider how food does this, although we shall also need to think about many other related matters. Indeed we cannot answer such a fundamental 'How' question without first finding out the answer to some simpler 'What' questions, such as what is food, what happens to it when it is stored, processed, preserved, cooked, eaten and digested. The answers to such questions can only be found out by experiment, and many different sciences play a part in helping to provide the answers. In recent years the study of food has been accepted as a distinct discipline of its own and given the name *food science*.

Food science, however, while embracing a number of distinct scientific disciplines is nevertheless a subject with its own outlook. It is, in a sense, a 'pure applied science' in that it exists not only to pursue academic knowledge, but also to promote the fulfilment of a basic human need – the need for a diet that will sustain life and health. Thus to be effective food science must be applied and this is the province of *food technology*.

The dividing line between the science and the technology of food is often blurred because the latter uses and exploits the knowledge of the former. The link between food science and food technology is well exemplified in considering how to solve what must be the foremost problem of our day; namely that of how to feed adequately the world's rapidly expanding population. The problems involved in determining what foods best meet the dietary needs of different countries, of what constitutes an adequate diet, of the nutritional merits of various new foods, of how to store and preserve food with minimum nutritional loss; these are the province of food science. But in order to use this information it must be applied – food must be grown, stored, processed, preserved and transported on a large scale, and this is the province of food technology.

The subject of food science and nutrition is an engrossing one. This is partly because of the inherent interest of the subject for we are all concerned about the food that we eat and as Samuel Johnson (p. vi) phrased it 'I look upon it that he who does not mind his belly will hardly mind anything else'. Its interest lies also in the fact that our knowledge of the subject is growing, leading to an unfolding of new perspectives about what is significant, while new techniques are being developed leading to new methods of food processing and of analysing nutrients, additives and possible contaminants in food.

We have already noted that one of the main functions of food is to keep us healthy, but in recent years the relation between food and disease has been much studied and has become a major preoccupation not only of scientists, nutritionists and doctors but also of the public and the popular press.

In the following pages we shall study the relation between food science, nutrition and health but first it is important to understand what we mean by the term food.

THE NATURE OF FOOD

Food is such an essential part of everyday life that it might be presumed that its nature would be universally understood. However, this is not so. To put it in the form of a paradox: food is what we eat, but not everything we eat is food. The explanation for this lies in the fact that food has a function – to keep us alive and healthy – and unless what we eat contributes to this function in some way, it should not strictly count as a food.

The difficulty of defining exactly the nature of food can be judged by the different definitions used. For example, Dorland's *Illustrated Medical Dictionary* defines it as 'Anything which, when taken into the body, serves to nourish or build up the tissues or to supply body heat; aliment; nutriment'. Bender's *Dictionary of Nutrition and Food Technology* states that foods are 'Substances taken in by mouth which maintain life and growth, i.e. supply energy, and build and repair tissue'.

The essence of these definitions is that unless what we eat fulfils the functions stated, it should not be classed as a food. Salt and pepper, for example, are used as condiments, but this in itself does not qualify them to rank as foods. Salt is a food because, in addition to being a seasoning agent, it acts as a regulator of body functions, but pepper has no function except as a flavouring agent and is therefore not a food.

Tea, coffee and cocoa are all widely used beverages which most people would classify as foods. The infusion obtained by adding boiling water to tea leaves has little more nutritional value, however, than the water itself. Tea and coffee are both esteemed on account of their flavour and mild stimulating action, the latter being due to the presence of *caffeine*. In fact they are drugs and not foods because they act through the nervous system and not the digestive system. The nutritional value of a cup of tea or coffee is almost entirely derived from the milk, sugar and water which it contains. Cocoa, on the other hand, contains the crushed cocoa bean and is therefore a true food because the nutrients of the bean are present in the beverage.

Alcoholic beverages contain *ethanol* (*ethyl alcohol*). This substance is both a drug and a food because it affects the nervous system and is also broken down inside the body with the liberation of energy. Alcoholic beverages are therefore properly classed as foods.

In *The Shorter Oxford English Dictionary* food is defined as 'What one takes into the system to maintain life and growth, and to supply waste; aliment, nourishment, victuals'. Thus, on this basis, an intravenous glucose drip and dietary fibre would both be regarded as foods. The glucose drip provides the body with energy while dietary fibre, although it is indigestible, would be considered a food because it does contribute to body waste in the form of faeces.

For the purpose of this book we shall regard as foods only those substances which, when eaten and absorbed by the body, produce energy, promote the growth and repair of tissues or regulate these processes. The chemical components of food which perform these functions are called *nutrients* and it follows that no substance can be called a food unless it contains at least one nutrient. Some particularly valuable foods, such as milk, contain such a variety of nutrients that they can fulfil all the functions of food mentioned above, while others, such as glucose, are composed entirely of a single nutrient and have only a single function. The study of the various nutrients in relation to their effect upon the human body is called *nutrition*.

Types of nutrient

Nutrients are of six types, all of which are present in the diet of healthy people. Lack of the necessary minimum amount of any nutrient leads to a state of *malnutrition*, while a general deficiency of all nutrients produces *under-nutrition* and, in extreme cases, starvation. The six types of nutrient are: *fats, carbohydrates, proteins, water, mineral elements* and *vitamins*. Water is not always included as a nutrient, but it seems advisable to do so because it is essential that our diet provides us with sufficient water required for many functions in the body. For example, it provides the medium by which the complex chemical processes which take place in the body can occur.

Apart from the nutrients already mentioned, the body also requires a continuous supply of oxygen. Oxygen is not normally regarded as a nutrient, however, because it is supplied from the air and passes into the body, not through the digestive system, but through the lungs.

Nutrients can be considered from two points of view – their functions in the body and their chemical composition. These two aspects are closely related, nutrient function being dependent upon composition, and in later chapters they will be considered in conjunction with each other.

The two basic functions of nutrients are to provide materials for growth and repair of tissues – that is to provide and maintain the basic structure of our bodies – and to supply the body with the energy required to perform external activities as well as carrying on its own internal activities. The fact that the body is able to sustain life is dependent upon its ability to maintain its own internal processes. This means that though we may eat all sorts of different foods and our bodies may engage in all sorts of external activities and even suffer injury or illness, the internal processes of the body should absorb and neutralize the effects of these events and carry on with a constant rhythm. This is only possible because the components of our bodies are engaged in a ceaseless process of breakdown and renewal – a theme to which we will return.

It is apparent that if the body's internal processes are to be maintained constant in spite of its ceaseless activity, and in the face of external pressures, some form of control must be exercised and, considering the complexity of the body's activities, it is evident that this control must be very precise. Thus nutrients have a third function, namely that of controlling body processes, a function which will be considered in the next chapter.

We have seen that food provides us with nutrients that perform three functions in our bodies. Although habits and patterns of eating may vary from person to person, and diets may be selected from hundreds of different foods, everyone needs the same six nutrients and needs them in roughly the same proportions. The relation between nutrients, their functions in the body and important foods that supply them is shown in Fig. 1.1.

Nutrients may also be considered according to their chemical composition. For example, though different oils and fats, such as

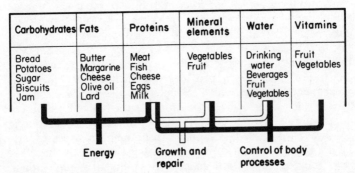

Carbohydrates	Fats	Proteins	Mineral elements	Water	Vitamins
Bread Potatoes Sugar Biscuits Jam	Butter Margarine Cheese Olive oil Lard	Meat Fish Cheese Eggs Milk	Vegetables Fruit	Drinking water Beverages Fruit Vegetables	Fruit Vegetables

Energy   Growth and repair   Control of body processes

Figure 1.1 The nutrients – showing their functions and representative foods in which they are found.

olive oil and palm oil, do not have identical compositions, they are chemically similar. In the same way different proteins (and carbohydrates) are constructed according to the same chemical pattern and are therefore conveniently grouped together. Vitamins are an exception to the method of classification according to chemical type; at the time of their discovery in food their chemical nature, which in most cases is complex, was unknown. They were grouped together because it was known that small quantities of them were essential to health. At first they were identified in terms of their effect on growth and health and distinguished by letters as vitamin A, B, C and so on. Their chemical composition is now known, and it has become apparent that they are not chemically related to each other. Nevertheless, it is still convenient to consider them together.

Food as a source of energy

Energy is required for sustaining all forms of life on the earth. The prime source of the earth's energy is the sun, without which there could be no life on this planet. The sun continually radiates energy, a fraction of which is intercepted by the earth and stored in various ways; plants and coal, for example, act as energy storehouses. Living plants convert the sun's energy into chemical energy and some plants of past ages have been converted, during many millions of years, into coal. Plants, by the process of *photosynthesis*, convert carbon dioxide and water into carbohydrate. Photosynthesis, which is discussed in Chapter 6 (p. 94), can only take place in daylight because solar energy is used in the process. A complex series of chemical changes occurs which can be represented by the following equation:

$$x\mathrm{CO_2} + y\mathrm{H_2O} \xrightarrow{\text{daylight}} \mathrm{C}_x(\mathrm{H_2O})_y + x\mathrm{O_2}$$

Carbohydrate

The formation of carbohydrate is, therefore, the method used by plants to trap and store a part of the sun's energy. Sugar-beet, which synthesizes carbohydrate in the form of the sugar *sucrose*, may be taken as an example:

$$12\mathrm{CO_2} + 11\mathrm{H_2O} \xrightarrow{\text{daylight}} \mathrm{C_{12}(H_2O)_{11}} + 12\mathrm{O_2}$$

Sucrose

When sucrose is formed from carbon dioxide and water, energy is absorbed and stored within the sucrose molecule.

Animals, unlike plants, cannot store the sun's energy directly and so must gain it secondhand by using plants as food; carnivorous animals and man take this process a stage further and also use other animals as food. In this way chemical compounds which have been photosynthesized and stored in plants are eaten by man and animals and the stored energy made available. For example, the energy that is stored within the sucrose molecule when it is synthesized by sugar-beet is liberated when sucrose reverts to carbon dioxide and water. This breakdown of sucrose into simpler units is brought about in the body by digestion and oxidation, but the overall reaction is simply the reverse of that represented above, namely:

$$C_{12}(H_2O)_{11} + 12O_2 \rightarrow 12CO_2 + 11H_2O$$

Sucrose

When sucrose is converted into carbon dioxide and water in this way, the energy stored during synthesis is evolved and made available for use by the body.

Sucrose may also be converted into carbon dioxide and water by burning it in air. The chemical reaction is the same as that represented by the equation above, exactly the same quantity of heat being liberated as when the oxidation occurs in the body. The difference in the two reactions concerns the speed at which they occur and their efficiency. Oxidation in the body takes place much more slowly than combustion in air as it takes place in a series of steps, so ensuring a slow, controlled and gradual release of energy to body tissues. The efficiency of combustion within the body is less than that in air because only about two-thirds of the energy of the sucrose becomes available as biological energy, one-third being 'lost' as heat which helps to maintain body temperature.

It can now be understood why the body is sometimes likened to a slow combustion stove and carbohydrate described as a fuel. It is clear that oxidation in the body is a most important process for it enables the energy stored in carbohydrates (and fats and proteins) to be liberated and made available for use by the body.

THE ENERGY VALUE OF FOOD   Energy may be measured in heat units called *calories*. A calorie is the amount of heat required to raise the temperature of 1 g of water by 1°C. As this is rather a small unit, energy derived from food may be expressed in units which are one thousand times larger and known as *kilocalories* (kcal). A kilocalorie is the amount of heat required to raise the temperature of 1 kg of water by 1°C.

The internationally recognised unit of energy is the *joule* (J), but like the calorie this is an inconveniently small unit with which to express the energy value of food so that the kilojoule (kJ), which is one thousand times larger than the joule, is usually used. Sometimes an even larger unit, the *megajoule* (MJ), is used. A megajoule is one thousand times larger than a kilojoule.

The relationship between these units may be expressed as follows:

$$1 \text{ kcal} = 4.19 \times 10^3 \text{ J} = 4.19 \text{ kJ} = 4.19 \times 10^{-3} \text{ MJ}$$

In the remainder of this chapter (and in Appendix I) both types of unit are given so that students become familiar with them and their relation to each other. Elsewhere, however, kilojoules (and megajoules) are used.

In order to compare the energy of different foods it is simplest to determine the amount of energy produced, calculated as heat, when one gram of the substance is completely oxidized by igniting it in a small chamber filled with oxygen under pressure. The result obtained represents the heat of combustion of food which is usually expressed as kcal or kJ per gram. If the calorie value of sucrose is expressed in this way, it is found to be 3.95 kcal/g. This means that when 1 g sucrose is completely oxidized the heat produced is sufficient to raise the temperature of 1000 g of water by 3.95°C. The average values of the heats of combustion of the energy-providing nutrients are shown in Table 1.1.

In order to express the energy value of nutrients in terms of the energy actually made available to the body it is necessary to calculate the available energy values. Such values are always lower than heats of combustion because of losses within the body. A small loss is due to incomplete absorption; such loss is suffered by all three nutrients and with protein there is an additional loss because protein, unlike carbohydrate and fat, is incompletely oxidized in the body. In addition a correction may need to be made for fibre. Insoluble fibre is not oxidized in the body and therefore cannot contribute to available energy. The magnitude of these energy losses may be appreciated from Table 1.1. It should be mentioned that there is some uncertainty about the size of losses within the body, and different nutritionists use different figures; the ones quoted may be taken as being sufficiently reliable for most purposes.

Although the available energy values given in Table 1.1 are only approximate values rounded off to the nearest whole number, they can nevertheless be used to calculate the energy value of any given diet. The available energy value of any food can be found using the

Table 1.1 Average energy value of nutrients (per gram)

Nutrient	Heat of combustion		Available energy value	
	kcal	kJ	kcal	kJ
Carbohydrate	4.1	17	4	17
Fat	9.4	39	9	37
Protein	5.7	24	4	17

average figures, provided that its composition in terms of carbohydrates, fats and proteins is known. The energy value of summer
milk, for example, can be calculated from its analysis, as shown in
Table 1.2. By similar calculations the available energy value of other
foods may be estimated; some average values are given in Table 1.3.

This table shows that foods, such as butter and lard, which contain
a high proportion of fat have the highest energy values. Carbohydrate
foods, such as those containing a high proportion of sugar (jam and
dates) or starch (bread and potatoes), are less concentrated sources of
energy. In spite of this, such foods supply a considerable proportion
of the energy in an average British diet. Indeed cereal foods contribute
no less than one-third of our total energy intake, which is a greater
proportion than that supplied by any other class of foodstuff. In
Eastern countries, starchy foods, often in the form of rice, supply an
even greater proportion of the total energy content of the diet.

The use of energy by the body

As we have already noted, energy is produced in the body from food
by a series of precisely controlled steps. Each step results in the release
of a small amount of energy which is used in the promotion of bodily
functions and may finally produce heat. Without attempting at this

Table 1.2   The energy value of milk

Nutrient	Amount in 100 g milk	kcal/g	kJ/g	Energy/100 g milk	
				kcal	kJ
Carbohydrate	4.7 g	4	17	18.8	79.9
Fat	3.8 g	9	37	34.2	140.6
Protein	3.3 g	4	17	13.2	56.1
	Total energy provided by 100 g milk			66.2	276.6

Table 1.3   The energy value of some foods (per 100 g edible portion)

Food	kcal	kJ	Food	kcal	kJ
Lard, dripping	900	3770	Eggs	147	612
Butter	735	3020	Potatoes	74	315
Cheese (Cheddar)	406	1682	Fish, white	77	324
Sugar	394	1680	Milk	65	272
Beef (average)	313	1296	Apples	46	196
Jam	260	1090	Tomatoes	15	60
Bread (white)	230	977	Lettuce	12	51
Dates	248	1056	Cabbage	25	100

stage any chemical explanation of the nature of the steps in the oxidation process, it is possible to gain a clear idea of how energy is used by the body.

Energy is required in the body for *basal metabolism*, for *thermogenesis*, for growth and for muscular activity.

BASAL METABOLISM   The term *metabolism* refers to the sum of all the chemical reactions going on in the body, and the energy needed to sustain the body at complete rest is known as the energy of *basal metabolism* or the *basal metabolic rate* (BMR).

The body requires a constant supply of energy to maintain its internal processes even when resting. Even during sleep, when the body is apparently at rest, energy is needed to ensure that essential internal processes continue. For example, energy must be supplied to maintain the powerful pumping action of the heart, the continual expansion and contraction of the lungs and the temperature of the blood. It is needed to maintain the ceaseless chemical activity of the millions of body cells and the tone of muscles. Living muscle must constantly be ready to contract in response to stimuli transmitted to it by nerves. Such a degree of readiness can only be achieved if energy is continually supplied to keep the muscles in a state of tension. Muscle tension decreases during sleep but does not become zero, so that a certain amount of energy, which ultimately appears as heat, is necessary to maintain it.

BMR is affected by many factors including shape, size, weight, age, sex and rate of growth. It is also affected by the amount of sleep a person has and even by climate and hormonal activity.

BMR values vary from person to person. Table 1.4 gives *average* values for different people. The table shows that BMR values are related to growth. The one-year-old infant has a higher BMR per unit weight than the eight-year-old child and adult because its growth rate is higher. On the other hand the BMR/day of an infant is less than that of the other categories mentioned because its weight is smaller. It can also be deduced from Table 1.4 that the BMR of men is greater per kg of body weight than that of women of the same weight. This is

Table 1.4   Some average values of basal metabolic rate

	Age	Weight (kg)	kcal/kg/day	kJ/kg/day	kcal/day	kJ/day
Infant	1 yr	10	50	210	500	2100
Child	8 yrs	25	40	170	1000	4250
Man	adult	55	25	100	1300	5400
Woman	adult	65	25	100	1600	6700

because, for a given body weight, women's bodies contain more fat than those of men and fat contributes little to BMR.

BMR also varies with age. As age increases, BMR falls; this fall is rather greater for men than for women. Climate also affects BMR, its value being reduced by 5–10 per cent in very cold and very hot climates. Active hormonal activity, especially of the thyroid gland, increases BMR.

BMR values are about 1600 kcal (6720 kJ) for an average man and 1500 kcal (6300 kJ) for an average woman. These values account for about two-thirds of total energy expenditure.

THERMOGENESIS   Thermogenesis, or more precisely *diet-induced thermogenesis*, refers to the increase in BMR that follows the eating of food. The size of the increase is proportional to the energy content of the meal, regardless of what nutrients provided the energy. It amounts to about ten per cent of the energy content of the meal and probably arises from the many metabolic processes – digestion, absorption, transport of nutrients round the body and metabolism – that follow after eating a meal. This energy involved in thermogenesis appears as heat.

GROWTH   During pregnancy some 30 000 kcal (336 000 kJ) of energy are required to produce the baby, increase the size of the placenta and reproductive organs, allow for the energy needed for the newly formed baby tissues and create additional stores of fat in the mother. Once the baby has been born lactation requires about 750 kcal (3150 kJ) per day.

Newly born infants grow at a remarkable rate and in the first three months of life 23 per cent of food energy is required for growth. This figure falls to six per cent by the time the infant is one year old and to two per cent by the fifth year. In general terms it has been estimated that the formation of 1 g of new tissue requires 20 kJ of food energy.

MUSCULAR ACTIVITY   Energy is needed to enable the body to perform external work. Muscular activity requires a supply of energy additional to that needed to maintain muscle tone and other internal processes. The simplest physical act, such as standing up, involves the use of many muscles, and the greater the degree of physical activity in daily life the greater is the energy requirement of the muscles. It is useful, therefore, to relate the degree of muscular activity to the energy that must be supplied by the diet.

The problem of equating muscular activity with energy requirement is complicated by the fact that the body is unable to convert energy that is supplied by food completely into mechanical work. The efficiency of conversion by the body, considered as a machine, is of the order of 15–20 per cent. If the higher value is taken it means that 100 units of energy supplied by food enable the body to perform physical work, for example, running, equivalent to 20 units. The other 80 units appear as heat and account for the fact that heat is lost from the body surface at an increased rate when physical work is done.

If allowance is made for this wasted heat energy it is possible to express the different daily energy requirements of people with various types of occupation in terms of BMR. This is done in Table 1.5. The energy required for specific occupations can be measured and some results are shown in Table 1.6 which again expresses energy expenditure in terms of BMR. It will be appreciated that such results are only average values for men and women, and that the results for an individual could vary considerably from those quoted.

Table 1.5 Average daily energy requirements when doing light, moderate and heavy work expressed as multiples of BMR

	Light	Moderate	Heavy
Men	1.55	1.78	2.10
Women	1.56	1.64	1.82

Table 1.6 Types of activity related to energy expenditure expressed as a multiple of BMR

Activity	Men	Women
Sleeping	1.0	1.0
Sitting	1.2	1.2
Walking (normal pace)	3.2	3.4
Housework (light cleaning)	2.7	2.7
Bricklaying	3.3	—
Mining (using pick)	6.0	—
Light activity (cricket, golf)	2.2–4.4	2.1–4.2
Moderate activity (swimming, tennis)	4.4–6.6	4.2–6.3
Heavy activity (athletics, football)	6.6 +	6.3 +

Energy requirements

For an individual, energy requirement can be expressed quite simply and precisely as follows: it is the level of energy intake required to match the level of energy expenditure, account being taken of the energy requirements of growth, pregnancy and lactation. Unfortunately it is much more difficult to estimate the energy requirements of groups of people because individual requirements vary so widely even for people with similar occupations and lifestyles. However, energy requirements for large groups of people are needed for government agencies to plan national food policies, and for these purposes the energy requirement for a group is taken as the average of individual requirements, no specific provision being made for the known individual variation in requirement.

In the United Kingdom the figures used for energy requirement are those published by the Department of Health and Social Security, and reproduced in Appendix I. When using this table it needs to be understood that the average body weights of men and women have been taken as 65 kg and 55 kg respectively and energy expenditure has been worked out by dividing the day into three parts – in bed (eight hours), at work (eight hours) and leisure (eight hours). The recommended intake of energy given in Appendix I is the same as the calculated energy expenditure, no allowance being made for individual variation.

Appendix I refers to three occupational grades for men and two for women. Typical 'sedentary' workers include office and shop workers and most professions, while 'moderately active' workers include those in light industry, postmen, bus conductors and most farm workers, and 'very active' workers include coal miners, steel workers and very active farm workers.

The recommended daily amounts of energy for babies and young children are not closely related to body weight and so ages are used instead; the weights referred to in Table 15.4 (p.316) are average weights at these ages.

CHAPTER 2

Enzymes and digestion

The human body is composed of some hundred thousand million cells, each of which is complete in itself. These cells are grouped together in the body to form tissues with specialized functions. Thus some cells comprise connective tissue and bind together the various organs of the body, others are concerned with muscular and nervous tissue, while others form the skeletal framework of bone that contributes strength and rigidity to the body.

Individual cells are so tiny that their internal structure can only be observed with the help of an electron microscope. The largest component of a cell is its *nucleus*, and this is surrounded by a watery fluid called *cytoplasm*. The cytoplasm contains a network of membrane-like material, the *endoplasmic reticulum*, which is studded with small dark bodies known as *ribosomes*. The cytoplasm also contains a number of bodies, among which are the egg-shaped *mitochondria* and the smaller *lysosomes*.

The complex activities needed to sustain life in the human body take place within the body's cells, and we may liken the activity of a cell to that of a chemical factory in which a great variety of raw materials are processed and converted into finished products. In a single cell many different raw materials are required, though they are largely composed of only four elements: carbon, hydrogen, oxygen and nitrogen. The processing stage, which is concerned with the conversion of these simple raw materials into the more complex substances required to carry out the many functions of the cell, involves thousands of different reactions. Each of these reactions comprises many steps which must be carried out in a definite sequence with the result that the chemical operations of a cell are much more complicated, and need much greater integration, than those of a chemical factory.

In order to sustain life the cell's activities must be controlled and organised into a self-regulating and self-renewing pattern. But how can such control be achieved, and how is it that although almost all human cells are built according to the same basic pattern, yet they are able to perform a multitude of different functions? The answer to these questions is to be found in the existence of a group of crucially important substances called *enzymes*. Their importance can be gauged from the fact that without them there could be no life. Without them the chemical reactions of the cell would get completely out of control.

Enzymes control all the chemical changes, that is the *metabolism*, which occur in living cells. They regulate the building up or *anabolic* reactions that result in the formation of complex substances such as proteins from single building units. They also regulate the breaking-down or *catabolic* reactions that result in release of energy. The role of enzymes can perhaps be appreciated more fully when it is realized that anabolic and catabolic processes involve very many steps, and that each step is controlled by its own enzyme. This control must be so carefully regulated that the life of the cell continues smoothly at all times, the whole metabolic process being kept carefully balanced.

The fact that different cells perform different functions is explained in terms of the enzymes that are present. Some thousand different enzymes have been recognised in the body, but in any one cell only a selection are present. Even so, most cells contain about 200 different enzymes, each of which is responsible for controlling a particular step. The complement of enzymes present in a cell automatically selects and controls those reactions which are to proceed.

THE CHEMICAL NATURE OF ENZYMES   One of the earliest known sources of enzymes was yeast, which is the simplest possible type of living organism. In fact this is how the name enzyme arose, for it means literally 'in yeast'. The fermentation of grape juice by yeast and the leavening effect of yeast in making bread have been known for many centuries. However, it is now appreciated that enzymes are of much more general significance than was originally realized, and that the chemical processes occurring in all living organisms are dependent on enzymes, those of the human body no less than those of yeast.

In spite of the importance of enzymes, a long time elapsed between the discovery of the first known enzyme in yeast and the isolation of an enzyme in the pure state. For many years it was believed that they were living organisms. This idea was only shown to be false at the end of the last century when the German chemist Buchner extracted from yeast cells a cell-free liquid which had a similar enzyme activity to the original living cells. So, although enzymes are made by living cells, they themselves are not living.

All known enzymes are proteins, the structure of which are described in Chapter 8. The properties of proteins – their ability to change their shape, their sensitivity to changes of conditions of temperature and acidity, their capacity to oppose changes of acidity that would upset the smooth working of the cell – make them peculiarly suited to control cell metabolism.

CLASSIFICATION OF ENZYMES   The substance upon which an enzyme acts is called the *substrate*, and enzymes are usually named after this substance. Thus the enzyme that acts on *urea* is called *urease* and that which acts on *maltose* is called *maltase*. It is a general rule that enzymes are named after the substrate upon which they act and given the suffix -*ase*. In company with most general rules, however, there are notable exceptions, mainly those enzymes which were named before the rule gained general acceptance. Some of these, such as *pepsin* and *trypsin*, will be encountered later.

Enzymes may be classified in a number of ways, but one of the most useful is to group them according to the type of reaction which they control. The five main groups of enzyme are shown in Table 2.1. Of these the first two are the most important in connection with what follows. *Hydrolases* control the *hydrolysis* of the substrate, that is its reaction with water and, as we shall see later in the chapter, this type of enzyme is of paramount importance in digestion. *Oxidases* control the *oxidation* of the substrate, and this usually takes the form of removal of hydrogen as indicated in the equation shown in Table 2.1.

CATALYTIC ACTION OF ENZYMES Enzymes are organic catalysts; they operate by speeding up a chemical process while appearing unchanged at the end of the reaction. In many respects their action is similar to that of the more familiar inorganic catalysts such as are often used in manufacturing processes. In the manufacture of margarine, vegetable oils are converted into solid fats by chemical reaction with hydrogen. In the absence of a catalyst the conversion of the oils into fats is very slow indeed, but the addition of small quantities of finely divided nickel produces a remarkable increase in the rate of the reaction; moreover, the nickel catalyst may be used time after time, for it is not used up in the process.

It is remarkable that only one part of nickel is needed to catalyse the conversion of several thousand parts of oil into fat, but this achievement appears quite insignificant when compared with the startling catalytic power of enzymes. One of the enzymes concerned with the breakdown of starch during digestion is amylase, produced by the pancreas. Only one part of amylase is needed to effect the conversion

Table 2.1 Classification of enzymes

Name	Reaction catalysed	General equation
Hydrolases	Hydrolysis	$AB + H_2O \rightarrow AOH + BH$
Oxidases	Oxidation	$ABH_2 \rightarrow AB + 2H$
Isomerases	Intramolecular rearrangement	$ABC \rightarrow ACB$
Transferases	Transfer of a group	$AB + C \rightarrow A + BC$
Synthetases	Addition of one molecule to another	$A + B \rightarrow AB$

of four million parts of starch into the sugar maltose. Where the efficiency of man-made catalysts is measured in thousands, that of nature's catalysts is measured in millions.

How do enzymes work? We can best approach this by considering first how an ordinary non-catalytic reaction proceeds. Suppose a reaction involves the conversion of reacting substances represented by A into products represented by B. The reaction will not start until A has received a 'push' in the form of energy, often supplied in the form of heat. The reason for this can be appreciated from Fig. 2.1.

Before A can react to form B it must surmount the energy hump shown by moving along path (i), and this requires an amount of energy ΔG^* (uncatalysed). When A has absorbed energy ΔG^* (uncatalysed), known as the *activation energy*, it is in an activated state and can decompose to form B.

This process can be likened to that of transferring a ball from one side of a hill to the other. If Fig. 2.1 represents a hill, the problem is that of transferring a ball from X to Y. If the only path lies over the summit of the hill, then it is necessary to push the ball up the hill – that is to work on it by supplying energy – until it reaches the top. Once there, it will run down the other side to Y of its own accord. The ball may now be at a lower level than it was at its starting point, as is the case in Fig. 2.1. This means that an amount of energy ΔG° has been released though this could not have been achieved without first pushing the ball to the top of the hill, which involved supplying it with energy ΔG^* (uncatalysed).

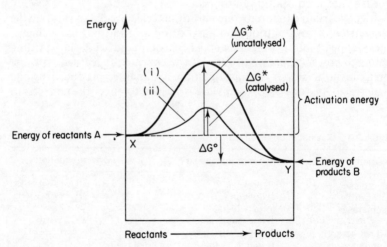

Figure 2.1   Reaction paths: (i) non-catalytic path; (ii) catalytic path.

In the cells of the human body, activation energy cannot be supplied in normal ways, such as heat, because this would damage the cells. The function of enzymes is to enable the reaction to proceed at a much lower activation energy than would otherwise be possible. In terms of Fig. 2.1 we must replace the reaction path (i) over the summit of the hill by one at a lower level such as (ii) involving a lower activation energy ΔG^* (catalysed).

Enzymes catalyse reactions by replacing a single-step of high-energy mechanism with a two- or multi-stage process, each step of which involves a low activation energy. If the enzyme catalyst is represented as E and the substrate as A we have:

$$A + E \rightleftharpoons AE$$
$$AE \rightleftharpoons B + E$$

Overall reaction: $A \rightleftharpoons B$

Enzymically catalysed reactions proceed by way of an unstable enzyme-substrate complex represented by AE. If both stages require little energy the reaction path is as represented by (ii) and the reaction will be a rapid and near spontaneous one with the enzyme E being regenerated. Thus we have a simple picture of how enzymes act as catalysts.

It should be noted that although enzymes speed up reactions, they cannot turn impossible reactions into possible ones. Neither can they affect the equilibrium position of a reversible reaction; this means that the *amount* of product in a reaction is the same whether an enzyme is involved or not. The presence of the enzyme merely reduces the time taken to reach the equilibrium position. In the absence of the enzyme the reaction may be so slow that for all practical purposes it does not proceed at all. In a cell thousands of different reactions are possible but the function of the enzymes present is to speed up particular ones, so that some reactions proceed rapidly while others proceed at a relatively insignificant rate. In this way cell metabolism is controlled and directed so that different cells are enabled to fulfil different functions.

SELECTIVITY OF ENZYMES Enzymes can be highly selective in choosing which reaction to catalyse. It is this characteristic which enables them to preserve order in living cells, for frequently one enzyme will catalyse only one cell reaction. Thus, although many other reactions may occur in a cell, the rate at which they proceed is insignificant compared with that of the catalysed reaction.

The selective power of enzymes is sometimes compared with the action of a key in a lock: the enzyme is the lock and only certain molecules can act as the key which exactly fits it. If reaction between two molecules is catalysed by an enzyme the lock can be imagined as having grooves into which the two molecules fit side by side; this leads to a brief union between the enzyme and the two molecules which are acting as keys, as is shown in stage 2 of Fig. 2.2. The key molecules are thus brought together and converted into an active state which enables them to react with each other. After reaction new molecules are formed, but as shown in stage 3 of the diagram, the enzyme remains unchanged and can catalyse further reaction.

In order to achieve a good fit between lock and key molecules, an enzyme frequently needs the help of another substance to be effective. Three main types of enzyme promotor have been distinguished, namely *coenzymes*, *cofactors* (or activators) and *prosthetic groups* (see p. 181). Coenzymes are smaller than enzyme molecules and are not proteins. They are not permanently bound to the enzyme but may become attached to it during enzyme reaction (Fig. 2.2, stage 1) only to be released later (once stage 3, Fig. 2.2, has been completed). Coenzymes are closely related to vitamins or are the vitamins themselves. One of the functions of the B group of vitamins seems to be to provide the body with suitable starting materials from which to make the coenzymes that it needs. The exact function of coenzymes is still only partly understood, but they certainly play an active and vital part

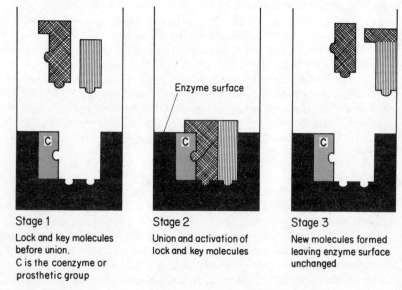

Stage 1
Lock and key molecules before union.
C is the coenzyme or prosthetic group

Stage 2
Union and activation of lock and key molecules

Stage 3
New molecules formed leaving enzyme surface unchanged

Figure 2.2 The lock and key theory of enzyme reaction.

in many reactions involving oxidizing enzymes. This is shown by the fact that if a coenzyme needed by an enzyme is absent, the enzyme can exert no catalytic effect.

In some cases it is found that metallic ions, such as magnesium, or non-metallic ions, such as chloride, are required to increase the activity of enzymes. Such substances are known as *cofactors* (or activators). Prosthetic groups are non-protein groups which are permanently bound to the enzyme.

Sometimes several molecules, which are similar to each other, can approximately fit the grooves of the same lock. In such cases the enzyme does not distinguish between them and acts as a catalyst to them all. In most cases, however, enzymes show great powers of discrimination as the following examples show.

The three enzymes, *maltase*, *lactase* and *sucrase*, are present in the small intestine, and during digestion these enzymes catalyse the hydrolysis of the sugars *maltose*, *lactose* and *sucrose* respectively. These three enzymes have considerable specificity, and in the case of lactase, complete specificity, for it will catalyse the hydrolysis of lactose and of no other substance, not even a similar sugar. Maltase and sucrase, however, catalyse the hydrolysis not only of maltose and sucrose but also that of certain other similar sugars.

A further example of enzymes which show a remarkable selectivity are those which catalyse the hydrolysis of proteins. The three enzymes, *pepsin*, *trypsin* and *chymotrypsin*, each select certain links of protein molecules and catalyse hydrolysis only at these links (see p. 179).

SENSITIVITY OF ENZYMES   Enzymes are very sensitive to effects of temperature and environment. All enzyme activity is destroyed on boiling because, being proteins, enzymes are denatured (see p. 169) and their nature changed by high temperatures. At low temperatures enzyme activity is greatly slowed down but as the enzymes are not denatured and, because enzyme reactions have a low activation energy, enzymes may retain some catalytic activity even at sub-zero temperatures. In general, plant enzymes work best at about 25°C and those in warm-blooded animals at about 37°C. An increase in temperature usually increases the rate of a chemical reaction, but in the case of an enzyme reaction it may also lead to inactivation of the enzyme.

Figure 2.3 shows the effect of temperature on the rate of catalysis. At 37°C the initial rate of reaction is rapid, but after a time the reaction rate slows down and stops, no further product being formed. This may be due to one of several causes. For instance, the reaction may be complete, all the substrate having reacted, or it may be that the products of reaction have made the environment unfavourable for

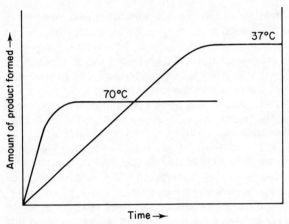

Figure 2.3 The effect of temperature on the rate of catalysis by an enzyme.

enzyme activity and the enzyme has been deactivated. If the temperature is raised to 70°C, the initial rate of reaction is increased. This is because the greater energy input increases the energy of substrate and enzyme molecules, and these molecules more rapidly gain the activation energy needed for reaction to take place. Although the initial rate of formation of products is rapid, it soon stops because the enzyme is rapidly inactivated at the higher temperature. The net result is that less product is formed at the higher temperature than at the lower. Enzymes catalyse reaction efficiently in man, the temperature being high enough to give rapid formation of products but low enough to avoid inactivation of the enzyme.

Enzyme activity is also dependent upon the acidity or alkalinity of the medium in which the enzyme acts. Most enzymes operate most efficiently in an environment which is nearly neutral, and if the medium becomes strongly acid or alkaline the enzyme becomes completely inactivated. Some enzymes, however, can only operate in an acid or alkaline solution. For example, the enzyme pepsin is present in gastric juice and during digestion it catalyses the initial hydrolysis of proteins. It can only act in strongly acid conditions such as are produced by the hydrochloric acid in the stomach. On the other hand, the enzyme trypsin which is present in pancreatic juice requires a slightly alkaline medium before it can catalyse protein hydrolysis. When food passes from the stomach into the small intestine the hydrochloric acid is completely neutralized and the medium becomes alkaline. Under these conditions pepsin becomes inactivated and trypsin carries on the digestion of proteins.

Cell metabolism

All activities that occur in cells are controlled by enzymes. However, these enzymes are not evenly distributed through the cell but are dispersed among the different parts of the cell so that each part has a distinct role in maintaining the life of the cell.

The nucleus of a cell contains genes and a small number of enzymes that together control cell growth and division while the surrounding cytoplasm contains water-soluble enzymes that control a variety of anabolic and catabolic processes. The mitochondria are important because they are the powerhouses of the cell and contain a number of oxidases responsible for the manufacture of high-energy materials used for energy production. The important job of protein synthesis in the cell is controlled by enzymes found in the endoplasmic reticulum. Finally, one of the most interesting bodies in the cell is the lysosome, also sometimes called the 'suicide bag'. Lysosomes contain a sufficient variety of hydrolases to destroy nearly all the components of the cell. Normally these 'suicide' enzymes are safely contained within the impermeable membrane that encloses the lysosome, but if the cell becomes injured or dies, the enzymes are released and the cell destroys itself.

Each cell, and each component of a cell, is surrounded by a membrane. The membrane allows only those raw materials needed by the cell to pass through it, all other substances being prevented from entering. This selection mechanism ensures that only those substances required for a particular job are available, and it also ensures that only those enzymes required to control this function are allowed through the membrane.

In later chapters we shall develop the theme of what happens to nutrients in metabolism in rather more detail, but enough has been said in this introductory survey to show that all the activities and functions of the body which constitute life are entirely dependent upon enzymes.

DIGESTION

It's a very odd thing –
As odd as can be –
That whatever Miss T eats
Turns into Miss T.

De la Mare

It is indeed a very odd thing – an extraordinary and remarkable thing – that no matter what we eat, the structure of the body, both flesh and blood, changes very little. There is no obvious similarity between the nature of the food we eat and the nature of our bodies. Yet within a few hours of being eaten, food is transformed into flesh and blood. This transformation is so complete that it cannot be accomplished before food has undergone a drastic breaking-down process known as *digestion*.

Digestion is both physical and chemical: the physical process involves the breakdown of large food particles into smaller ones while the chemical process involves the breakdown of larger molecules into smaller ones. Foodstuffs are mainly complicated, insoluble substances that must be converted into simpler, soluble, more active ones before they can be used by the body. Not all nutrients need digesting, however, for there are some such as water and simple sugars (e.g., glucose) and many vitamins and mineral salts which do not need to be broken down. Whether or not nutrients need to be broken down by digestion they cannot be utilized by the body until they have passed into the bloodstream, a process which is known as *absorption*. Once in the bloodstream nutrients are distributed to all the cells of the body where they sustain the complex processes of metabolism.

The role of enzymes in digestion

The chemical processes involved in digestion are brought about by enzymes. The chemical breakdown of food molecules, which in the absence of enzymes would be very slow indeed, is thereby speeded up so that digestion is completed in a matter of hours. Thus in the space of some three or four hours a remarkable change in the nature of the food has occurred. Substances such as starch, which may contain as many as 150 000 atoms in a single molecule, have been converted into molecules containing only 24 atoms (simple sugars such as glucose). The breakdown of protein molecules is almost equally spectacular, for an average protein molecule is split up into about 500 amino acid molecules during digestion. These two examples perhaps make clearer the magnitude of the chemical task performed by the enzymes of the digestive system.

Each stage of digestion involves hydrolysis and is catalysed by a hydrolysing enzyme or hydrolase. The hydrolysis can be represented: $AB + H_2O \rightarrow AOH + BH$. The equation shows how water is involved in splitting up a molecule AB into two smaller molecules AOH and BH. In some instances (e.g., sucrose), a single step involving the breakdown of a molecule into two parts is sufficient to produce a smaller soluble molecule that can be absorbed. In other instances (e.g., proteins), a very large number of hydrolytic steps is required before breakdown is complete.

The digestive process involves a fairly small number of different enzymes which catalyse the chemical breakdown of proteins, carbo-

Table 2.2 Hydrolases involved in digestion

Name	Substrate	Product
Amylases	Starch	Maltose
Maltases	Maltose	Glucose
Lipases	Fats	Fatty acids and glycerol
Peptidases	Proteins	Amino acids

hydrates and fats. The names of the hydrolases that catalyse the hydrolysis of these different nutrients are shown in Table 2.2. Unfortunately there is no general agreement about these names and different authors use different names according to fancy and fashion. The names given, however, are descriptive of the main changes brought about by hydrolases in digestion. *Peptidases* can be conveniently subdivided into *exopeptidases*, which split off amino acids from the ends of protein molecules, and *endopeptidases*, which attack and split the inside of protein molecules.

We have already noted the high selectivity of some enzymes such as peptidases. Amylases, and enzymes which break down sugars, show a similar high degree of selectivity. It is therefore evident that a whole series of such enzymes is required to achieve the stepwise breakdown of proteins and carbohydrates, Lipases, on the other hand, are relatively unselective, so that only a few lipases are required to break down fats.

Stages of digestion

The digestive system is separate from the body system proper and can be regarded very simply as a series of tube-like organs which pass through the body from the mouth at one end to the anus at the other. Food enters the system at the mouth, passes down the oesophagus into the stomach and then through the small intestine and large intestine, being gradually digested and absorbed in the process. Any that remains leaves the body at the other end of the system. In what follows the stages of digestion are described in a very simple way, to give an overall picture of the process. The chemical details will be filled in later after a discussion of the chemical nature of the nutrients concerned. The main parts of the digestive system are illustrated in Fig. 2.4 and the digestive process is summarized in Fig. 2.5.

DIGESTION IN THE MOUTH   When food is chewed, the size of the individual pieces is reduced and saliva is secreted by the salivary glands. The secretion of saliva takes place in response to various sorts of stimuli; the sight of a well-cooked meal, an appetizing smell or even the thought of a good meal may cause the salivary glands to pour forth saliva. This fluid becomes well mixed with food during mastication, lubricating it and so making it easier to swallow. Saliva is a dilute aqueous solution having a solid content of only about one per cent. Its main constituent is a slimy substance called *mucin* which assists lubrication. It also contains the enzyme *salivary amylase*, and various inorganic salts, the most abundant being sodium chloride which furnishes chloride ions that activate the enzyme. The initial hydrolysis of cooked starchy food is catalysed by salivary amylase in the mouth, and this catalytic action is continued as the food moves down the oesophagus and into the stomach. The enzyme soon becomes inactivated in the stomach, however, because it cannot tolerate a strongly acid environment.

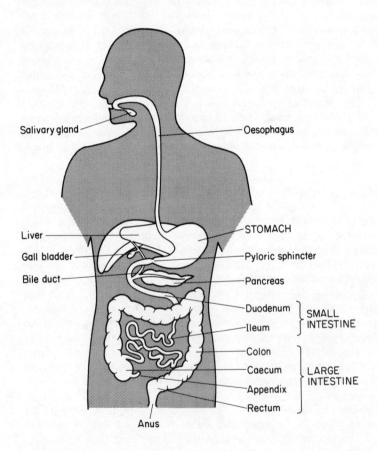

Figure 2.4   The digestive system.

Food is carried down the oesophagus by gentle muscular action called *peristalsis*. The muscles contract, producing a peristaltic wave and this moves down the oesophagus, carrying the food with it.

DIGESTION IN THE STOMACH   The stomach may be regarded as a reservoir in which food is prepared for the main stage of digestion in the small intestine. This does not mean that no digestion takes place there, however, for cells in the lining of the stomach produce a fluid called *gastric juice*. The two essential constituents of this dilute aqueous solution are its enzymes and its acid content. The main enzyme is pepsin, secreted as the inactive pepsinogen and becoming activated when it comes into contact with the hydrochloric acid which forms the acid constituent of the gastric juice.

Some 20 minutes after starting to eat a meal, vigorous muscular movements begin in the lower region of the stomach. Muscular contraction produces an inward pressure and this moves down the stomach wall as a peristaltic wave, so moving food through the stomach and causing it to become intimately mixed with the gastric juice. In this way the acidity of the semi-fluid food mixture called *chyme* increases, until the endopeptidase pepsin is able to catalyse the conversion of part of the protein into slightly simpler molecules called *peptones*. The other enzyme in the gastric juice is *rennin*, which also acts in an acid medium and brings about the coagulation or clotting of milk. The acidity of the gastric juice also causes some bacteria, which enter with the food, to be killed.

A copious flow of gastric juice is necessary during a meal and its production is stimulated both by psychological and chemical means. The former is the more important and is governed by involuntary nervous action which may be brought about by the appearance, smell and taste of food. The mere thought of food may be sufficient to stimulate gastric secretion; on the other hand, the flow of gastric juice may be inhibited by such factors as excitement, depression, anxiety and fear. Certain foodstuffs act as chemical stimulants to secretion. *Meat extractives*, for instance, which are dissolved out of meat when it is put in boiling water, are particularly potent in this respect. Soups and meat dishes in which the extractives have been preserved are therefore valuable aids to digestion in the stomach.

Peristaltic action moves the chyme into the lower region of the stomach which is separated from the upper region of the small intestine, called the duodenum, by the pyloric valve. The valve opens at intervals, so allowing small portions of chyme to leave the stomach. This process continues for about six hours after eating a meal until no chyme remains in the stomach.

DIGESTION IN THE SMALL INTESTINE   The main stage of digestion occurs during the passage of chyme through the long small intestine.

As soon as food enters the duodenum, digestive juices pour forth. There are three sources: the liver secretes *bile* which is then stored by the gall bladder, and the pancreas secretes *pancreatic juice* – these two secretions enter the small intestine through a single duct situated a short way down the duodenum; the third secretion is produced in the lining of the small intestine and is called the *intestinal juice*. They are all produced at the same time and as they are alkaline they neutralize the acidity of the chyme. Under these conditions the enzymes of the three secretions are able to exert their catalytic influence.

The pancreatic juice contains enzymes which enable it to help in the digestion of the three main types of nutrient. The endopeptidases trypsin and chymotrypsin, among others, carry on the degradation of proteins begun by pepsin in the stomach; they complete the breakdown of proteins into peptones. *Pancreatic amylase* is another enzyme present in the pancreatic juice; its capacity for catalysing the hydrolysis of large amounts of starch and converting it into maltose has already been mentioned. Finally, *pancreatic lipase* brings about

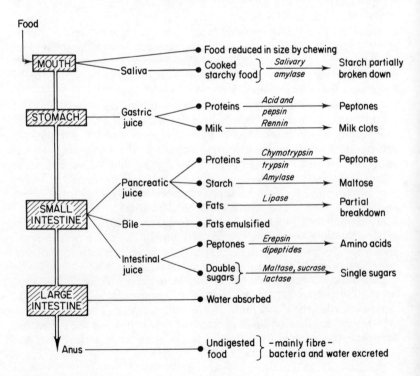

Figure 2.5 Summary of the digestive process.

the partial hydrolysis of some fat molecules, converting them into simpler substances which can be absorbed.

The bile has no enzyme action, but contains bile salts which convert fats (liquefied by the warmth of the stomach) into a fine emulsion of tiny oil droplets which may then be acted upon by the lipase of the pancreatic juice.

The intestinal juice contains a number of enzymes, three of which have already been mentioned, namely maltase, lactase and sucrase, which break down the double sugars maltose, lactose and sucrose respectively into simple sugars that can be absorbed. In addition to these, a group of exopeptidases, called *erepsin*, continue the breakdown of proteins begun by the endopeptidases pepsin, trypsin and chymotrypsin. The exopeptidases attack the ends of the chain-like peptone molecules until they are broken down into small units called dipeptides containing only two amino acids. Finally, another group of enzymes called *dipeptidases* break down the dipeptides into free amino acids which can be absorbed.

Apart from these chemical changes, muscular activity continues, so causing the various substances to move slowly down the small intestine.

ABSORPTION IN THE SMALL INTESTINE   The digestive process is almost complete after the food material has been in the small intestine for some four hours. The most complicated of all nutrients, the proteins, have been converted by stages into amino acids; all carbohydrates, except dietary fibre, have been broken down into simple soluble sugars, while fats have been emulsified and partly split into simpler substances called *fatty acids* and *glycerol*. However, before digested nutrients can be used by the body they must pass through the walls of the digestive tube into the blood, a process known as *absorption*.

The walls of the long small intestine are folded into finger-like projections called *villi* (like the pile of velvet) which contain both blood capillaries and a lymph vessel and which have a very large surface area. Non-fatty nutrients, such as the products of protein and carbohydrate digestion, namely amino acids and simple sugars, are absorbed directly through the villi into the blood. Products of fat digestion, namely fatty acids and glycerol, pass through the villi into the lymph vessel where they are resynthesized into fat molecules. However, in the process of being re-formed the fatty acids are rearranged so that the resulting fat molecules are more suitable for use by the body.

THE LARGE INTESTINE   About 7–9 hours after a meal has been eaten any food that has not been digested and absorbed in the small intestine passes through the ileocaecal valve into a wider and shorter tube called the large intestine. No new enzymes are produced by the body during this stage but the large intestine is a rich source of bacteria. These may attack undigested substances such as dietary fibre with their own enzymes and partially break them down. In addition, vitamin K and certain vitamins of the B group are synthesized, i.e. built up by the bacteria. Such bacterial action is not on a large scale but the small molecules so formed, if able to be absorbed, pass through the walls of the large intestine into the blood.

The main function of the large intestine is to remove water from the fluid mass; this process continues as the fluid passes along, so that by the time it reaches the end of the tube it is in a semi-solid form known as *faeces*. In a day, between 100 and 200 grams of moist faeces may be produced containing undigested food material, residues from digestive juices, large numbers of both living and dead bacteria, and water. After having been in the large intestine for about 20 hours these materials are passed out of the body.

TRANSPORT IN THE BODY   Food, after digestion and absorption, provides nutrients that are the raw materials of body metabolism. But this process is not complete without an efficient transport system capable of carrying nutrients to the cells that require them. We have already seen that nutrients, during absorption, pass into the blood and it is the constant circulation of blood through the body system that enables these nutrients to be transported to where they are needed. Blood, which is four-fifths water, contains many substances, such as nutrients and hormones, in solution. Other substances, such as the red blood corpuscles which transport oxygen, are present as cells in the blood and are carried round with it in suspension.

Most digested nutrients, being water-soluble, are easily transported dissolved in blood but the transport of fat is more complicated. Fat is transported in an emulsified form as minute droplets of *lipoprotein*. Lipoproteins are composed of clusters of fat molecules which have been coated with protein and *phospholipids* (mainly *lecithin*) and so converted into an easily transportable emulsion.

The heart pumps blood through the arteries and into successively smaller tubes, the smallest of which are capillaries. In the capillaries, nutrients and oxygen from the blood diffuse into the surrounding cells, while waste products from the cells diffuse into the blood. Blood carrying the waste material passes into a network of veins, carbon dioxide being removed by the lungs, while soluble substances are removed by the kidneys. There is also free diffusion of water between the blood and tissue fluid, which enables the fluid bathing the cells to be continually renewed.

CHAPTER 3

Food, health and disease

Today, in Western countries, most people are able to choose what they eat. Most people are also able to choose how much they eat. It was not always so however. Since earliest times man has had a strong instinct for survival and this primary driving force caused him to seek out animals and discover edible plants that would enable him to survive. Thus man's diet – that is his 'habitual pattern of consumption of food and drink' – was determined qualitatively by what was available and quantitatively by how much was available. At this early stage of man's development choice was limited, being mainly restricted to an instinctive selection of food that would not cause ill effects when eaten. As time went by man extended the range of foods available both by breeding animals and by cultivating those crops which proved to be useful sources of food.

Even in Britain as recently as 100 years ago the diet of ordinary people was very restricted and most people simply lived on the food that was available to them. Food choice amounted to selecting food that could be afforded. People in these circumstances were not concerned with whether food was 'good' or 'bad' for health but only with whether they could find enough food to stave off hunger. Considerations of availability and cost meant that the average diet was restricted to a few foods only. At the beginning of the twentieth century the average Englishman fed largely on a very limited diet, the principal elements of which were bread and tea.

There is no doubt at all that the diet in Britain today is both much healthier and more attractive than it was a century ago. At that time the general level of health was much inferior to current standards and infectious diseases such as cholera, tuberculosis, typhoid, smallpox and scarlet fever were commonplace and the major causes of death. It was not until the nineteenth century that the origin of infectious diseases was discovered, and the discovery and appreciation that they were caused by harmful bacteria was a landmark in man's understanding of the nature of disease. It demonstrated that infectious diseases result from the *presence* of something harmful in the environment, whether the air we breathe, the water we drink or more generally the insanitary nature of our living conditions.

The knowledge that infectious diseases were caused by harmful bacteria in the environment allowed preventive measures to be developed to exclude them. For example, untreated water supplies were a common source of infection, including cholera, but the treatment of water by sterilizing it before use caused the destruction of all harmful bacteria. Today in Britain all water for domestic consumption is carefully treated before use so that it is not injurious to health; a subject treated in more detail in Chapter 10.

DEFICIENCY DISEASES   The past prevalence of infectious diseases was not directly due to any deficiency of diet, though inadequate diet and consequent poor health made people susceptible to the harmful bacteria causing them. However, there was another type of disease that came to be linked directly with an inadequate diet and this type of disease became known as a *deficiency disease*. *Scurvy*, *rickets* and *beriberi* are well-known examples of deficiency diseases and they have been responsible for causing a great many deaths.

From the time of Pasteur onwards the part played by the presence of harmful bacteria in causing infectious diseases was fully appreciated. However, the opposite possibility – namely that disease could also be caused by the *absence* of something – was more difficult to comprehend.

Scurvy is a very ancient disease and is particularly associated with long sea voyages. At the end of the sixteenth century Admiral Hawkins noted that from his own experience he knew of 10 000 seamen who had died of scurvy. In the 1750s Captain Cook, during a long naval expedition to Australia, added citrus fruits and fresh vegetables to his sailors' diet with the result that scurvy was either prevented or cured. At about the same time (1753) the naval surgeon James Lind published his *Treatise of the Scurvy* in which he produced the first unequivocal experimental evidence linking diet and disease. He demonstrated, by means of the first controlled clinical trial ever carried out, that adding oranges and lemons to the sailors' diet cured scurvy.

Lind's treatise on scurvy established that the absence of citrus fruit in the diet could cause scurvy and that scurvy could be cured if such fruit were added to the diet of scurvy sufferers. Thus the concept of a deficiency disease came to be accepted. It was not until 1912, however, in a paper published by Gowland Hopkins, that the nature of *vitamins* and their vital role in maintaining health was established.

The discovery of vitamins and the consequent understanding of, and ability to cure, deficiency diseases has been hailed as the greatest advance in nutritional science in the twentieth century. The knowledge that the absence of vitamins from the diet could cause diseases such as scurvy and rickets brought about a fundamental change in man's attitude to food. It also focused attention on diet and established it as a major factor in causing disease as well as in maintaining health.

Concept of a balanced diet

As we saw in Chapter 1, our bodies will only attain and maintain a healthy condition if they receive a supply of food that provides adequate amounts of all the nutrients, and an appropriate amount of energy. Such a diet is known as a *balanced diet*. A healthy diet will also be one that is wholesome, i.e. one that does not contain anything harmful.

The concept of a balanced diet makes it easy to develop an analytical and chemical approach to food and diet. Such an approach is based on using food analysis tables and calculating the amount of different nutrients supplied by eating a given quantity of food. Thus it is easy to devise a balanced diet that supplies the recommended amount of each nutrient as specified in the table in Appendix I. Similarly, values for the energy content of foods can be used to ensure that the diet provides an appropriate amount of energy.

The balanced diet concept arose out of a desire to ensure that a sufficiently wide variety of foods should be eaten to prevent a deficiency of any particular nutrient. This approach was a relevant one up to the beginning of the twentieth century when malnutrition in general, and deficiency diseases in particular, were commonplace in Britain. However, at the present time malnutrition in Britain and other Western countries is rare and deficiency diseases are confined to specific circumstances. For example, immigrant children whose traditional diets have been replaced by inadequate Western-style diets may show vitamin deficiency, especially of vitamin D, and consequently suffer from rickets.

In the West today infectious and deficiency diseases are no longer a real problem. Most people eat a wide variety of foods and have plenty (or more than plenty) to eat. In spite of this a range of new diseases has appeared during the first part of the twentieth century, diseases such as coronary heart disease (CHD) and cancer, which have become known as *diseases of affluence* because they particularly afflict the more affluent parts of the world.

The prevalence of diseases of affluence in the Western world is believed to have some relation to diet and these diseases have developed in spite of the supposedly healthy, balanced diets being eaten by most people. It is evident that eating a wide variety of foods has not helped in preventing the spread of the diseases of affluence. Indeed, it is now believed that a different approach is needed; one in which careful consideration is given to the proportion of particular types of food eaten and to reducing the intake of certain types of food.

There is now a consensus of nutritional opinion about the *general* sort of changes that are required in the diets of affluent countries to make them more healthy. These changes will be considered in much more detail later in the book, but it may be helpful to give here the broad outlines of what is believed to constitute a healthy diet.

Until the 1960s protein was considered to occupy a key place in a healthy diet. Great emphasis was given to the need for sufficient protein and to its quality. Animal protein in particular was considered to be especially desirable on account of its high biological value. Today much less emphasis is placed on the need to eat plenty of high quality protein. This is partly because the body's need for protein is less than was previously considered and partly because if the diet contains enough 'energy foods', i.e. carbohydrate and fat, it will almost certainly also contain enough protein. In other words, if the carbohydrate and fat in the diet are right, the protein will look after itself.

Fat is the most compact source of energy in the diet and diets low in fat tend to be considered unappetizing. The quality and quantity of fat in a healthy diet is the subject of much research and debate and the matter is a complex one which is considered in more detail in Chapter 5. In general, however, there is agreement that the quantity of fat in the diet needs to be reduced and that the quality needs to be adjusted in favour of vegetable oils and fats (or more accurately in favour of *polyunsaturated* oils and fats) at the expense of animal (or saturated) fats.

One result of promoting the balanced diet concept, which nutritionists were doing until the late 1970s, was that food was thought to be as good as the nutrients that it contained. That part of food which did not contain nutrients was written off as *roughage*. Roughage is the indigestible part of food; chemically it consists mainly of *cellulose* while physically it consists of the cell walls of plants. *Refined* foods are those which have been processed so as to remove part or all of the cell walls of the natural foods from which they are derived. Sugar, for example, is a completely refined food. It contains no trace of the cell walls of the sugar-cane or sugar-beet from which it is derived; it is 'pure' sucrose. White flour is a refined food in the sense that a considerable proportion of the cell walls of the wheat from which it is made has been removed. Nevertheless, it is not completely refined, it still contains some 'roughage'.

It was during the 1960s that the trend towards eating increasing amounts of refined foods started to be questioned. Dr T. L. Cleave, Dr Denis Burkitt and Dr Hugh Trowell challenged the conventional wisdom of the day by promoting 'whole foods', that is foods which contain all the cell walls of their plant origin. What had been denigrated as 'roughage' was now acclaimed as *dietary fibre*.

We now acknowledge that food is more than the nutrients it contains and that dietary fibre is a necessary part of a healthy diet. At present there is an emphasis on the need for having plenty of unrefined carbohydrate in the diet while reducing the amount of refined carbohydrate. This can be achieved by eating whole foods rather than processed foods. Apart from protein, fat and carbohydrate (including dietary fibre) the diet must also supply adequate amounts of vitamins and mineral elements. As already noted, examples of vitamin deficiency are rare nowadays in Western countries. As for mineral elements, the consensus is not that diets are deficient in minerals but that one mineral salt, namely sodium chloride, should be restricted as a diet high in salt may predispose to hypertension (high blood pressure).

Diet and diseases of affluence

The rise in the incidence of diseases of affluence such as CHD and cancer in the Western world has become a major preoccupation of both medical and nutritional science. It is known that these diseases, unlike infectious, and deficiency diseases, do not have a single origin, but that a number of different factors are involved – they are *multifactorial* in origin. The factors implicated include age, sex, weight, lifestyle (including, for example, exercise and stress), smoking, blood pressure and *diet*.

It is the inclusion of diet among these factors that is our main concern, for it highlights a possible link between nutrition and disease that is highly significant. It is important to emphasize that there is no proof that diet *causes* any of these diseases (unless obesity is included among them). On the other hand, there is considerable evidence that diet does play some part, and the nature of this evidence will be examined in more detail in later chapters.

Coronary heart disease (CHD) has been more intensively studied than other diseases of affluence, not least because it is a major cause of death in developed countries. In the UK, which has a particularly high death rate from CHD, over half the deaths of men aged 45–54 are caused by strokes and CHD. CHD and other diseases of affluence, having no single cause, are difficult to investigate. *Epidemiology*, the study of disease patterns, is often used for this purpose. It is known that CHD is more prevalent in some countries – the developed ones – than in others, and that the disease occurs more in men than in women, mainly in old age but increasingly in middle age. From these observations common factors are looked for that might correlate with the distribution of the disease. Up to the present such correlation studies have been most useful in identifying certain factors, known as *risk factors*, that appear to have some connection with the disease.

Correlation studies of CHD have identified a number of risk factors that relate to the Western way of life. Among these risk factors are a number that are diet-related, though the three main risk factors have been identified as *smoking, raised blood pressure* and *high blood cholesterol*. In a long-term study of CHD carried out in Framingham, USA, the nature of these risk factors was investigated. Taking men in the 30–62 age group it was found that smoking more than 20 cigarettes a day almost doubled the risk of CHD, that very high blood pressure more than doubled the risk and that very high blood cholesterol increased the risk by a factor of two and a half.

The particular interest of the Framingham study is this; it demonstrates that the risk of CHD increases with each risk factor and this is illustrated in Fig. 3.1 which relates to men aged 45. The chart shows that where all three major risk factors are present the risk of CHD is nearly four times greater than the average. This study also showed that people who live stressful lives – those who are aggressive, competitive and harassed – are more likely to be at risk from CHD than those who lead less stressful ones.

While correlation studies of this sort provide valuable circumstantial evidence, a risk factor is simply a statistical concept and so cannot demonstrate cause. The importance of a risk factor needs to be evaluated in practical ways. *Clinical trials* are used to try and assess the importance of a particular risk factor while *intervention studies*, in

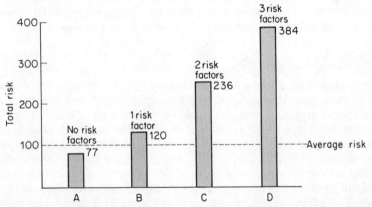

Figure 3.1   How the number of risk factors increases the risk of CHD. **A**: Men who do not smoke and have normal blood pressure and blood cholesterol. **B**: Men who smoke and have normal blood pressure and blood cholesterol. **C**: Men who smoke, have normal blood pressure but very high blood cholesterol. **D**: Men who smoke, have very high blood pressure and very high blood cholesterol.

which groups of people change their habits in a particular way, can show how changing a particular habit or factor can reduce the incidence of the disease.

The link between CHD and diet has been, and is, under intense scrutiny. Many different aspects of diet have come under suspicion, including excessive intake of food in general (leading to obesity), excessive intake of particular foods (including fat, especially saturated fat, sugar and even coffee) and insufficient intake of polyunsaturated fat and fibre.

There is need to emphasize again the fact that there is no proof that any of these dietary factors causes the disease but, on the other hand, there is considerable evidence (though much of it is contradictory) that at least some of these dietary factors do play a part. While proof concerning these dietary factors is lacking, the weight of evidence has caused national bodies in the UK such as the National Advisory Committee on Nutrition Education (NACNE Report, 1983), and the Committee on Medical Aspects of Food Policy (COMA) of the DHSS (Report 1984 on 'Diet in Relation to Cardiovascular Disease') to issue nutritional guidelines suggesting ways in which diet should be altered in order to reduce the incidence of CHD and other diseases of affluence. These guidelines are considered in more detail in Chapter 15.

The fact that such nutritional guidelines have been published confirms the view of nutritional and governmental bodies that there is *some* link between diet and diseases of affluence. The fact that many investigators remain unconvinced about the diet–disease link confirms the difficulty of obtaining reliable evidence. Whatever the truth of the matter it is certain that the proposed nutritional guidelines can do no harm and it seems probable that they will do some good.

Concept of health

A balanced diet will prevent malnutrition, under-nutrition and starvation; it will prevent deficiency diseases. The use of 'dietary goals' based on nutritional guidelines may help combat diseases of affluence but of themselves they cannot make us healthy, for health is more than the absence of disease. Health is 'the attainment and maintenance of the highest state of mental and bodily vigour of which any individual is capable'.

Our present understanding of nutrition releases us from the narrow chemical view (we write as chemists who, we trust, have come to a position of some humility concerning a chemical approach to food) that food is nothing more than the nutrients (chemicals) that it contains. Nutrients *are* important, but so is the crunchy texture of an apple, a raw carrot, or a nut, the chewiness and aroma of freshly baked wholemeal bread, the satisfying feeling that comes from eating a bowl of muesli and dried fruit, the mouth-watering appearance and smell of fresh strawberries.... Food should be *fun* as well as *functional*.

It is unhelpful only to see our bodies as slow combustion machines and food as fuel, just as it is unhelpful to see food as a sort of medicine that can reduce the possibility of disease. Food is for enjoying as well as doing us 'good'. Health comes not merely from eating the 'right foods' but from our attitudes, approach to life and our lifestyle. Stress and anxiety may contribute as much, or more, to our lack of health as a poorly chosen diet.

FOOD-RELATED DISEASES

While much prominence has been given to the diseases of affluence discussed in previous pages, other diseases that are clearly related to diet, though less dramatic in their effects, have increasingly come to be recognized as important in our modern society. It is strange that two of these diseases, *obesity* and *anorexia nervosa*, result from opposite tendencies, namely the eating of too much food and the eating of too little respectively. Other important food-related diseases are *coeliac disease* and forms of *food intolerance* including *allergy* diseases.

Obesity

Obesity is the state in which a person accumulates an excessive amount of body fat. There is about 12 per cent fat in the body of an average adult male and about 25 per cent in the body of an average adult female. Men and women are considered obese if these figures exceed 20 per cent and 35 per cent respectively. Most body fat occurs just below the skin and it is possible to estimate total body fat from the thickness of skin folds. However, it is much more usual to assess body fat from measurements of weight even though such a method only gives a crude estimate of the body's fat content.

The guidelines for body weight shown in Table 3.1 give the acceptable weight range for men and women of different heights. The acceptable weight range is based on actuarial analyses that indicate the weight range for each height associated with the lowest mortality rate in an insured population. A person is considered to be obese if their body weight is 20 per cent above the upper limit of the acceptable weight range. Such a definition of obesity is no more than a rough-and-ready guide, for on the one hand it ignores any consideration of age, and on the other hand it ignores differences in muscular development.

Although few comprehensive studies of the prevalence of obesity

Table 3.1 Guidelines for body weight (based on Royal College of Physicians Report, 1983)

Height (m)	Acceptable weight range (kg)		Obesity (kg)	
	Men	Women	Men	Women
1.48	–	42–54	–	65
1.52	–	44–57	–	68
1.56	–	45–58	–	70
1.60	52–65	48–61	78	73
1.64	54–67	50–64	80	77
1.68	56–71	52–66	85	79
1.72	59–74	55–69	89	83
1.76	62–67	58–72	92	86
1.80	65–80	–	96	–
1.84	67–84	–	101	–
1.88	71–88	–	106	–
1.92	75–93	–	112	–

have been carried out, it is recognized in the UK to be one of the most common nutritional disorders and a major health problem. It is believed that about one-third of adults in Western countries are obese. In the report *Obesity* by the Royal College of Physicians (1983), it is estimated that in the UK up to 30 per cent of adults and about five per cent of children are obese at some time.

EFFECTS OF OBESITY   Obesity is considered to be a major threat to health. In particular, obesity has an adverse effect on the cardiovascular system. In the Framingham study already referred to, and in subsequent studies, it has been found that obesity increases the risk of high blood pressure, angina pectoris (which produces severe chest pains) and coronary heart disease.

Obese people are also more likely to suffer from diabetes than slim people and this may lead to kidney failure and blindness. Obese women are more likely to be infertile than slim women and if they do become pregnant they risk having toxaemia and difficulty in childbirth. Obese people (especially women) are more likely to develop gall-bladder disease than slim ones because they produce excessive amounts of cholesterol which predisposes to the development of gallstones. It is also possible that obesity increases the risk of osteoarthritis, especially of the knees, hips and spine.

There is considerable evidence, much of it obtained by the American Cancer Society, that increasing obesity gives rise to an increased risk in men of cancer of the colon, rectum and prostate and an increased risk in women of cancer of the breast, cervix and uterus.

The catalogue of diseases associated with obesity given above needs to be seen in proper perspective in order that the risk of being overweight is not exaggerated. For example, the risk to health arising from smoking is so much greater than the risk arising from obesity that it is less hazardous to be an overweight non-smoker than a smoker of normal weight.

The effects of obesity in children are of concern because excessive weight gain during early infancy is believed to be related to obesity in later years. American studies suggest that about 80 per cent of obese children turn into obese adults, while British studies (presented in the report *Obesity* already referred to) indicate that although overweight in infancy does not *necessarily* lead to overweight in adults, it does increase the *likelihood* of becoming overweight in adult life. Obesity seems to be a familial condition so that children of overweight parents are particularly at risk and such children should be encouraged to develop eating and exercise patterns which will prevent weight gain. There is evidence that children who are prone to weight gain have a lower energy output than average and so require to eat less energy foods, such as sugar and fats, than other children.

Enough has been said to substantiate the claim that obesity is a major hazard to health; indeed in the USA it has been described as the greatest dietary problem. Ways in which overweight and obesity can be prevented or cured are considered in Chapter 15.

Anorexia nervosa

Anorexia nervosa is a diet-related disease, the term anorexia suggesting a loss of appetite. It is also known as the 'slimming disease'. However, neither of these terms is completely accurate as the disease results from fears of the results of satisfying appetite rather than loss of appetite itself. It is a psychological illness where cause is often difficult to diagnose. It mainly affects adolescent girls and while some genetic factors may be involved, environmental factors are much more significant.

The disease results in severe loss of weight and may lead to the cessation of menstrual periods. If unchecked anorexia may result in death. Treatment is principally by psychotherapy but also involves dietary help. The anorexic needs to be encouraged to embark on a gradually increasing intake of food and while no special diet is required it should be nutritionally adequate but without an excessive energy content. Although recovery may be slow, taking as much as two or three years, there is the possibility that it may be complete.

Coeliac disease

Coeliac disease is caused by malabsorption of food resulting in loss of weight and deficiency of mineral elements and vitamins. Coeliacs are sensitive to the protein *gliadin* which is part of the gluten (see p. 139) contained particularly in wheat and rye, but only in small quantities in other cereals such as barley, rice and oats. The disease normally originates in childhood when cereals containing gluten are first introduced into the diet.

The cause of the disease is unknown, though it may be due to the absence of an enzyme (see Chapter 2) required for the digestion of gliadin. Whatever the cause, sensitivity to gliadin leads to irritation, and eventually destruction, of the villi (see p. 27) in the small intestine which are responsible for the absorption of nutrients and their transfer into the bloodstream. As a result some nutrients are not absorbed but are excreted from the body, usually as fatty diarrhoea.

Children suffering from coeliac disease do not grow properly, usually have a characteristic 'pot belly' and may suffer from anaemia and rickets. In the past the disease could prove fatal but now it can be cured simply by eliminating all foods containing untreated wheat flour, rye and barley. Such a diet involves avoiding not only the obvious sources of gluten such as bread and pasta but also manufactured foods to which wheat flour is added such as sausages and sauces. Some manufacturers of baby foods produce gluten-free products, this fact being indicated by a symbol on the label. Gluten-free foods, such as biscuits, bread and flour, are made from wheat which has had the gluten removed and such products are widely available.

Food intolerance

Some people are hypersensitive, that is their bodies produce unpleasant physical reactions, to pollen or dust in the air, to drugs or to food or food additives. Such adverse reactions are usually described by the term intolerance, and when food is involved by the term *food intolerance*.

Food intolerance may be defined as an adverse reaction to food that is not psychologically based. It is hard to determine how prevalent this condition is, though it has been estimated that up to 20 per cent of children suffer from some form of dietary intolerance.

It is believed that the condition may be inherited, though fortunately infants who suffer from food intolerance often grow out of it before they are three years old.

Food intolerance may be caused in a number of ways, the principal ones being as follows.

1. *Specific substances in food.* Some foods contain substances that trigger off reactions that produce unpleasant symptoms. For example, strawberries and tomatoes can cause the release of

histamine in some people, producing an irritating rash. *Amines* present in wine, cheese, yeast extracts and bananas can cause unpleasant effects as can *caffeine* contained in tea and coffee. Adverse reactions only occur, however, after consuming fairly large quantities of the offending food.

2. *Inability to digest certain foods.*   Some people suffer from *lactose intolerance*. Most people digest milk with ease but in some babies and children and also in some adults especially in Africa, Asia and South America, this ability is lacking. Milk contains the sugar *lactose* which is digested in the small intestine with the help of the enzyme *lactase*. Those who suffer from lactose intolerance produce insufficient lactase to digest lactose with the result that lactose passes into the large intestine where it is converted by bacteria into lactic acid causing diarrhoea.

3. *Reaction to certain chemicals and drugs.*   Some people react to additives found in food. For example, some artificial colours such as the yellow dye *tartrazine* found in some drinks, and some preservatives such as *benzoic acid* added to pickles and soft drinks produce allergic reactions in certain people. In addition, some people are susceptible to certain drugs such as *salicylates* found in aspirin but also in many foods including fruits (especially dried fruit), vegetables and herbs and spices.

Food aversion is distinguished from food intolerance because it has a psychological origin and is often associated with depression or anxiety.

FOOD ALLERGY   Food allergy is a form of food intolerance involving a breakdown in the body's immune system. Normally the immune system protects the body against substances such as harmful viruses and bacteria but in food allergy it also reacts to certain harmless substances contained in food. The substance producing an allergic reaction is known as an *allergen* and it is either a protein or bound to a protein (technically a *glycoprotein*, see p. 168).

In normal metabolism proteins are broken down during digestion as explained in Chapter 2. In people who show an allergic reaction to a particular food, the protein contained in it is not digested but passes into the bloodstream and reacts with antibodies which are part of the immune system. The consequence is that *histamine* and other powerful and irritating substances are released into the bloodstream producing different symptoms depending on whether the antibodies are found in the skin, the digestive tract or the respiratory system. The symptoms most commonly associated with food allergy are urticaria (nettle rash), eczema, asthma, vomiting and diarrhoea.

The treatment of food intolerance is easiest where there is an immediate reaction to food consumed. In this instance it is relatively easy to isolate the food causing the trouble and eliminate it from the diet. Sometimes reaction to allergens does not develop for several hours after eating and the cause may be difficult to discover. The most reliable, but time-consuming, test for allergy is the use of an *elimination diet*. This involves feeding the patient on a limited diet containing foods unlikely to contain allergens and then adding foods containing allergens one-by-one until an allergic reaction is produced. To be effective the suspect foods should be added to the diet without the patient's knowledge so that psychological aversion to particular foods can be avoided.

The foods and food additives most commonly associated with food intolerance and aversion are cows' milk, eggs, fish, shellfish, wheat and other cereals, flour, yeast, chocolate, pork, bacon, tenderized meat, coffee, tea, preservatives and artificial colours.

SUGGESTIONS FOR FURTHER READING

British Medical Association (1986). *Diet, Nutrition and Health*. BMA, London.

British Nutrition Foundation (1986). *Energy Balance in Man*. BNF Briefing Paper No. 8.

Department of Health and Social Security (1984). Committee on Medical Aspects of Food Policy (COMA), *Diet and Cardiovascular Disease*. HMSO, London.

Eagle, R. (1986). *Eating and Allergy*. Thorsons, Wellingborough.

Franklin, A. and Lingham, S. (Eds) (1987). *Food Allergy in Children* . Parthenon Publishing Group, Carnforth, Lancs.

Garrow, J. S. (1981). *Treat Obesity Seriously*. Churchill Livingstone, Edinburgh.

National Advisory Committee on Nutrition Education (NACNE) (1983). *Proposals for Nutritional Guidelines for Health Education in Britain*. Health Education Council, London.

Royal College of Physicians and British Nutrition Foundation (1984). *Food Intolerance and Food Aversion*. BNF, London.

Royal College of Physicians (1983). *Obesity*. RCP, London.

CHAPTER 4

Lipids and colloids

The term *lipid* is a general one that is used to describe a large group of naturally occurring fat-like substances. They are organic compounds which all contain carbon, hydrogen and a small amount of oxygen. They form a diverse group of compounds that have little in common except that they are soluble in organic solvents such as chloroform and alcohols but are not soluble in water. Most of them are also derivatives of fatty acids. *Oils and fats*, *waxes* and *phospholipids* are examples of lipids that are fatty acid derivatives. *Steroids* are also classified as lipids, though they are not fatty acid derivatives and they are quite dissimilar in structure to the rest of the group.

Oils and fats are of major importance in food and nutrition. Chemically they belong to a class of substances known as *esters*, which result from the reaction of acids with alcohols. Fats are esters of the trihydric alcohol glycerol. The three hydroxyl groups of the glycerol molecule can each combine with a fatty acid molecule and the resulting ester is called a *triglyceride*. The simplest type of triglyceride results when the three acid molecules are all the same. For example, if three molecules of stearic acid react with a molecule of glycerol, the fat formed is *tristearin*:

$$
\begin{array}{ccc}
& \mathrm{HOCH_2} & \mathrm{C_{17}H_{35}COOCH_2} \\
& | & | \\
\mathrm{3C_{17}H_{35}COOH} + \mathrm{HOCH_2} & \rightleftharpoons \mathrm{C_{17}H_{35}COOCH} + \mathrm{3H_2O} \\
& | & | \\
& \mathrm{HOCH_2} & \mathrm{C_{17}H_{35}COOCH_2}
\end{array}
$$

Stearic acid   Glycerol   Tristearin

Chemically speaking, oils and fats are the same; the only distinction between them is that at normal air temperatures oils are liquids while fats are solids. However, this distinction is rather vague, as a 'normal' temperature cannot be accurately defined, and some oils, such as palm oil, are usually solid at the prevailing temperatures of the British climate.

Phospholipids are important substances concerned with the transport of lipid in the bloodstream. *Lecithin* is one of the most significant phospholipids; it is made up of a group of substances that play a major role in the digestion of fats and as such it has already been encountered in Chapter 2. Phospholipids are similar to fats in that they have a structure based on glycerol but instead of the three hydroxyl groups being combined with a fatty acid (see below) only two are so combined, the third being combined with phosphate linked to choline. Lecithin is found in some foods, notably egg-yolk. It is not an essential part of the diet and it is manufactured by the body.

Steroids, as already indicated, are very different in structure to other lipids but include the important substances *cholesterol, cortisone* and *sex hormones*. Cholesterol is a white fat-like substance which is present in body tissues and is found in a variety of animal foods, notably brain, kidney, liver and egg-yolk. Meat and fish contain small amounts and dairy foods contain a little. The body of an adult contains 140 g cholesterol found in all parts of the body, but particularly in the membranes of every cell, especially nerve cells. Cholesterol is also important in the body as the source of raw material from which bile acids and several hormones – including corticosterone from the adrenal gland and sex hormones – are made. The body obtains some cholesterol from the diet (except strict vegetarian diets) but some is also made in the body, especially by the liver, although all body cells can make cholesterol.

Waxes are made from fatty acids combined with any alcohol *except* glycerol. Animal waxes are often esters of cholesterol. Waxes are of no particular importance in food science and will not be encountered again.

OILS AND FATS

Oils and fats are of great importance in food science for they are used in their own right, for example, in cooking, in salad oils and as spreads, and as ingredients in many manufactured and cooked foods. They are important in nutrition as the most compact energy source available and they play an important role in body metabolism. In addition, their role in promoting health and disease, a subject considered in the next chapter, is currently under intensive scrutiny.

As already mentioned, oil and fat molecules are triglycerides formed by reaction of one molecule of glycerol with three fatty acid molecules. The nature of the fatty acids involved plays an important part in determining the character of oils and fats, and this aspect will now be considered in some detail.

Fatty acids

Fatty acids are those organic acids which are found in fats chemically combined with glycerol. Fatty acids are known as *carboxylic* acids because they contain the carboxyl group $-COOH$. Over 40 different fatty acids are found combined as part of triglycerides. It should be noted that where fatty acids are referred to as being part of fats this means that they occur *combined* with glycerol and not as *free* acids. Natural fats never consist of a single triglyceride but are mixtures of triglycerides.

Fatty acids consist of a chain of carbon atoms (each with hydrogen atoms attached) with a carboxyl group at the end. The length of the carbon chain varies, and the number of carbon atoms is an even number between four and 24. The commonest fatty acids contain 16 or 18 carbon atoms (see Table 4.1). For example, stearic acid, $CH_3(CH_2)_{16}COOH$, contains 18 carbon atoms. When combined with glycerol it forms tristearin. There are three types of fatty acid which may be classified as follows.

1. *Saturated fatty acids* in which the carbon atoms are linked together by single bonds. Thus the carbon chain consists of repeated CH_2 groupings: $-CH_2-CH_2-$. For example, stearic acid whose formula is noted above, is a saturated fatty acid.
2. *Monounsaturated acids* in which there is *one* double bond in the carbon chain (see below). Thus the chain contains two unsaturated carbon atoms each joined to only one hydrogen atom: $-CH=CH-$. For example, oleic acid, $CH_3(CH_2)_7CH=CH(CH_2)_7COOH$, is a monounsaturated fatty acid.
3. *Polyunsaturated fatty acids* (PUFA) in which there are two or more double bonds in the carbon chain: $-CH=CH-CH_2-CH=CH-$. For example, linoleic acid, $CH_3(CH_2)_4CH=CHCH_2CH=CH(CH_2)_7COOH$, contains two double bonds in its carbon chain.

THE NATURE OF SINGLE AND DOUBLE BONDS   The distinction between saturated and unsaturated fatty acids is that the former contain carbon atoms linked by single bonds whereas the latter contain at least two carbon atoms linked by double bonds.

A carbon atom contains four electrons in its outer shell, and when it combines with other elements such as hydrogen it does so by sharing these electrons in order to gain the stable structure conferred by eight electrons in the outer shell. The combining together of elements by electron sharing is known as *covalence*; the bond so formed is a *covalent bond*. A covalent bond consists of a shared pair of electrons where each of the two atoms involved provides one electron for sharing. When carbon atoms link together to form a chain consisting of single bonds each carbon atom can combine with two hydrogen atoms:

$$\begin{matrix} & H & H & \\ & | & | & \\ -C & - & C- & \\ & | & | & \\ & H & H & \end{matrix} \qquad \text{or} \qquad -CH_2-CH_2-$$

Each dash between atoms represents a shared pair of electrons, i.e. a single covalent bond. In order to gain a stable octet of electrons each carbon atom forms four covalent bonds with adjacent atoms.

Carbon atoms can also link together so as to form chains containing double bonds. Four electrons are involved in the formation of a double bond – two from each of the carbon atoms concerned – so that less electrons are available for forming bonds with hydrogen. Each carbon atom can now only combine with one hydrogen atom in order to gain its stable octet of electrons. A double bond is represented by a double dash:

$$\begin{matrix} & H & H & \\ & | & | & \\ -C & = & C- & \\ \end{matrix} \qquad \text{or} \qquad -CH=CH-$$

the more double bonds that a carbon-hydrogen chain possesses the greater is its *degree of unsaturation*.

The degree of unsaturation of a fat is important in determining its properties. All natural fats contain both saturated and unsaturated fatty acids (combined with glycerol), but the greater the proportion of the latter the lower is the melting point of the fat. Fats which are high in unsaturated fatty acids (e.g., olive oil, sunflower seed oil) are therefore liquid at room temperature (i.e. they are oils) while those rich in saturated fatty acids are solid at room temperature (e.g., butter) (see Table 4.2).

The degree of unsaturation of an oil or fat can be measured by its *iodine* value. When iodine (in practice the more reactive iodine monochloride is used) is added to a triglyceride formed from an unsaturated fatty acid, it reacts with the double bonds in the molecule, and the degree of unsaturation may be calculated from the amount of iodine absorbed:

$$-CH=CH- + I_2 \rightarrow -CHI-CHI-$$

One molecule of iodine is used to saturate each double bond. The result is usually expressed as the iodine value which is the number of grams of iodine needed to saturate 100 grams of oil. For example, the iodine value of butter is 26–38 while the iodine value of olive oil is 80–90.

The fatty acids most commonly found as part of edible oils and fats are shown in Table 4.1. Oleic acid is the most commonly occurring fatty acid in nature, and in most fats it forms at least 30 per cent of the total fatty acids. Palmitic acid is also widely distributed and is found in every natural fat, usually accounting for 10–50 per cent of the total fatty acids.

Table 4.1 uses a convenient shorthand method for describing fatty acids. Butyric acid $C_{4:0}$ means that it contains four carbon atoms but no double bonds while linolenic acid $C_{18:3}$ contains 18 carbon atoms and three double bonds.

Table 4.2 shows the contribution made by fatty acids in combination with glycerol to some important oils and fats. The total fatty acid content is taken as 100, the figures for the individual combined acids are average values, as oils from different places vary in composition. Some other fatty acids not shown in the table occur in these oils and

Table 4.1   Fatty acids most commonly found as part of triglycerides

Saturated acids		Monounsaturated acids		Polyunsaturated acids	
Butyric	$C_{4:0}$	Palmitoleic	$C_{16:1}$	Linoleic	$C_{18:2}$
Caproic	$C_{6:0}$	Oleic	$C_{18:1}$	Linolenic	$C_{18:3}$
Caprylic	$C_{8:0}$	Erucic	$C_{22:1}$	Arachidonic	$C_{20:4}$
Capric	$C_{10:0}$				
Lauric	$C_{12:0}$				
Myristic	$C_{14:0}$				
Palmitic	$C_{16:0}$				
Stearic	$C_{18:0}$				

Table 4.2   The per cent combined fatty acid content of oils and fats

Oil or fat	Myristic $C_{14:0}$	Palmitic $C_{16:0}$	Stearic $C_{18:0}$	Oleic $C_{18:1}$	Linoleic $C_{18:2}$	Other PUFA
Butter	11	26	11	30	2	1
Lard	1	24	18	42	9	0
Margarine	5	23	9	33	12	1
Margarine (polyunsaturated)	1	12	8	22	52	1
Fish oil	5	15	3	27	7	43
Olive oil	0	12	2	73	11	1
Palm oil	1	40	4	45	9	0
Rapeseed oil	0	3	1	24	15	10
Corn oil	0	12	2	31	53	2
Soya bean oil	0	10	4	24	53	7
Sunflower seed oil	0	6	6	33	58	0
Safflower seed oil	0	7	2	13	74	0

Table 4.3 Sources of saturated and polyunsaturated fats

High in saturated fats	Dairy products	*Butter, cream, milk, cheese*
	Meat	*Liver, lamb, beef, pork*
	Others	Coconut oil, palm kernel oil, palm oil, hard margarine, *lard*
High in polyunsaturated fats	Vegetable oils	Corn (maize) oil, soya bean oil, safflower seed oil, sunflower seed oil
	Nuts	Most, except coconut and cashew nuts
	Margarines	Many soft varieties especially soya bean and sunflower seed

fats. In particular butter contains 13 per cent $C_{4:0}$ to $C_{12:0}$ fatty acids, rapeseed oil may contain as much as 33 per cent erucic acid (though rapeseed oil used for margarine manufacture is low in erucic acid and contains about two per cent) while the 43 per cent PUFA in fish oil is arachidonic acid.

It is often stated that animal fats are saturated while vegetable oils are unsaturated. While there is a considerable basis of truth in this generalization, it cannot be taken as a rule, as Table 4.3 demonstrates. This table shows the more important dietary sources of saturated and polyunsaturated fats, those of animal origin being shown in italics. From the above generalization it might be presumed that when the label on a packet of margarine declares that it is made from blended vegetable oils, it would contain a high proportion of PUFA. However, if it is a cheap brand, this may well not be so as it will almost certainly be made from palm and coconut oils and will therefore contain a high proportion of saturated fatty acids.

Essential fatty acids

The polyunsaturated fatty acids (PUFA) are of particular interest in human nutrition because certain of them, although essential to the body, cannot be made by the body and so must be supplied by food. The most important essential fatty acid (EFA) is *linoleic acid*, which is found in large amounts in oil made from corn, soya beans and sunflower seeds. *Arachidonic* acid is often referred to as an EFA although strictly speaking this is not true as it can be made in the body from linoleic acid.

Linolenic acid is also classed as an EFA although it is not certain whether this is correct because the amount required by the body is small and it is possible that these small amounts can be made from other fatty acids within the body. Linolenic acid occurs in small amounts in some vegetable oils such as rapeseed oil and soya bean oil.

The amount of EFAs required by the body is small, amounting to only about 10 g per day and deficiency in human nutrition is rare. EFAs have two important roles in the body: they provide the raw materials from which the hormones known as *prostaglandins* are made and they also form part of the structure of all cell membranes.

Cis and *trans* fatty acids

When two carbon atoms are joined by a double bond there is no freedom of rotation about the axis of the double bond. Consider oleic acid (Fig. 4.1). In one form the two parts of the hydrocarbon chain are on the *same* side of the double bond (*cis* form) while in the other they are on *opposite* sides of the double bond (*trans* form).

Naturally occurring unsaturated fatty acids have *cis* forms while the corresponding *trans* forms lack EFA activity. Enzymes can recognize the difference between *cis* and *trans* forms, acting on the former but not the latter.

The distinction that is drawn between *cis* and *trans* fatty acids may seem like an unnecessary complication but it will be seen later (p. 87) that it has some relevance when considering the relationship between the types of fat we eat and our health.

The physical nature of oils and fats

Having considered the chemical nature of oils and fats it is now necessary to consider their physical nature.

The physical characteristics of oils and fats are important in many food applications, such as their use in cake and pastry mixes and in mayonnaise and ice-cream. Unlike pure chemical compounds fats do not melt at a fixed temperature but over a range of temperatures. In this range they are *plastic*, that is they are soft and can be spread, but they do not flow. In other words their properties are intermediate between those of a solid and a liquid.

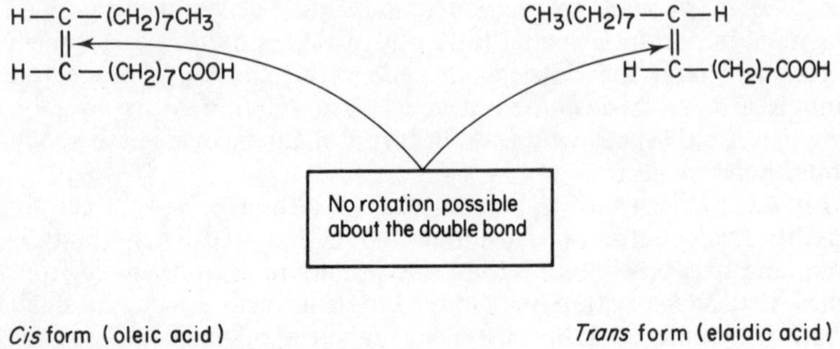

$$H - C - (CH_2)_7CH_3$$
$$H - C - (CH_2)_7COOH$$

$$CH_3(CH_2)_7 - C - H$$
$$H - C - (CH_2)_7COOH$$

No rotation possible about the double bond

Cis form (oleic acid)   *Trans* form (elaidic acid)

Figure 4.1   *Cis* and *trans* forms of a fatty acid.

The plasticity of a fat results from its being a mixture of a number of different triglycerides, each triglyceride having its own melting point. If a large proportion of the triglycerides are below their melting point the mixture will be solid and will consist of a network of minute crystals surrounded by a smaller quantity of liquid triglycerides. The solid network is not rigid, however, and the crystals can slide over one another, so giving rise to the plastic character of the fat. If the temperature of the fat is raised an increasing proportion of triglycerides melt, the solid network gradually breaks down and the plasticity of the mixture increases until it becomes liquid, when all the triglycerides have melted.

The melting point of fats is also affected by the fact that many triglycerides can exist in several crystalline forms; that is they are *polymorphic*. Each crystalline form has its own melting point and when oils are cooled different mixtures of these separate crystalline forms, and therefore of different melting point, may be obtained depending upon how the cooling was carried out. The way in which an oil is cooled therefore affects the texture and consistency of the product formed. Such considerations are important in commercial methods of fat manufacture.

Animal fats

The two most important animal fats are lard and butter, the latter being considered in Chapter 5 together with other dairy products. Lard is prepared by melting pig's fat and is virtually 100 per cent fat. As can be appreciated from Table 4.2 lard is relatively low in PUFA. Natural lard is a low-melting fat with good properties as a shortening agent, an acceptable white colour and bland flavour, but it suffers from the disadvantage that it has poor creaming properties and is therefore not a desirable fat for cake making.

The creaming properties of lard, i.e. its ability to incorporate air when beaten, may be improved by *interesterification* which, as the name implies, involves a rearrangement of the combined fatty acids between the triglyceride ester molecules. This regrouping process results in a more random distribution of fatty acids. To understand why this should be so it is necessary to consider the arrangement of combined fatty acids in the triglycerides of lard. Lard is peculiar among fats in that the saturated fatty acids, mainly palmitic, are found predominantly in the middle position of the triglyceride molecules. It is this factor which causes lard to form large coarse crystals that prevent it from creaming well.

During interesterification the fatty acids change position, thus reducing the proportion of triglyceride molecules with a saturated fatty acid in the centre position. Interesterification is carried out by heating lard at about $100°C$ in the presence of a catalyst such as sodium ethoxide, $NaOC_2H_5$. This gives a product with greatly improved creaming properties and so more suitable for incorporation in margarine and cooking fats.

Marine oils

Whale and fish oils are characterized by their high content of polyunsaturated fatty acids containing 20 and 22 carbon atoms and up to six double bonds (see Table 4.2). This large proportion of highly unsaturated fatty acids is reflected in the high iodine value which is in the range 100–140. Certain fish oils, such as herring oil, contain an even greater proportion of highly unsaturated fatty acid units and consequently have high iodine values, which may reach about 200.

The highly unsaturated nature of whale and fish oils makes them prone to spoilage and they are therefore unsuitable for use until they have been processed.

Vegetable oils and fats

Vegetables constitute the most important source of edible oils and fats. Most vegetable oils are liquid at $20°C$, though there are a few notable exceptions such as palm oil, palm kernel oil and coconut oil which melt above this temperature. The nature of the combined fatty acids in vegetable oils can be appreciated from Table 4.2, which shows the relatively high proportion of mono and polyunsaturated fatty acids they contain compared with animal fats.

Soya beans are grown extensively in China and the USA and soya bean oil is now the most important edible vegetable oil. The residue left when the oil has been extracted from the bean constitutes a valuable source of protein and as such is discussed on page 201. Soya bean oil is the major vegetable oil used in margarine manufacture and large quantities are also used in cooking fats.

Vegetable oils are normally extracted from seeds, kernels and nuts, either by mechanical pressure or by solvents. The latter method involves the use of a liquid solvent of low boiling point in which the oil is soluble. After the seed or nut has been ground, it is shaken with the solvent, the oil is extracted and a solid residue remains. When the liquid mixture is heated, the low boiling solvent evaporates, leaving the oil. Groundnut oil, for example, occurs in groundnuts (or peanuts) to the extent of 45–50 per cent, most of which may be extracted by means of a screw-press which squeezes out the oil. In some modern methods a two-stage process is used. After an initial extraction with the press, an oil solvent is used to extract the remaining oil. The residue is not wasted, in fact its high protein content makes it a valuable substance, and is used as a cattle food.

Another useful source of edible oil is the palm tree, which is not to be confused with the more slender coconut palm. The fruit of the palm tree grows in large bunches which may contain over a thousand fruit. Each fruit is rather like a plum, having a thin orange to dark red skin covering a fleshy interior in which is embedded a hard shell containing the kernel. Palm oil is extracted from the fleshy part of the fruit and palm kernel oil from the kernel. Both these oils are edible, and palm oil is extensively used in Britain for the manufacture of margarine and cooking fats.

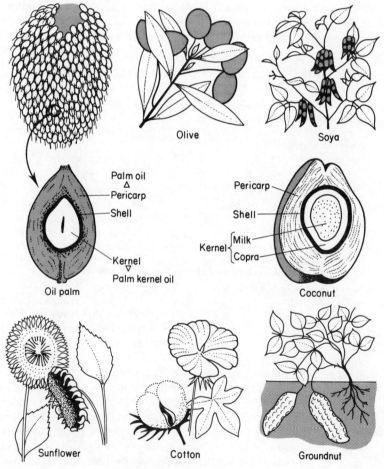

Figure 4.2 Important sources of vegetable oils.

The olive is also valuable as a source of oil, over a million tons of olive oil being produced annually. Olive oil is notable for the large proportion of oleic acid it contains (see Table 4.2) and for its purity. The finest olive oil does not need purification and is used for all manner of purposes in the Mediterranean countries, while in Britain it is used as a salad oil.

Refining of crude oils

Crude olive oil is exceptional in that it can be used for edible purposes without refining. Most vegetable oils, however, contain a number of impurities such as moisture, free fatty acids, colouring matter, resins, gums and sometimes vitamins. These impurities affect flavour, odour and clarity, and are removed during refining. The refining process is carried out in a number of stages which may be considered in turn.

1. *Degumming.* Crude oils often contain impurities in suspension which in the presence of water form gums. The impurities are removed by adding hot water to the warm oil, which is then transferred to a centrifugal separator. The separator revolves at a high speed and the gum particles, which have a higher density than the oil, are thrown to the bottom of the vessel, leaving an upper layer of clarified oil.

2. *Neutralizing.* Owing to spoilage all crude oils contain a small proportion of free fatty acids and low-grade oils may contain considerable quantities. The acids are removed by neutralizing the oil with a solution of caustic soda, which converts the fatty acid into an insoluble soap. The soap is then removed by allowing it to settle to the bottom of the neutralizing tanks. If the acid impurity is palmitic acid, for example, insoluble *sodium palmitate* is formed:

$$C_{15}H_{31}COOH + NaOH \rightarrow C_{15}H_{31}COONa\downarrow + H_2O$$

Palmitic acid   Sodium palmitate

3. *Washing and drying.* In order to remove the last traces of soap from the oil, it is washed with warm water. Two layers form and the lower water layer is run off, leaving the oil layer, which is then vacuum dried.

   In modern plants these separate stages are being replaced by a continuous automatic process in which the neutralizing stage is carried out very much more quickly in a centrifugal separator.

   The oil is now clear and free from acid, but it is usually yellowish in colour and still has a distinct odour. It is, therefore, bleached and deodorized.

4. *Bleaching.* The oil is warmed and *fuller's earth* and activated carbon are added. Both these materials have a large capacity for

adsorbing coloured matter. The mixture is stirred and a partial vacuum is maintained. When all the coloured matter has been adsorbed, the oil–earth mixture is passed through filter presses, from which the oil emerges as a clear, colourless liquid.

5. *Deodorizing*. The oil is heated under vacuum in a tall tank and steam is injected so that the liquid mixture is violently agitated. In one method it is sprayed upwards as an umbrella-shaped fountain, so that a large surface area of liquid is continually exposed, and the volatile odoriferous substances and remaining free fatty acids are stripped from the oil.

The oil is now pure and ready for use, or, as is usually the case, ready for blending. It is desirable that the oil should not come into contact with air once it has been refined, as this leads to deterioration due to oxidation. In some modern plants, therefore, the oil is stored under an inert atmosphere of nitrogen.

Hydrogenation of oils

Hydrogenation is the process whereby an oil is converted into a fat, i.e. by which it is hardened. This important process has resulted in a considerable increase in the use of hardened vegetable oils at the expense of animal fats which were once such a staple feature of the diet.

Hydrogenation is simply the addition of hydrogen to the double bonds of unsaturated fatty acids combined with glycerol in an oil. During hydrogenation one molecule of hydrogen is absorbed by each double bond:

$$-CH=CH- + H_2 \rightarrow -CH_2-CH_2-$$

The most commonly occurring unsaturated fatty acids found in combination with glycerol in vegetable oils, namely oleic, linoleic and linolenic acids, contain one, two and three double bonds respectively. As they all contain 18 carbon atoms, complete hydrogenation converts them all into stearic acid. Stearic acid has a much higher melting point ($70°C$) than any of the other three, so the hydrogenated oil is harder than the original. The equation below illustrates the conversion of liquid triolein into solid tristearin which takes place when a molecule of triolein absorbs three molecules of hydrogen.

$$
\begin{array}{ll}
CH_2OCOC_{17}H_{33} & CH_2OCOC_{17}H_{35} \\
| & | \\
CHOCOC_{17}H_{33} + 3H_2 \rightarrow & CHOCOC_{17}H_{35} \\
| & | \\
CH_2OCOC_{17}H_{33} & CH_2OCOC_{17}H_{35} \\
\text{Triolein} & \text{Tristearin}
\end{array}
$$

Hydrogenation only proceeds at a reasonably fast rate in the presence of a catalyst, finely divided nickel being used industrially. The nickel is usually made by reducing finely divided nickel carbonate or nickel formate. The catalyst is added in small quantities to the oil, which is contained in large closed steel vessels called converters, operating at a temperature of about 170°C and high pressure. The oil is stirred and hydrogen gas is pumped in. The oil is heated to start the reaction, but as the reaction is exothermic, further heating is not necessary. After hydrogenation the oil is cooled and filtered to remove the nickel which can be reused.

The way in which a catalyst affects a reaction was considered in Chapter 2 and it will be recalled that catalysts reduce the energy required for a reaction to proceed, i.e. the energy of activation; reference to Fig. 2.1 on page 16 will serve as a reminder of this. Nickel, like enzymes, catalyses a reaction by providing a surface upon which the reaction can take place, and converts a single-step high energy mechanism into one involving several low energy stages.

The first stage of hydrogenation is the adsorption of reactants, in this case hydrogen and oil, onto the nickel surface. Adsorption takes place only at certain preferred parts of the surface known as *active centres* and results in the adsorbed hydrogen and oil molecules being brought close to each other. Exchange of energy between the nickel and the reacting molecules weakens the internal bonds of the latter and activates them sufficiently to provide them with the necessary energy of activation. Reaction occurs, after which the hydrogenated oil molecules are desorbed, i.e. they leave the nickel surface which is then available to catalyse further reaction. These stages of hydrogenation are illustrated in Fig. 4.3.

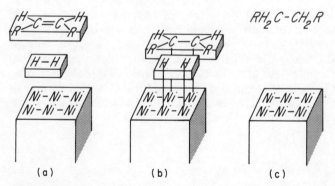

Figure 4.3   The catalytic action of nickel in the hydrogenation of an oil. (a) Oil and hydrogen before reaction. (b) Molecules adsorbed and activated on nickel surface. (c) Hydrogenated oil after reaction.

Surface catalysts, such as nickel, are easily poisoned by substances which are adsorbed in preference to the hydrogen and oil molecules. Carbon monoxide and sulphur compounds, if present in even minute quantities, poison the catalyst owing to their strong adsorption by the nickel. This means that the hydrogen gas, which is often produced from water gas (a mixture of carbon monoxide and hydrogen), must be most carefully purified before use and that the oil to be used must be carefully refined before hydrogenation.

Hydrogenation is a selective process, some triglycerides becoming saturated more rapidly than others. The most unsaturated triglycerides are partially hydrogenated before the less unsaturated ones react, so that in terms of the fatty acids combined in the triglycerides, more linolenic acid is converted into linoleic in a given time than linoleic into oleic. The relative rates of reaction of oleic, linoleic and linolenic are in the ratio 1:20:40. This fact enables the hydrogenation to be controlled, and food oils are only partially saturated with hydrogen.

Hydrogenation of oils not only converts unsaturated fatty acids into saturated ones, but it also converts naturally occurring *cis* forms of unsaturated fatty acids into their *trans* forms. Such *trans* forms are treated by the body in the same way as saturated fatty acids.

Careful control of the hydrogenation process is important. For health reasons it may be desirable to retain a degree of unsaturation in the partially hardened oil. In addition, complete hydrogenation would make fats too hard and lacking in plasticity for use in food preparation. Apart from its hardening effect, hydrogenation is advantageous in that it also bleaches the oil and increases its stability.

Rancidity

Oils and fats are liable to spoilage which results in the production of unpleasant odours and flavours; such spoilage is usually described by the general term *rancidity*. Different types of oil and fat show varying degrees of resistance to spoilage; thus most vegetable oils deteriorate only slowly whereas animal fats deteriorate more rapidly and marine oils, which contain a relatively high proportion of combined highly unsaturated fatty acids deteriorate so rapidly that they are useless for edible purposes unless they have been refined and hydrogenated.

Spoilage may occur in many ways, but two important types of rancidity may be distinguished, namely *hydrolytic rancidity* and *oxidative rancidity*.

HYDROLYTIC RANCIDITY   Hydrolytic rancidity occurs as a result of hydrolysis of triglyceride molecules to glycerol and free fatty acids and it is brought about by the presence of moisture in oils. The rate of hydrolysis in the presence of water alone is negligible, but it is hastened by the presence of enzymes and microorganisms. Oils and fats that have not been subjected to heat treatment may contain lipases which catalyse hydrolysis. They may also contain moulds, yeasts and bacteria present in the natural oil or they may become contaminated with them during processing. Such microorganisms hasten hydrolytic breakdown.

The nature of the unpleasant flavours and odours produced by hydrolysis depends upon the fatty acid composition of the triglycerides. If the triglycerides contain combined fatty acids of low molecular weight containing 4–14 carbon atoms, hydrolysis yields free acids having characteristically unpleasant odours and flavours. For example, hydrolysis of butter yields the rancid-smelling butyric acid, while palm kernel oil yields considerable amounts of lauric and myristic acids. Oils containing combined fatty acids with more than 14 carbon atoms are not liable to hydrolytic rancidity as the free acids are flavourless and odourless.

OXIDATIVE RANCIDITY   Oxidative rancidity is the most common and important type of rancidity and it results in the production of rancid or 'tallowy' flavours. It is caused by the reaction of unsaturated oils with oxygen and its occurrence does not depend, therefore, on the presence of impurities or moisture in an oil; it can consequently affect pure and refined oils. The actual mechanism of the oxidation is complex and not completely elucidated, but the main features are known and are as follows.

The oxidation of oils takes place by means of a *chain reaction* which is a type of reaction that is characterised by extreme speed. A chain reaction takes place in three stages known as *initiation*, *propagation* and *termination*. In the initiation stage, which is slow to occur, a hydrogen atom is removed from an unsaturated triglyceride molecule with the production of a free radical (R·). Free radicals, which are groups containing an 'unpaired' electron, are extremely unstable and immediately react with another molecule to form a more stable substance. This initiation stage only occurs under the influence of catalysts in the form of minute traces of metals, particularly copper, and in the presence of heat and light.

The slow initiation stage is followed by a fast propagation stage in which free radicals from the initiation stage combine with atmospheric oxygen to form an unstable peroxy free radical which reacts with molecules of unsaturated oil to form another free radical and an unstable hydroperoxide. It can be seen from the sequence shown below, that for every free radical R· used up another is generated with the result that the reaction is self-generating. The site of the reaction in the unsaturated oil is a methylene ($-CH_2-$) group adjacent to a double bond and we can represent such molecules by

RH, H being the hydrogen atom of the methylene group adjacent to a double bond.

Initiation   Initiator $\xrightarrow[\text{catalyst}]{\text{energy}}$ R·   (free radical)

Propagation ┌→R· + O_2 → RO_2·   (peroxy free radical)

│ RO_2· + RH   → R· + ROOH (hydroperoxide)

└────────────←────────┘

As the reaction proceeds hydroperoxide is continually formed and, being unstable, it breaks down to form ketones and aldehydes, which are responsible for the off-flavours of rancid fats. The reaction continues either until all the oxygen (or oil) is used or until the free radicals which are responsible for maintaining the reaction are removed. Many fats, particularly vegetable oils, contain natural substances, such as vitamin E, known as *antioxidants* which help to retard rancidity by reacting with peroxy free radicals (RO_2·) so that they are no longer available for the propagation stage. Synthetic antioxidants are also added to fats to control rancidity, a subject which is considered further in Chapter 16.

COLLOIDAL SYSTEMS

The physical nature of a solution is familiar enough, and an aqueous sugar solution, for example, is known to consist of sugar molecules dispersed through water, the whole system being homogeneous. If we consider what happens when we add to water a substance made up of relatively large molecules, such as a starch or a protein, we find that the system formed is *not* homogeneous but consists of two distinct parts or *phases*. The large molecules dispersed through water form one phase, known as the *disperse phase* and the water forms the other phase, known as the *continuous phase*, the complete system being described as *colloidal*.

A colloidal solution is usually called a *sol* and contains particles – consisting of single large molecules or groups of smaller molecules – that are intermediate in size between small molecules and visible particles. Such systems have properties that are intermediate between those of true solutions and suspensions of visible particles.

All types of colloidal systems are similar to sols in that they contain two distinct phases and in that the disperse phase contains particles intermediate in size between small molecules and visible particles. The disperse phase may be solid, liquid or gaseous, but in every case the properties of colloidal systems depend upon the very large surface area of the disperse phase. The types of colloidal systems which are important in food are summarized in Table 4.4, and will be considered in the following pages.

Table 4.4 Types of colloidal systems

Disperse phase	Continuous phase	Name	Examples
Solid	Liquid	Sol	Starch and proteins in water
Liquid	Liquid	Emulsion	Milk, mayonnaise
Gas	Liquid	Foam	Whipped cream, creamed fat, beaten egg-white
Gas	Solid	Solid foam	Ice-cream, bread
Liquid	Solid	Gel	Jellies, jam, starch paste
		Solid emulsion	Butter, margarine

Emulsions and emulsifying agents

When oil is added to water it forms a separate layer above the water; the oil and water do not dissolve in each other and are said to be *immiscible*. If oil and water are shaken vigorously the two liquids become dispersed in each other and an *emulsion* is said to be formed. Such an emulsion is unstable, however, and on standing it reverts to the original two layers. Emulsions are described as being either oil-in-water (o/w) or water-in-oil (w/o) emulsions. An o/w emulsion is one in which small oil droplets form the disperse phase and are dispersed through the water (Fig. 4.4), whereas a w/o emulsion is one in which small water droplets are dispersed through the oil.

Although food emulsions are described as being either o/w or w/o, these terms may be misleading in that the oil and water may contain other substances. Thus, in addition to triglycerides the oil phase may contain other lipids and fat-soluble materials and the water phase may, for example, be vinegar or milk. Table 4.5 gives details of some important food emulsions.

Why is it that once water drops have been dispersed in oil they come together again and form a continuous water layer? The answer is to be found in the fact that the action of dispersing water through oil in the form of drops increases the area of oil and water in contact. In order to do this, work must be done against the force of surface tension which causes a liquid to assume minimum surface area. The natural tendency is, therefore, for the water drops to coalesce, for by so doing the interfacial area is decreased and a more stable system is produced.

In order for oil and water to form a stable emulsion a third substance called an *emulsifying agent* or *emulsifier* must be present. Although the complete mechanism by which emulsifiers facilitate the

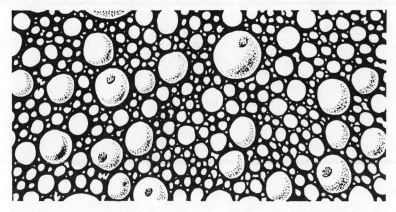

Figure 4.4 Diagram of an oil-in-water emulsion based on a photomicrograph. The diameter of the oil droplets (light shading) is in the range 10^{-4}–10^{-6} mm.

formation of stable emulsions is complex, variable and incompletely understood, an outline of the main factors can be given.

As the surface tension between oil and water, known as *interfacial tension*, is great, it is difficult for a stable emulsion to be formed. Emulsifiers lower interfacial tension by becoming adsorbed at the oil–water interface and forming a film one molecule thick round each droplet. The adsorbed film prevents the droplets from coalescing and, in some instances, may form a film which, by virtue of its mechanical strength, imparts stability. For example, protein emulsifiers are notable for the mechanical strength of the adsorption film which they produce. If an emulsifier contains charged groups the adsorption process gives rise to charged droplets which repel each other. Such droplets will not coalesce and this factor therefore promotes emulsion stability.

Emulsifying agents are substances whose molecules contain both a *hydrophilic* or 'water-loving' group and a *hydrophobic* or 'water-hating' group. The hydrophilic group is polar and is attracted to the

Table 4.5 Examples of food emulsions

Example	Type	Main emulsifiers present or added
Milk, cream	o/w	Proteins (casein)
Butter	w/o	Proteins (casein)
Mayonnaise, salad cream	o/w	Egg-yolk (lecithin), GMS, mustard
Margarine	w/o	Proteins (casein), lecithin, GMS
Ice-cream	o/w	Proteins (casein), GMS, plus stabilizers (gelatin, gums, alginates)

water while the non-polar hydrophobic group, which is frequently a long chain hydrocarbon group, is attracted to the oil. Thus in a w/o emulsion the emulsifier is adsorbed in such a way that the polar 'heads' of the emulsifier molecules are in the water and the non-polar 'tails' stick out into the oil as shown in Fig. 4.5.

The type of emulsion formed by an oil–water system depends upon a number of factors including the composition of the oil and water phases, the chemical nature of the emulsifying agent and the proportions of oil and water present. If the polar group of an emulsifier is more effectively adsorbed than the non-polar group, adsorption by the water is greater than by the oil. The extent of adsorption at a liquid surface depends upon the surface area of liquid available and increased adsorption of emulsifier by water is favoured by the oil–water interface becoming convex towards the water, thus giving an o/w emulsion.

The relative proportions of oil and water also help to determine whether an o/w or w/o emulsion is formed. If more oil than water is present the water tends to form droplets and a w/o emulsion is formed. On the other hand if more water than oil is present an o/w emulsion is favoured.

Many substances show some activity as emulsifying agents and among naturally occurring ones phospholipids, proteins and complex carbohydrates such as gums, pectins and starches are important. Natural food emulsions are often stabilized by proteins and milk, for example, is stabilized by the casein and other proteins present. Artificial emulsifiers are added during the preparation of many emulsions, though in Britain only those on a permitted list (see p. 377)

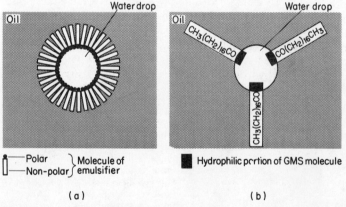

Figure 4.5   (a) Molecules of emulsifier adsorbed at a water–oil interface forming a complete protective film round a water droplet. (b) The adsorption of glyceryl monostearate in a w/o emulsion.

may be used. Of these *glyceryl monostearate* (GMS) is the most important and will serve as an example.

GMS is a monoglyceride which is formed when one hydroxyl group of glycerol is esterified with stearic acid as shown in the following equation.

$$
\begin{array}{lll}
CH_2OH & CH_2OH & \\
| & | & \\
CHOH + CH_3(CH_2)_{16}COOH \longrightarrow CHOH & + H_2O \\
| & | & \\
CH_2OH & CH_2O- \boxed{CO(CH_2)_{16}CH_3} &
\end{array}
$$

Hydrophobic portion
of glyceryl monostearate

One part of the GMS molecule is hydrophilic because it contains hydroxyl groups and the rest of the molecule, as indicated in the formula, is hydrophobic. When GMS is added to a w/o emulsion, the hydrophilic parts of the molecules are adsorbed onto the surface of the water droplets and the lipophilic parts are adsorbed onto the surface of the oil round the drops, as illustrated in Fig. 4.5b.

Commercial GMS is not a single substance and in addition to glyceryl monostearate it contains some di- and triglycerides. It is widely used in food manufacture and is added, for example, to margarine, mayonnaise, salad dressing and ice-cream.

The phospholipid *lecithin* is an important natural emulsifier which favours o/w emulsions and which is present in egg-yolk and many crude oils, particularly vegetable oils. Lecithin is extracted from vegetable oils, notably soya bean oil, and added to some manufactured products.

Sometimes *stabilizers* are added to products in addition to emulsifiers, their function being to maintain an emulsion once it has been formed. Such substances improve the stability of emulsions mainly by increasing their viscosity. As viscosity increases, the freedom of movement of the dispersed droplets of the emulsion is reduced and this lessens the chance of their coming into contact and coalescing. Stabilizers are high molecular weight compounds, usually proteins such as gelatin, or complex carbohydrates, such as pectins, starches, alginates and gums. For example, starch or flour may be added to thicken gravies, sauces and salad cream and several stabilizers, such as gelatin and gums, are added to ice-cream.

Uses of emulsifying agents

A number of natural foods are emulsions (milk is the prime example) and are stabilized by emulsifiers which are present as constituents of the food. Here we are concerned, however, not with natural foods but with manufactured foods to which emulsifiers are added. Emulsifiers are added to several products containing fat, such as margarine, cooking fats, salad dressings and ice-cream.

MAYONNAISE AND SALAD CREAM   Emulsifiers play an important part in making salad cream and mayonnaise, which are o/w emulsions. The term salad cream means, according to the legal definition, 'any smooth, thick stable emulsion of vegetable oil, water, egg or egg-yolk and an acidifying agent with or without the addition of one or more of the following substances, namely vinegar, lemon juice, salt, spices, sugar, milk, milk products, mustard, edible starch, edible gums and other minor ingredients and permitted additives'. The minimum proportions of vegetable oil and egg yolk solids allowed in Britain are 25 per cent and 1.35 per cent respectively.

In Britain the legal standards for mayonnaise are the same as those for salad cream and this gives rise to the confusing situation that two products with different names can be identical. In practice, mayonnaise is normally thicker than salad cream and contains a higher proportion of both oil and egg-yolk (and thus less carbohydrate and water). Indeed, in many countries the oil content of mayonnaise must be greater than that of salad cream. In America, for example, mayonnaise must contain at least 65 per cent oil (compared with 30 per cent for salad cream) and in certain countries as much as 80 per cent is required.

The finest oil for making salad cream is undoubtedly olive oil though, because of its high cost, other vegetable oils are normally used. Such oils must be of high purity, light colour and bland flavour and they are therefore refined, bleached and deodorized before use. As they must also be liquid, vegetable oils such as groundnut oil, soya bean oil, cottonseed oil and maize oil are employed.

Salad cream should be viscous and have a creamy consistency, which can only be achieved if the o/w emulsion is stable. In order to produce such a product emulsifying agents must be present, the chief one being lecithin contained in egg-yolk. In addition, mustard and often GMS are added and these are both effective emulsifiers. The stability of the emulsion formed is increased by the addition of stabilizers which increase viscosity. Such an increase in viscosity becomes more important the smaller the oil content, and is brought about by the addition of starches and gums.

ICE-CREAM   Present day ice-cream is one of the triumphs of food technology and is noteworthy in that air is a major ingredient. Without air ice-cream would simply be a frozen milk ice, but with air it becomes a highly complex colloidal system. It consists of a solid foam of air cells surrounded by emulsified fat together with a network of minute ice crystals that are surrounded by watery liquid in the form of a sol.

Ice-cream is made from fat, non-fat milk solids, sugar, emulsifiers, stabilizers, flavourings and colour. A typical ice-cream contains 12 per cent fat, 11 per cent non-fat milk solids, 15 per cent sugar and about one per cent minor ingredients, the rest being water. During processing air is incorporated and accounts for about half the volume of the final product; as ice-cream is sold by volume this latter point is not unimportant. In Britain ice-cream must contain a minimum of five per cent fat and $7\frac{1}{2}$ per cent non-fat milk solids and to bc classified as a 'dairy' ice-cream all the fat must be milk fat. 'Non-dairy' ice-cream contains suitable vegetable fats, such as hydrogenated coconut oil.

In the manufacture of ice-cream the ingredients are mixed together, pasteurized (see p. 354) and homogenized. The latter treatment, together with the added emulsifiers, produces a stable o/w emulsion. The emulsion is cooled to a temperature which partially freezes the mixture; at the same time air is whipped in. A solid film forms on the walls of the freezer and this is continuously scraped off. The temperature is subsequently reduced further and this causes the rest of the water to freeze and the product to harden.

Fat plays an important role in determining the texture of ice-cream, and also flavour if milk fat is used. Fat converts the ice-cream mix into an o/w emulsion and when air is incorporated into the mixture it aids the formation and stabilization of a foam by forming a stable fat film round the air bubbles. The main components of the aqueous phase of the emulsion are sugar and non-fat solids. The latter are important because they contain proteins which are natural emulsifiers in ice-cream. During pasteurization the milk proteins become denatured (see p. 169) and form a solid film of considerable strength around the oil droplets. This prevents the oil droplets from coalescing and thus stabilizes the emulsion.

An important function of synthetic stabilizers added to ice-cream is to increase viscosity rather than aid emulsification. Gelatin, gums and alginates are all used and help to promote the firm texture, smooth taste and good-keeping qualities associated with modern ice-cream. These stabilizers also assist the formation of a weak network of hydrated molecules through the ice-cream and this gives a firm-textured product which is slow-melting and resistant to the formation of large ice crystals. The use of gelatin exemplifies this.

Gelatin molecules are relatively large, have a thread-like shape and are hydrophilic. When gelatin is added to the ice-cream mix its long, thin molecules are dispersed through the emulsion, and water molecules are attracted and held at the large surface area of gelatin exposed. In this way the water molecules lose their freedom of movement, and melting, which occurs when the freedom of movement of molecules suddenly increases, is made more difficult. The gelatin molecules, by forming a three-dimensional mesh or *gel*, also give added firmness to the structure.

When an ice-cream emulsion is frozen, ice crystals are formed. If ice-cream is to give a sensation of smoothness when it melts in the mouth the crystals must be very small. Because gelatin molecules are dispersed through the emulsion and because of their hydrophilic nature, they ensure that the water molecules are also dispersed. In this way the formation of large ice crystals, which can only occur when large numbers of water molecules come together, is prevented.

Margarine

Margarine is a manufactured food which was invented by a French scientist, Mège-Mouriés, in 1869. The nineteenth century saw a rapid rise in the population of Europe so that the increasing demand for the two most common fats − butter and lard − soon out-stripped their production. Mège-Mouriés' intention was to make a fat that would resemble butter as closely as possible. His recipe was extremely odd; he obtained an oil called *oleo oil* from beef tallow and mixed the warm oil with chopped cow's udder, milk and water and stirred them until a solid mixture called *oleo margarine* was obtained.

Today's margarine is, happily, made by a very different process to that employed by Mèges-Mouriés. His method was based on animal fat but as such fat became increasingly scarce it was necessary to find an alternative source for making margarine. Although vegetable and fish oils were available these were liquids and the problem was how to turn them into solid fats resembling butter. The problem was solved by the invention of hydrogenation at the beginning of the twentieth century.

Margarine is now made from a water-in-oil emulsion, the aqueous phase being fat-free milk and the oil phase being a blend of different oils. The two phases are mixed together and, with the aid of suitable emulsifiers, a stable emulsion is formed. The emulsion is processed until it forms a solid product having the desired consistency.

1. *The oil blend.* Several different oils are blended together in preparing the oil phase. The blend may include animal, vegetable and marine oils, the actual oils being selected at any time depending upon cost and availability as well as the type of product it is desired to make. Today the major vegetable oils used include those derived from soya, sunflower, palm, rapeseed, safflower, maize, cotton seed, coconut and groundnut. Marine oils include whale oil while animal fats include beef fat.

It is important that the oil phase should have a bland taste and a wide plastic range; to achieve the former oils are carefully refined and to achieve the latter some are hydrogenated. It has already been pointed out that in order to have suitable plasticity both liquid and solid triglycerides must be present. In margarine the desired liquid:solid ratio is obtained by selective hydrogenation of the oils used. After hydrogenation oils are refined again as shown in Fig. 4.6 and pass to the blending tank in which they are heated until they are all liquid.

2. *The aqueous blend* The skimmed milk is matured after pasteurization by adding a 'starter' in the form of lactic acid bacteria, and the ripening and souring is allowed to continue until the desired flavour is produced. Small amounts of other materials are added to

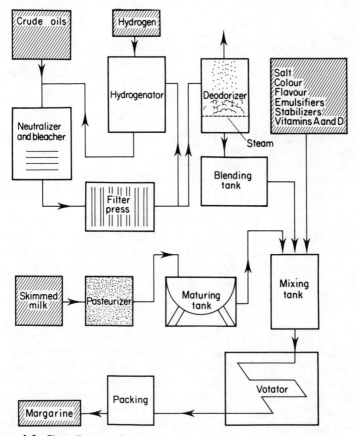

Figure 4.6 Flow diagram of margarine production.

the ripened milk and these have an important influence on the nature of the final product. Artificial flavouring and colouring agents, vitamin A and D and salt are all added.

3. *Emulsification.* The oil and aqueous phases together with emulsifying agents, such as lecithin and GMS, are now mixed together in large cylindrical tanks. These are fitted with two sets of paddles which rotate in oposite directions and mix the fluids until they form a stable emulsion, which has the appearance of a thick cream.

4. *Processing the emulsion.* The emulsion is now passed on to a roller which slowly revolves and which is cooled internally. When the emulsion comes into contact with the cold surface of the roller it is rapidly cooled and converted into a solid film. The film is scraped off the roller, and after being stored for about a day it is worked until it gains a smooth even texture. This involves breaking up the structure of the flakes of fat, by passing it through rollers and by kneading it.

In the more modern continuous process, using a machine known as a 'votator', the emulsifying and processing stages follow one another directly and take place in closed machines. This enables a great saving in time to be made, and prevents the margarine from coming into contact with air.

All that remains to be done is to pack the margarine by weighing it into blocks, wrapping and labelling it. The label must state that the product is margarine and give the vitamin content and the butter content (if butter has been added).

The term margarine no longer describes a single product but a whole range of products that provide a wide variety of different blends of oils and flavours to meet every need and taste. Hard margarines are available as an alternative to butter for table use and also for baking. Soft margarines are available in abundance resulting from the use of different oil blends including those, such as sunflower oil brands, which are designed to be high in PUFA (and low in cholesterol).

Early varieties of margarine contained oils and fats that had been completely hydrogenated and as a result contained no combined essential fatty acids (EFA). Modern methods of manufacture, involving careful selection of vegetable oils rich in PUFA including linoleic acid, and selective hydrogenation that allows some oils rich in PUFA to escape hydrogenation, produce margarines containing EFA.

Margarine produced by early methods was significantly inferior to butter in its vitamin content but all table margarine now manufactured contains added vitamin A and D. The vitamin content is controlled by law so that it contains 800–1000 μg vitamin A/100 g (as retinol equivalent) and 7–9 μg vitamin D/100 g. This means that the vitamin content of margarine is equal to that of summer butter.

A typical margarine contains 81 per cent fat and has an energy value of about 3000 kJ/100 g. This compares with an average value of 82 per cent fat for butter and an energy value of 3140 kJ/100 g. The water content of margarine is controlled by law and it must not exceed 16 per cent.

LOW FAT SPREADS   Owing to the considerable body of opinion which believes that we should, for health reasons, reduce our intake of fat, a number of low fat spreads is now available. Low fat and reduced fat spreads contain 40–80 per cent fat compared with margarine which must contain a minimum of 80 per cent fat. Such products may be based either on butter or margarine. They cannot be called margarine because they do not comply with the legal requirements for that product.

Consumption of margarine and low fat spreads in the UK now accounts for 70 per cent of total solid fats eaten and butter for the remaining 30 per cent.

Cooking fats

Cooking fats, or *shortenings* as they are sometimes called, differ from margarine in that they are pure fat products rather than solid emulsions. They are made from an oil blend which may contain animal, vegetable and marine oils. The actual blend will depend on the nature of the product required and on the availability and cost of suitable oils. The oil blend is partially hydrogenated so as to give a product having the required plasticity and after refining and blending, the fat blend is cooled and processed in a 'votator'-type of machine similar to that used in making margarine. After the oil blend has been cooled down it is a near-white solid but, during processing, air is sometimes incorporated and it is transformed into a pure white thick creamy liquid. While in a liquid condition it may be forced under pressure through a texturizing valve which ensures that on cooling the product sets into a smooth-textured soft mass.

SUPERGLYCERINATED FATS AND THEIR USE IN CAKE MAKING   In commercial practice cooking fats are often required for specific purposes, such as cake making, and these incorporate an emulsifier. Such fats are called *superglycerinated* or *high ratio* fats. The emulsifying agent most frequently used is glyceryl monostearate. The efficiency of such a fat is measured in terms of its creaming, emulsifying and shortening powers. The use of high ratio fats allows the use of a higher ratio of sugar to flour than with ordinary cooking fats.

The creaming power of a fat is measured by its capacity for incorporating air bubbles when it is beaten up. This power is dependent upon the plastic properties of the fat which enable it to entrap air bubbles within its structure without loss of mechanical strength, which would certainly occur if it were in a liquid condition. In the first stage of making a rich type of cake, fat is warmed to increase its plasticity and then the softened fat is creamed with sugar by beating the two together until sufficient air becomes trapped in the mixture. This entrapped air assists the action of baking powder in the aeration of the cake during baking and has an important influence on the volume and evenness of texture of the final product.

When creaming of the fat and sugar is complete, eggs are usually beaten into the mixture and mixing is continued until the whole is light and foamy. The result is an emulsion, usually of the o/w type, because egg-yolk tends to emulsify fat as an o/w emulsion. The emulsion is stabilized by the lecithin of the egg-yolk and the GMS of the superglycerinated fat. The presence of the GMS greatly increases the stability of the emulsion and enables a large amount of water to be emulsified with the fat. As the amount of sugar that can be used depends on how much water there is to dissolve it, the use of high ratio fats enables cakes of high sugar and moisture content to be produced. Looking at it in another way, a cake having a certain sugar and moisture content can be produced using less of a superglycerinated fat than of a normal cooking fat. This has led to the term 'fat extender' being used to describe such a fat.

After eggs have been beaten into the mixture, flour and baking powder (and milk if it is used) are added. A foam is created, each tiny air bubble being surrounded by an oil film which, at this stage, is not very strong. When the cake mixture is put into the oven, the air and water vapour trapped within the bubbles expand due to the rise in temperature. If the cake is to rise evenly during baking to give it the desired lightness of texture, the foam must be stable. Rupture of the oil films will produce the familiar 'sad' or 'sinking' effect. The protein of the egg-white acts both as a *foaming* agent and a foam *stabilizer*. The protein becomes adsorbed onto the interface between the air and the oil film, and as the temperature rises and the bubble expands, the protein stiffens, or coagulates, so forming a rigid wall around the bubble. This prevents the bubble from bursting and also prevents the production of very big bubbles which would spoil the evenness of the texture. If the ratio of fat to egg is too high, the foam structure is weak and during baking some bubbles break and the cake sinks.

As already indicated, a fat must not only act as a creaming and emulsifying agent but also as a 'shortener'. This function is of primary importance in the production of biscuits and shortbread but is also an important factor in cake making. Fat coats the starch and gluten of the flour with an oily film so breaking up the structure and preventing the formation of a tough mass. This leads to a cake which has a tender and 'short' crumb, whereas the use of too little fat produces a cake which is tough and of poor keeping quality. The greater the proportion of fat in the mixture the greater the shortening effect.

If a cooking fat is to be an effective shortening agent it must have good plasticity because this enables it to spread over a large area of flour, coating the surface with a film of oil. Such a fat must be neither too hard, in which case it will have poor spreading power, nor too liquid as with an oil, in which case it tends to form globules rather than a film. In other words, a soft fat is required; this may be achieved by suitable blending of the oils used in manufacture. Blends which contain a high proportion of PUFA have better shortening power than those composed of mainly saturated fats. The presence of monoglycerides, such as the GMS added to superglycerinated fats, and diglycerides, also improves shortening power.

CHAPTER 5

Dairy products

Dairy products constitute an important group of foods, as reference to Table 5.1 shows. Apart from the nutrients mentioned in the table, they also make an important contribution to vitamin intake. In the UK dairy products provide 23 per cent of the thiamine, 40 per cent of the riboflavin and 14 per cent of the nicotinic acid in an average diet. Although traditionally dairy products have played an important role in the British diet, their reputation has suffered recently on account of doubts expressed about the quantity and nature of fats in the diet. The animal fats and cholesterol present in dairy products have both been linked with modern 'diseases of affluence', a subject which is discussed later in this chapter.

Milk

Milk is a food of outstanding interest. In the first place, it is designed by nature to be a complete food for very young animals. Its extremely high nutritional value is a consequence of this and cow's milk – which is the only type to be discussed – is not only a complete food for young calves but is also an excellent food for babies and young children and a valuable food for adults. In the second place, milk is an interesting and complex colloidal system, the properties of which are of great practical importance in making butter and cheese and in other ways of processing milk. The complexity of this colloidal system can

Table 5.1 Per cent contribution of dairy products to diet in the UK

	Liquid milk	Other milks and cream	Cheese	Butter	Total dairy products
Energy	10	2	3	5	19
Protein	15	2	6	–	23
Fat	12	2	5	11	30
Carbohydrate	6	1	–	–	7
Calcium	37	5	14	< 1	56

be judged from the fact that, in spite of much research, there is still much that is not fully understood.

Milk is an oil-in-water emulsion containing 3.5–4 per cent fat. In addition to milk fat the fat phase contains fat-soluble vitamins, phospholipids, carotenoids and cholesterol while the aqueous phase contains proteins, mineral salts, sugar (lactose) and water-soluble vitamins. The composition of different specimens of milk may show some variation with such factors as breed of cow, the nature of its food and the season of year. The figures which are given in Table 5.2 are, therefore, average values which refer to fresh summer milk. Winter milk contains only about two-thirds as much vitamin A as summer milk. The table also shows the percentage contribution made to the nutritional allowances of a man and a child by one pint of fresh milk each day, assuming that proteins contribute to energy. The figures given indicate the importance of milk as a source of calcium and riboflavin; they also show why milk is regarded as such a valuable food, for it makes contributions to every class of nutrient. The main nutrients in which fresh milk is deficient are iron, nicotinic acid and vitamin D, while milk as received by the consumer may also be deficient in ascorbic acid.

THE FAT OF MILK is in the form of minute droplets, most of which have diameters of 5–10 μm (1 μm is 10^{-4} cm). The oil droplets are so small that a single drop of milk contains several million of them and the fact that milk is so highly emulsified makes it particularly easy to

Table 5.2 Composition of fresh summer milk and its contribution to the diet

Nutrient	Amount in 100 g milk	Per cent contributions to nutrient allowances by 1 pint of milk daily	
		Man with daily expenditure of 12.6 MJ	Girl 3–4 years with daily expenditure of 6.2 MJ
Energy	272 kJ	13	25
Protein	3.3 g	22	34
Carbohydrate	4.7 g	4	8
Fat	3.8 g	6	12
Water	88 g	–	–
Calcium	103 mg	86	69
Iron	0.1 mg	1	2
Vitamin A	56 μg	17	29
Thiamine	50 μg	17	33
Niacin	90 μg	4	9
Riboflavin	170 μg	48	95
Ascorbic acid	1.5 mg	55	75
Vitamin D	0.1 μg	–	3 (max)

digest: it is digested more easily than any other fat. Milk fresh from the cow contains uniformly distributed oil droplets but when it is allowed to stand the oil droplets, being lighter than the aqueous phase, tend to rise to the surface and form a layer of cream. As the oil droplets rise they coalesce and form larger droplets but the emulsion does not break down.

Milk may be *homogenized* by forcing it through a small hole under pressure. This breaks up the oil droplets and reduces their size to 1–2 microns. This may not sound like a dramatic reduction in size but it increases the number of droplets by a factor of about 500 and results in a tremendous increase in surface area of fat. Such treatment has a considerable effect on the properties of milk; for example, it prevents the cream from separating out and it gives the milk a higher viscosity and richer taste. Homogenized milk coagulates more easily and has greater proneness to off-flavours and rancidity than unhomogenized milk.

Milk, whether homogenized or not, is a stable emulsion and the fat–water interface is stabilized by adsorbed natural emulsifiers present in the milk. The main emulsifier is protein which is adsorbed around each oil droplet forming a protective monolayer, but other emulsifiers such as phospholipids (e.g., lecithin) and vitamin A also play a part.

The triglycerides of milk fat contain many different combined fatty acids. However, relatively few are present in significant amounts, chief of which are the saturated fatty acids shown in Table 5.3. Milk fat contains a very low proportion of PUFA – less than three per cent.

Table 5.3　The more important fatty acids in cow's milk

Fatty acids		g fatty acid/100 g total fatty acids
Saturated		61.1
$C_{4:0}$	Butyric	3.2
$C_{12:0}$	Lauric	3.5
$C_{14:0}$	Myristic	11.2
$C_{16:0}$	Palmitic	26.0
$C_{18:0}$	Stearic	11.2
	Others	6.0
Monounsaturated		31.9
$C_{18:1}$	Oleic	27.8
	Others	4.1
Polyunsaturated		2.9
$C_{18:2}$	Linoleic	1.4
$C_{18:3}$	Linolenic	1.5

THE PROTEINS OF MILK consist of molecules which are relatively speaking so large that single molecules constitute colloidal particles dispersed through the aqueous phase of the emulsion, i.e. as a sol. The most important proteins in milk are *casein* (2.6 per cent), which is precipitated under acid conditions and *lactalbumin* (0.12 per cent) and *lactoglobulin* (0.3 per cent) which are both whey proteins that remain in solution after acidification (see cheese making, p. 79).

Casein is not a single substance, but a family of phosphorus-containing proteins that bind the calcium and other minerals present. The colloidal particles of casein are stabilized by a positive charge owing to the presence of bound calcium and magnesium ions. The charged particles are sensitive to changes of pH and to changes in concentration of surrounding ions. For example, during digestion milk becomes solid owing to the coagulation or 'clotting' of casein. This is brought about by the enzyme *rennin* which, at the low pH prevailing in the stomach, converts casein into a coagulated form. The coagulated casein reacts with calcium ions to give a three-dimensional gel which is in the form of a tough clot called *calcium caseinate*.

The making of junket is also an example of clotting. When milk is warmed to body temperature (37°C) and *rennet* added, it slowly coagulates into a white solid and exudes a slightly yellow liquid called *whey*. The change in the milk which occurs on making junket is the same as that which occurs in the stomach. Rennet is obtained from a calf's stomach but its essential constituent is rennin which is responsible for the clotting of milk.

Lactalbumin and lactoglobulin are not coagulated by rennin but they are more easily coagulated by heat than casein. Thus, when milk is heated, lactalbumin and lactoglobulin coagulate and form a skin on the milk surface. This skin is responsible for the way in which milk so easily 'boils over', for expanding bubbles are trapped beneath the skin and build up a pressure which eventually lifts the scum and allows the milk to boil over the sides of the pan.

THE CARBOHYDRATE OF MILK consists of the disaccharide lactose, also called milk sugar. Lactose is the only sugar manufactured by mammals and is distinguished by its lack of sweetness compared with other sugars.

As is well known, milk readily becomes sour when it is stored. This is because milk contains bacteria, called *lactic bacilli*, in which are enzymes that bring about the breakdown of lactose into the sour-tasting lactic acid:

$$C_{12}H_{22}O_{11} + H_2O \xrightarrow{\text{lactic bacilli}} 4CH_3CH(OH)COOH$$

Lactose $\qquad\qquad\qquad\qquad$ Lactic acid

The pH of fresh milk is 6.4–6.7, the value being maintained within this narrow range by proteins, phosphate and citrate which act as buffers. As milk turns sour the pH drops and when it reaches 5.2 the milk *curdles* and the casein is precipitated in the form of flocculent curds. It will be noted that curdling and clotting are not the same chemically, for in curdling casein is merely precipitated, while in clotting a tough mass of calcium caseinate is formed. Although milk which has been kept for a period curdles naturally because of the presence of lactic acid, any acid will produce the same effect. Curdling is hastened by warmth and it is for this reason that care must be taken when preparing such dishes as tomato soup where the acidity of the tomato juice may be sufficient to curdle the hot milk. If conditions are sufficiently acid, curdling may occur in the cold, for example, when milk is added to acid fruit such as rhubarb.

THE MINERAL ELEMENTS OF MILK are either in the form of mineral salts or occur as constituents of the organic nutrients. Some are in solution and some are colloidally dispersed either as sol particles or combined with protein. As milk is the sole food of a young calf it contains all the mineral elements required by the animal for growth. It is particularly rich in calcium and phosphorus, both of which are needed to build bone and teeth. These mineral elements are found in milk combined together in the form of calcium phosphate, which is not soluble but is held in suspension in the form of fine particles. Combined phosphorus is also found in casein and in phospholipids such as lecithin.

Many other mineral elements are present in small quantities; chlorine and iodine, for example, occur as soluble chlorides and iodides. Milk is an important source of iodine because in winter it is added to cattle feed to prevent goitre and stillbirth. Winter milk contains about 40 μg iodine per 100 g milk while in summer, when cattle are grazing, milk contains only about 5 μg iodine per 100 g milk. Iron is present in milk, but in such small quantities as to make it the one important mineral element in which milk is seriously deficient for human nutrition.

THE VITAMINS OF MILK are found either dissolved in fat globules or in aqueous solution. Vitamins A and D are both found in milk fat, the former in appreciable amounts, the latter in very much smaller quantities. The amount of both vitamins present in milk is greatest in summer when the cows are feeding on grass and receiving what sunshine our summers afford. The vitamin A value of milk is in part due to carotene and it is this substance which gives to milk its creamy colour.

Milk is a valuable source of riboflavin (see Table 5.2). It also contains useful amounts of thiamine and ascorbic acid and a small amount of niacin. The actual amounts of these vitamins in milk when it reaches the consumer depends upon the treatment it has received. Both ascorbic acid and thiamine are destroyed by heat treatment, while exposure to light destroys ascorbic acid and riboflavin.

TYPES OF FRESH MILK   The nutrient composition of whole milk is given in Table 5.2. By law it must contain at least three per cent fat. Such 'ordinary' milk is distinguished from milk from Jersey, Guernsey and South Devon breeds of cow, which has a higher fat content (an average of 4.8 per cent) than 'ordinary' milk. By law such milk must contain at least four per cent fat.

Average milk consumption in Britain is now about half a pint per day, and an increasing proportion is in the form of skimmed or semi-skimmed milks from which a proportion of fat has been removed. The nutritional value of skimmed milk is as good as that of whole milk, except for the loss of fat and the fat-soluble vitamins A and D which occurs when the cream is skimmed off. By law skimmed milk must contain less than 0.3 per cent fat; in practice it contains, on average, only 0.1 per cent fat. Semi-skimmed milk must by law contain 1.5–1.8 per cent fat. The use of skimmed or semi-skimmed milk is helpful for those who wish to reduce their fat intake but such milks are unsuitable, because of their reduced vitamin content, for babies and children under five.

Cream

Cream, like milk, is an oil-in-water emulsion. The way in which it is manufactured from milk is illustrated in Fig. 5.1. The milk is first heated to $50°C$. This makes easier subsequent separation by centrifugal action into an upper cream layer and a lower layer of skimmed milk. The fat content of the cream produced can be adjusted by means of a pressure valve up to a maximum of 70 per cent fat.

The fat content of different creams is regulated by law, the minimum values for fat content being as follows: half cream, 12 per cent; single cream, 18 per cent; whipping cream, 35 per cent; double cream, 48 per cent; clotted cream, 55 per cent. The nutrient content of different types of cream is shown in Table 5.4.

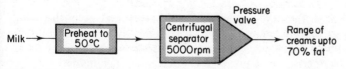

Figure 5.1 The manufacture of cream.

Table 5.4   The nutrient composition of creams per 100 g

Type	Energy (kJ)	Protein (g)	Fat (g)	Carbohydrate (g)	Sodium (mg)	Calcium (mg)
Half cream	568	2.8	12.3	4.1	55	96
Single cream	813	2.6	19.1	3.9	50	91
Whipping cream	1536	2.0	39.3	3.0	42	62
Double cream	1847	1.7	48.0	2.6	39	50

Cream can be heat treated in a number of different ways including pasteurization, sterilization or ultra heat treatment. Pasteurized frozen creams and UHT aerosol creams are available.

Yoghurt

Yoghurt is made from heat treated, homogenized milk which is inoculated with a culture containing equal amounts of *Streptococcus thermophilus* and *Lactobacillus bulgaricus* bacteria. The essential change produced by these bacteria is that the lactose in the milk is converted into lactic acid. In the initial stages of the fermentation the streptococci are most active, converting lactose into lactic acid and also producing *diacetyl* until the acidity increases to pH 5.5. Thereafter the *lactobacillus* continue the production of lactic acid until the acidity increases further to pH 3.7–4.3. At the same time acetaldehyde is produced and this gives yoghurt its characteristic flavour.

The fermentation process is continued until 0.8–1.8 per cent lactic acid is present and the product thickens. At this stage, if desired, flavours or fruit and sugar may be added. All such yoghurt contains live bacteria unless it is heat treated after fermentation, in which case this will be mentioned on the carton label.

The nutrient content of various yoghurts is shown in Table 5.5. Yoghurt is a nutritious and easily digested food though it does not have the somewhat magical health properties suggested for it at the beginning of this century and still believed by some people. In general the nutrient value of yoghurt is that of the milk it contains and of any substances added to it during manufacture.

Butter

Butter, no less than milk, is a complex colloidal system and although the conversion of cream into butter is an ancient art, complete understanding of the mechanism of the process is still wanting. Butter is made from cream by churning. The cream used contains 35–42 per cent milk fat and may be used fresh or allowed to go sour, a process

Table 5.5 Nutrient content of low-fat yoghurts per 100 g

Nutrient	Plain	Flavoured	Fruit
Water (g)	85.7	79.0	74.9
Lactose (g)	4.6	4.8	3.3
Other sugars (g)	1.6a	9.2b	14.6b
Protein (g)	5.0	5.0	4.8
Fat (g)	1.0	0.9	1.0
Minerals (g)	0.8	0.8	0.8
Energy (kJ)	216	342	405

a = galactose   b = mainly sucrose

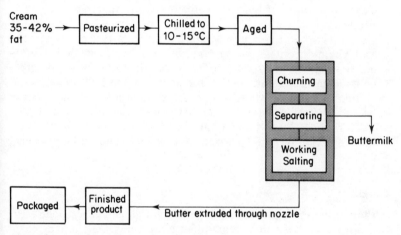

Figure 5.2 Flow diagram of butter production.

known as *ripening*, in which lactose is converted into lactic acid and flavour developed. After pasteurization the cream is chilled and agitated or churned. During churning the cream becomes increasingly viscous and eventually granules of solid butter appear. In modern creameries churning is carried out in a continuous buttermaker as shown in Fig. 5.2.

The conversion of cream into butter involves 'breaking' the o/w cream emulsion and turning it into a w/o emulsion, a process which is known as *inversion*. The detailed mechanism of this change is not known but a simple outline can be given. At an early stage in churning air becomes trapped in the cream emulsion and a foam is formed. As already noted, natural emulsifiers present in cream stabilize the emulsion by being adsorbed at the oil–air interface. When air is trapped in the emulsion these emulsifiers become desorbed and spread out onto the surface of the air bubbles. The oil droplets lose their stability and coalesce into larger drops. Continued churning brings about collapse of the foam structure, and the coalescing fat particles appear as visible granules and separate out from the aqueous phase.

Although both o/w emulsion and foam structures have collapsed, further churning brings about dispersion of a small amount of water in the fat and a w/o emulsion is formed. The nature of the final product is affected by the proportion of solid to liquid triglycerides present in the fat phase. Some liquid oil droplets are dispersed in the continuous fat phase, though continued churning reduces the proportion of such dispersed oil. If the temperature of churning is lowered the proportion of crystalline triglycerides in the fat increases and a correspondingly harder butter is formed.

The composition of butter is variable. An average butter contains the following: 82 per cent fat, 0.4 per cent protein, 15 per cent water, 2 per cent salt, and vitamins A and D, particularly the former. The actual vitamin content varies considerably, being much higher in summer when the cows feed on grass than in winter when no fresh food is available. Summer butter may contain up to 1300 μg vitamin A (as retinol equivalent) whereas the average vitamin A content of butter is about 1000 μg per 100 g. Summer butter contains about 0.8 μg vitamin D per 100 g and about half this in winter. By law butter must contain a minimum of 80 per cent fat and a maximum of 16 per cent water.

Cheese

Although there are more than 400 different named cheeses the basic principles governing their manufacture are the same. Milk is coagulated and the solid formed is cut into small pieces to allow the whey to drain off. The solid curd is dried, salt is added and the cheese pressed or moulded and allowed to ripen. Cheddar cheese, which is a typical and popular hard cheese, will serve as a convenient example for discussing the main stages of cheese manufacture which are summarized in Table 5.6.

The essence of cheese making is the coagulation of milk and its conversion from a colloidal dispersion into a gel known as curd, and the subsequent release of water in the form of whey. The loss of moisture from a gel is known as *syneresis* and results in a fall in water content from 87 per cent in milk to less than 40 per cent in mature cheddar cheese. The control of this water loss constitutes a major part of the art of cheese making. The rate at which water is lost depends upon three factors, namely temperature, pH and the way the curd is cut, and in practice all three are controlled so as to give rapid syneresis. Reduction of water content is most important as it determines the hardness and keeping quality of the cheese. The

Table 5.6 Stages in making Cheddar cheese

Stage	Description of physical and chemical changes
1. Pasteurization	Whole milk is pasteurized.
2. Ripening or souring	Lactic acid bacteria starter added. Lactose converted into lactic acid with consequent fall in pH.
3. Clotting or coagulation	Rennet added to sour milk at 30°C. Casein precipitated as a tough gel or clot of calcium caseinate known as curd.
4. Cutting	Curd is cut into small pieces.
5. Scalding and pitching	The temperature is raised to about 40°C, pH continues to fall and cutting is continued. Pieces of curd matt together and whey is run off.
6. Cheddaring and piling	Curd is cut into blocks and piled up. Whey drains off and curd forms solid mass with a firm, soft texture.
7. Milling and salting	The dry curd is milled into small pieces and salt is added. More whey is lost.
8. Pressing	The soft cheese is put into moulds, pressure is applied and more whey expressed.
9. Maturing	After removal from the mould, the cheese is allowed to mature for three months or longer.

changes in water content and pH during cheese making are shown in Fig. 5.3.

The chemical changes which occur during maturing of cheese are still not completely understood, but are certainly brought about by enzymes. Lactic acid bacteria thrive in the immature acid cheese and the enzymes present in them bring about a number of chemical reactions which are responsible for the development of flavour and aroma. A week after manufacture is started all the lactose has disappeared, having been converted into lactic acid. Apart from lactose breakdown, maturing mainly involves breakdown of protein and fat. Protein is broken down by enzymatic hydrolysis brought about by rennin and other peptidases. Proteins are progressively broken down into smaller molecules such as peptones and ultimately into amino acids. Such soluble and low molecular weight nitrogen compounds certainly contribute to cheese flavour and in addition bring about physical changes in the cheese, causing it to become softer and creamier. Fat, like protein, is broken down by enzymatic hydrolysis and is converted into glycerol and free fatty acids. Milk fat is relatively rich in low molecular weight fatty acids such as butyric, caproic and capric, which are released on hydrolysis and, being volatile and strong-smelling, contribute to cheese flavour.

Amino acids and fatty acids produced by breakdown of protein and fat may be further broken down by enzymes yielding low molecular

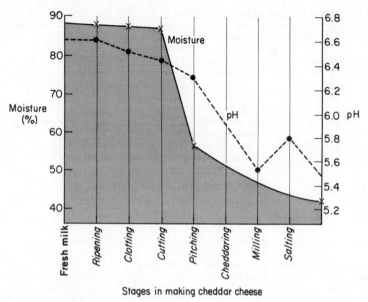

Figure 5.3 The variation of water content and pH during cheese making.

weight molecules such as amines, aldehydes and ketones which are volatile and strong-smelling, and so contribute to the flavour of mature cheese.

It is evident that the flavour of cheese is due to a very large number of different substances and much more research is needed before they all become known. After Cheddar cheese has been stored for about three months it has developed its full flavour, although it may be stored for a year or longer.

The very large number of cheeses available makes it impracticable to do more than describe the main types and varieties, and this is done in Table 5.7. The main distinction is between soft cheeses, which are not pressed and therefore have a high moisture content, and hard cheeses, which are pressed and therefore have a lower moisture content and better keeping qualities. Soft cheeses have an open texture and provide suitable conditions for development of moulds, which require air for successful growth. The best known of the British mould-ripened cheeses is Stilton, the blue-green veins of which are caused by moulds. White Stilton is merely ordinary Stilton which has not been matured long enough for mould to develop.

Cheese has a high nutritional value, as would be expected from the fact that a pint of milk produces only about 56 g cheese. Certain water-soluble nutrients of milk are lost in the whey, but most of them

Table 5.7 Some types and varieties of cheese

Type	Varieties	Milk used	Water (%)	Nature
Very hard	Parmesan	Skimmed ripened milk	<25	Dry cheese excellent keeping qualities, matured >1 year
Hard	Cheddar	Whole ripened milk	35	Mature, non-crumbly
	Cheshire	Whole ripened milk	38	Immature, crumbly, mild flavour
Semi-hard	Pecorino	Unripened sheep's milk	35–40	Soft, mild cheese (Italian)
	Edam	Unripened skimmed milk	35–40	Mild, firm, red colour (Dutch)
Soft	Cambridge	Unripened whole milk	>40	Immature cheese, curd is not cut
Cream	Cream	Ripened cream	45–50	Soft, mild, rich flavour
Internal mould	Stilton	Ripened whole milk	33–35	Blue mould, mellow flavour, cured 4–6 months
External mould	Camembert	Ripened whole milk	45–55	Soft, creamy consistency (French)

are retained in the curd. A hard cheese, such as Cheddar, consists of roughly one-quarter protein, one-third fat and one-third water. It is a rich source of calcium, phosphorus and vitamin A, and also contains useful quantities of other nutrients, as shown in Table 5.8. It is a much more concentrated food than milk but is less complete because of its lack of carbohydrate. Soft cheeses retain a higher percentage of moisture than hard ones and, therefore, have a lower percentage of other nutrients. Cream cheeses, which are made from cream, are rich in fat but contain much less calcium, phosphorus and vitamin A than Cheddar. The amount of water and milk fat in cheese is regulated by law and Cheddar cheese, for example, must not contain more than 39 per cent water and at least 48 per cent milk fat expressed on a dry basis.

The actual vitamin content of a cheese is very variable depending upon the quality of milk used in its production. Thus, Cheddar cheese which is rich in vitamin A contains on average 363 μg per 100 g of cheese, although the amount in a particular sample may differ greatly from this figure.

Hard cheeses are largely free of additives, except for the use of the vegetable dye *annatto* used to colour some varieties such as Red Leicester. Processed cheese, on the other hand, contains a variety of additives such as dried milk powder, emulsifiers and flavours. A preservative such as sodium nitrite may also be used.

Table 5.8   A comparison of the composition of cheese and summer milk

Name	Energy (kJ) per 100 g	Protein (g)	Fat (g)	Carbohydrate (g)	Calcium (mg)	Vitamin A (µg)	Vitamin D (µg)
Cheddar	1680	26	34	0	800	363	0.3
Stilton	1930	26	40	0	350	450	0.3
Cream cheese	1840	3	47	0	100	450	0.3
Summer milk	272	3.3	3.8	4.7	103	56	0.03

LIPIDS IN THE DIET

Lipids in the diet mainly occur in the form of fats, but also include cholesterol. Fat is present in two forms, the 'visible' fats like butter and margarine and 'invisible' fats which occur in foods like milk and eggs where the fat is in a highly emulsified form. Some foods, such as meat, contain both forms of fat. The amount of fat in selected foods is shown in Table 5.9.

Fat has four functions in the diet. Firstly, it serves as a source energy. As fat has a higher energy value than either carbohydrate or protein, foods containing a high proportion of fat form a compact energy source. A man doing heavy work expends about 14 MJ per day. If he obtained all his energy from fat he would need to eat only 378 g – less than a pound – of fat. A less fatty diet containing a large amount of carbohydrate would be more bulky.

Secondly, fat makes diets more palatable. Without any fat a diet would be completely unattractive. The amount of fat in diets varies considerably, the higher the standard of living the higher is the proportion of fat in the diet. For example, in the poorest countries fat may only contribute about ten per cent of the total energy in the diet, whereas in the wealthiest industrialized nations, fat may contribute

Table 5.9   The fat content of selected foods

Food	Fat (%)	Food	Fat (%)
Lard	99	Beef, rump steak	14
Margarine	81	Eggs	10.9
Butter	82	Milk	3.8
Cream cheese	47	Bread, white	1.7
Cheddar cheese	34	Rice	1
Pork sausage	32	Haddock	0.6
Herring	19	Potatoes	0

Table 5.10 Cholesterol content of some foods (mg/100 g food)

Egg-yolk	1260	Cheddar cheese	70
Whole egg	450	Chicken, white meat	69
Kidney, pig	410	Beef	65
Liver, pig	260	Milk	14
Butter	230	Vegetable oil	0

40 per cent of the total energy of the diet. Moreover, in wealthier countries a greater proportion of the fat in the diet is animal fat.

Thirdly, fats provide the body with essential fatty acids and, finally, they provide the body with fat-soluble vitamins.

Cholesterol is another lipid that is provided by the diet, although only about 20–25 per cent of body cholesterol comes from food, the rest being made by the body itself. A person of average weight will have about 140 g of cholesterol in his or her body and the body makes about 1 g of cholesterol each day. The actual amount varies with the individual and also with the amount received from food. If the amount supplied in the diet is reduced the body compensates by making more for itself. The body normally has more cholesterol than it needs and some of the excess is made into bile salts which help with the digestion of fat. It is this excess of cholesterol in the body which some consider harmful in relation to coronary heart disease. Table 5.10 shows the cholesterol content of some foods.

Lipids in the body

Lipids in the body have three main functions.

1. In the fat depots of the body – the adipose tissue – they provide the chief store and source of energy for the body.
2. In all body tissues lipids form the main part of the structure of cell membranes.
3. Lipids provide the raw materials from which many hormones are made.

When fatty foods are eaten they pass through the digestive system in the manner described in Chapter 2. In the small intestine they are partly hydrolysed by lipase, one fatty acid molecule being split off from a fat molecule at a time, so that the result is a mixture of free fatty acids and mono- and diglycerides. As the hydrolysis is not complete, little free glycerol is produced. During absorption of fats, fatty acids and glycerol pass through the villi into the lymph vessel where they are resynthesized into fat molecules, being rearranged in the process so that the new fat molecules formed are more suitable for use by the body.

Fat, as it is insoluble in water, cannot be carried round the body dissolved in blood. Instead it is transported in the form of complex *lipoproteins* which are made up of two parts: a *lipid* component of triglyceride, cholesterol and phospholipid together with a *protein* component. Thus fat is carried by the blood in emulsified form as tiny droplets of lipoprotein. The lipoprotein complex responsible for transporting lipids from the small intestine is in the form of *chylomicrons*. It is the chylomicrons that are responsible for the 'milky' appearance of blood plasma that occurs after a meal rich in fat has been eaten.

The lipoproteins may be divided into four types, distinguished from one another by their different densities. They vary from chylomicrons with a high triglyceride content and correspondingly low density to high density lipoproteins with a low triglyceride content. These differences are shown in Table 5.11.

Chylomicrons carry triglyceride to both muscle and adipose tissue where lipoprotein lipases rapidly hydrolyse them. Some of the free fatty acids produced plus the remains of the chylomicrons (minus the triglyceride component), together with glucose, pass to the liver where they are converted to triglycerides and then transported as very low density lipoproteins (VLDL) to muscle and adipose tissue.

Most body fat lies under the skin as adipose tissue. It was once supposed that such fat formed an inert store that was drawn upon when the body required energy. However, it is now known that adipose tissue is a source of considerable metabolic activity. Triglycerides (as part of both chylomicrons and VLDL) are constantly being broken down and re-formed within adipose tissue. Glucose derived from carbohydrate food is also converted into fat in adipose tissue, this process being controlled by the hormone *insulin*. The constant breaking down and reforming of triglycerides involves the interchange of fatty acids between triglyceride molecules and the formation of many different triglycerides. Essential fatty acids must be provided by the diet, however, as these cannot be produced by the body's metabolic processes.

When the body requires energy from its reserves triglycerides from adipose tissue can be used, and they are broken down to fatty acids which are carried by the blood to muscles and other tissue. Glycerol is

Table 5.11　Per cent composition of blood plasma lipoproteins

Lipoprotein	Abbreviation	Protein	Triglyceride	Cholesterol	Phospholipid
Chylomicrons	–	2	85	4	9
Very low density	VLDL	10	50	22	18
Low density	LDL	25	10	45	20
High density	HDL	55	4	17	24

also produced and it is broken down to carbon dioxide and water and energy by a process involving *adenosine triphosphate* (ATP) and the Krebs cycle (see p. 111).

The production of energy from fatty acids involves, firstly, an activation step, the energy for which is provided by ATP – an energy-rich molecule – which is considered in more detail in Chapter 7. In this step a molecule of *coenzyme A* (containing the B vitamin pantothenic acid) is added to the fatty acid. This active form is then acted upon by about four enzymes (which contain B vitamins) in a series of steps, each of which involves the removal of a fragment containing two carbon atoms. Each two-carbon fragment is in the form of acetyl coenzyme A, which is then oxidized in a complex cycle of reactions known as the citric acid cycle (or Krebs cycle). It appears that the final oxidation of both carbohydrate and fat follows the same pathway; also both nutrients finally yield the same products, namely carbon dioxide, water and ATP.

If energy is not required immediately, molecules of glycerol and fatty acid recombine and are deposited again as fat. There is, therefore, a dynamic equilibrium between the breakdown and rebuilding processes.

KETOSIS During normal fat metabolism a small amount of acetoacetic acid, CH_3COCH_2COOH, and beta-hydroxy butyric acid, $CH_3CH(OH)CH_2COOH$, is produced by combination of two-carbon fragments. Acetone, CH_3COCH_3, is also produced from breakdown of acetoacetic acid. These three substances are known as *ketone bodies* and if they accumulate in the blood *ketosis* is said to occur. Ketone bodies can be used as an energy source by most tissues, including the brain, in place of glucose

Ketosis occurs when fat rather than carbohydrate is the prime energy source, as happens in starvation or in a fat-rich diet or in severe diabetes. Provided that ketosis is controlled it is probably beneficial, particularly as it allows ketones to act as an energy source for the brain when carbohydrate energy sources have been exhausted. On the other hand, in a diabetic, severe ketosis may be fatal, partly because of the toxicity of the ketone bodies themselves and partly because the accumulation of acetoacetic acid and beta-hydroxy butyric acids reduces the pH of blood (i.e. increases acidity). If the pH is reduced from its normal value of 7.4 to a value of 7.2 or lower *acidosis* is said to occur. This condition may result in a coma, which, if untreated, may be fatal.

Lipids and health

Lipids are an essential part of the diet, but in recent years there has been a growing controversy as to the quantity and quality of lipids that are desirable to ensure a healthy diet. In wealthy countries, as already noted, fat contributes about 40 per cent of the energy content of the diet, while in poorer countries it only contributes some ten per cent. Moreover, in affluent countries a much larger proportion of the fat eaten is derived from animals.

In the Western world during the twentieth century certain 'diseases of affluence' have appeared as described in Chapter 3. Among these diseases *coronary heart disease* (CHD) has become prominent as the most frequent cause of death. The study of this disease has been, and continues to be, intense. It has been established that there is no single cause, while correlation studies (see p. 33) indicate that the three main *risk factors* are smoking, raised blood pressure and high blood cholesterol. Of all the factors implicated, diet is the one that has been most studied and is the one of most concern in this book. Although we do not know the causes of CHD we do know that it is associated with the build-up of fatty material, especially *cholesterol*, in the blood. Much research therefore is directed towards trying to discover the way in which diet affects blood cholesterol. Firstly, however, it is necessary to explain the nature of CHD.

CHD is related to a reduced flow of blood to the coronary arteries which supply blood to heart muscle. A healthy coronary artery has the features shown in Fig. 5.4. Blood flows through the artery which is surrounded by the *endothelium*, a smooth lining to the artery which allows blood to pass through it without resistance. This in turn is surrounded by a layer of the protein *elastin*, which has an elastic character, and a thick layer of smooth muscle and connective tissue called the *media*.

With increasing age there is some 'hardening of the arteries', or

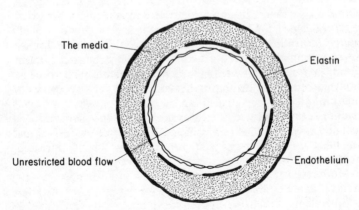

Figure 5.4   Cross-section through a healthy artery.

atherosclerosis, which involves the formation of a swelling between the elastin and endothelium layers and consequent narrowing of the channel through which blood flows. The swelling (or *atheroma*) involves accumulation of a mixture of cholesterol and other fatty substances together with complex carbohydrates, blood, calcium and fibrous material. The thick rough area of the arterial wall containing this material is called a *plaque*.

If the growth of plaque restricts blood flow unduly symptoms occur and the greater the amount of plaque formed the more severe the symptoms become. Undue restriction of blood flow produces *angina* in the form of stabbing pains which normally occur during exercise and which subside following rest. If the plaque grows large enough to form a clot or thrombus which cuts off the flow of blood to heart muscle, a heart attack results.

It has already been mentioned that CHD, which includes both angina and heart attack, is related to raised levels of cholesterol and other fatty material in the blood. Dietary theories about CHD which involve fat are based on the belief that such raised levels in the blood are related to the amount of fat and cholesterol in the diet. The main theories involving fat are summarized below.

DIET/FAT THEORIES OF CHD

1. *Fat theory*.   Epidemiology has been much used in seeking to establish risk factors in CHD (see Chapter 3). Such an approach compares the prevalence of CHD in different groups of people and seeks to correlate this with particular factors. If these factors seem to be significant they are referred to as risk factors. One of the earliest studies of this sort was carried out by Ancel Keys in 1953 when he compared the incidence of CHD in six countries with a number of factors and found that the best correlation was with *total fat* intake.
2. *Saturated fat theory*.   Additional epidemiological studies involving seven countries, carried out by Keys in 1970, produced further significant correlations, namely that levels of blood cholesterol in the body correlated well with intake of *saturated* fat and also with incidence of CHD.

   It was also demonstrated that intake of *polyunsaturated* fat (and protein and energy) did *not* correlate with incidence of CHD. This finding has been the basis for a surge of dietary advice advocating the replacement of saturated fats in the diet by polyunsaturated ones. More recently it has been proposed that it is *cis* polyunsaturated fats that are desirable from a health viewpoint rather than the *trans* equivalents, because the latter are treated by the body in the same way as saturated fats.

3. *Cholesterol theory*. The basic argument is simple: CHD is associated with high levels of blood cholesterol, therefore by reducing the amount of cholesterol in the diet, blood cholesterol levels will be reduced and hence the risk of CHD. Originally blood cholesterol was measured as LDL as this is rich in cholesterol (see Table 5.11) and is the form in which cholesterol is delivered to all body cells. In more recent experimental studies HDL has also been measured and it has been found that *low* levels of HDL correlate with high risk of CHD. On this basis it would seem possible that high levels of LDL may be a risk factor while high levels of HDL may be a protective factor.

The cholesterol theory has gained some support, but it needs to be appreciated that while some cholesterol in the blood comes from dietary sources, additional amounts are synthesized by cells within the body. Cholesterol is transported round the body as part of a lipoprotein. It passes to cells as LDL and the cells use what cholesterol they need from the LDL. The cells regulate the level of cholesterol by adjusting the amount they produce. If the supply of cholesterol to the cells (in the form of LDL) is lower than the demand within the cell, then the cell makes more cholesterol. When the supply of cholesterol (as LDL) to the cells increases (because more cholesterol has been consumed in the diet) to an amount greater than the demand within the cell the amount produced by the cells is reduced.

It needs to be understood, therefore, that reducing cholesterol intake from food will, at best, have only a very small effect on total cholesterol in the body. In particular, consumption of low-cholesterol margarines rather than butter will have no appreciable effect on total body cholesterol. Butter and margarines with the highest content of cholesterol contain only about 0.2 per cent cholesterol so that the amount of cholesterol we receive from such foods is insignificant compared with the cholesterol we receive from other foods (see Table 5.10) – and even less significant when compared with the amount of cholesterol made by the body.

MODIFICATION OF DIET AS A MEANS OF REDUCING RISK OF CHD
Medical and scientific opinion is sharply divided about the lipids – that is fats and cholesterol – in our diet, and their relation to health in general and CHD in particular. There are those who consider that the heart–diet relationship is not valid and that CHD is in no way related to diet. At the other extreme there are those who believe that diet is a major – if not *the* major – risk factor in CHD, and who consequently advocate modification of diet as the best way of reducing incidence of the disease. Some recommend a low-fat, low-cholesterol diet while others favour a diet low in cholesterol and saturated fat but

moderately high in polyunsaturated fats. Between these two opposing positions there are those who believe that we are not yet in a position to make general recommendations about modifications to our diet because the evidence to date is inconclusive and, indeed, often conflicting.

One major method of collecting evidence about the diet–CHD relationship is the use of *intervention trials* in which selected people are divided into two groups. In one group, the intervention group, diet (and possibly some habits such as exercise or smoking) are modified, while in the control group no such changes are made. Health in general and incidence of CHD in particular can then be studied in the two groups over a period of years and the results compared.

Where such intervention studies have involved the general population rather than selected high-risk groups the results have proved to be confusing and contradictory. Where diet has been modified by the adoption of low-fat diets in which most dietary fat is polyunsaturated, blood cholesterol levels have been somewhat reduced. On the other hand, even where death from CHD has also been reduced, total deaths in the intervention and control groups has remained the same because lower deaths from CHD have been balanced by more deaths from cancer.

However, where intervention trials have been conducted with high-risk groups of people rather more positive results have been reported. One such trial is the *Oslo Study*, the details of which are summarized in Table 5.12. In this study men with high cholesterol levels were chosen. The intervention group reduced both total fat and saturated fat intake and also reduced smoking. The results after five years showed a convincing reduction in death from CHD compared with the control group.

Table 5.12 Detail of the Oslo study, showing results after five years

	Intervention group	Control group
At start of trial:		
Age in years	45	45
Cases of CHD	0	0
Smokers as %	70	70
Serum cholesterol as mg/100 ml	328	329
After 5 years:		
Dietary fat as % of energy intake	28	40
Saturated fat as % of energy intake	8	18
Serum cholesterol as mg/100 ml	263	341
Deaths from CHD per 1000	31	57

From the evidence gained from intervention trials it seems fair to conclude that for middle aged men who are in the high-risk category by virtue of high levels of blood cholesterol or because there is a family history of CHD, there is enough positive evidence to recommend a reduction in the quantity of fat in the diet and a change to a diet involving less saturated fat and more polyunsaturated fat.

Countries with a high incidence of CHD have naturally been concerned to give advice that would lower the risk of CHD. In view of the conflicting evidence available, this is not an easy matter. Nevertheless, since the 1960s there has been a flow of official and semi-official reports giving advice concerning fats in the diet. Until the mid-1970s a majority of such reports offered fairly sweeping advice advocating that national diets should be changed by a reduction in total fat consumed and that some saturated fats should be replaced by unsaturated ones.

More recently there has been increasing caution as to the advice given and a more conservative attitude towards recommending major changes in diet for whole populations when the evidence is not completely convincing. Nevertheless, most nutritionists in the UK now seem to accept that some modifications to diet involving fat is desirable. This view is reflected in two recent reports. The report issued by the National Advisory Committee on Nutrition Education (NACNE) in 1983, based its findings on seven earlier reports issued in the UK and one by the World Health Organization (WHO), advocates that in the long-term fat intake should be reduced by 25 per cent to provide not more than 30 per cent of the total energy intake compared with 40 per cent at present. It also recommends that intake of saturated fats should provide no more than ten per cent of total energy intake, rather than the present figure of 17 per cent.

In 1984 the DHSS issued a report from its Committee on Medical Aspects of Food Policy (COMA) regarding the relation between diet and CHD. This recommends that in the UK total fat intake should be reduced to 35 per cent of total energy intake and that no more than 15 per cent of the total energy intake should be provided by saturated and *trans* unsaturated fatty acids (as these are treated by the body in the same way as saturated fatty acids).

Although the recommendations of the NACNE and COMA reports appear rather different they are in fact very similar when the different methods of assessing total energy intake and the inclusion of *trans* acids in the COMA report are taken into account.

NON-DIETARY RISK FACTORS AND CHD   We have already noted that CHD appears to have no single cause but that a number of factors are involved. While diet is considered by many people to be one major risk factor there are a number of others.

Some factors associated with CHD are non-environmental and therefore cannot be changed. For example, CHD is very rare in infancy and youth but increases with *age*. It is apparent that incidence of CHD does vary with *sex*. Up to the age of about 45, men are more likely to have CHD than women. After the menopause, however, this distinction disappears. *Heredity* is a third factor. There is clear evidence that family history is important and that people are more likely to develop CHD if there is a history of the disease in the family.

CHD has already been referred to as a 'disease of affluence', indicating that the disease is, in a general way, related to prosperity. It is evident, therefore, that environmental factors are involved. However, there are so many differences between the lives of the rich and the poor that it is not an easy matter to pinpoint the factors that are relevant to CHD. One factor that has been identified is *smoking*. It has been demonstrated with some conviction that the more people smoke the greater is the risk of developing CHD. Other factors that may relate to CHD include *personality* and *exercise*.

It has been suggested that people with an aggressive, dominant personality are more prone to CHD than those with a more relaxed approach to life. The former are more likely to live under stress and it is possible that this may be a risk factor. Similarly, it is possible that people who live active lives are less likely to develop CHD than those who are inactive. The importance and effects of exercise have been much studied but remain incompletely understood. Generally speaking, a moderate amount of exercise seems to be desirable and good for the heart. In particular, exercise raises the level of HDL cholesterol in the body and this is a protective factor against CHD. In addition, exercise can help in regulating weight, which is itself a risk factor. This risk arises because overweight people are more likely to develop diabetes than people of normal weight and they may also have high levels of blood cholesterol and high blood pressure, the last two also being risk factors in CHD.

CHD: SOME CONCLUSIONS While a number of risk factors have been identified in respect to CHD, it seems probable that all the known hereditary and suspected environmental factors do not fully account for the prevalence of CHD, and further, unknown factors need to be discovered before we can fully explain the causes of the disease.

While recommendations about desirable changes in diet to reduce the risk of CHD have been issued in a number of countries, it remains uncertain as to how effective these can be. For people who are not at high risk it is difficult to give precise advice. However, in general terms some guidelines can be given and these would include adopting a moderate diet both in terms of quantity of food eaten so as to prevent overweight, and quality in terms of restricted intake of fat, especially saturated fat. (In addition an active life-style involving a moderate amount of exercise and no smoking are desirable.)

For people who are at high risk from CHD, such as middle aged men with high levels of blood cholesterol and/or a family history of CHD, there is much to be said for more specific dietary action involving both the quantity and quality of fat in the diet. Such action may have to be quite drastic to be effective. It may also be wise to adopt other specific changes to diet. For example, there is fairly strong evidence to suggest that heavy coffee drinking is a risk factor in CHD. In a careful study of 1130 men aged between 19 and 35 years it was found that heavy coffee drinking (more than five cups per day) produced raised blood cholesterol levels and increased the risk of CHD in mainly non-smoking men by a factor of 2–3.

Although much remains to be done before the causes and remedies for CHD are fully understood, one encouraging trend can be noted by way of conclusion. In the USA and certain other countries, although not in Britain, the incidence of CHD is decreasing slightly. While the reasons for this decline are not understood, it does give hope that the spread of CHD as a major 'disease of affluence' can be reversed.

SUGGESTIONS FOR FURTHER READING

ALFA-LAVAL (1980). *Dairy Handbook*. Alpha-Laval Co., London.

CHRISTIAN, G. (1977). *Cheese and Cheesemaking*. Macdonald Educational, London.

DEPARTMENT OF HEALTH AND SOCIAL SECURITY (1984). Committee on Medical Aspects of Food Policy (COMA), *Diet and Cardiovascular Disease*. HMSO, London.

EDWARDS, G. (1968). *Vegetable Oils and Fats*, 2nd edition. Unilever Booklet, Unilever, London.

EVERETT, D.H. (Ed.) (1988) *Basic Principles of Colloid Science*. Royal Society of Chemistry, London.

FAO/WHO (1980). *Dietary Fats and Oils in Human Nutrition*. HMSO, London.

GURR. M.I. (1984). *Role of Fats in Food and Nutrition*. Elsevier, Amsterdam.

HYDE, K.A. AND ROTHWELL, J. (1973). *Ice-cream*. Churchill Livingstone, Edinburgh.

KON, S.K. (1972). *Milk and Milk Products in Human Nutrition*, 2nd edition. FAO, Rome.

MEYER, A. (1973). *Processed Cheese Manufacture*. Food Trade Press, Orpington.

MOORE, E. (1971). *Margarine and Cooking Fats*. Unilever Booklet, Unilever, London.

MULDER, H. AND WALSTRA, P. (1974). *The Milk Fat Globule. Emulsion Science as Applied to Milk Products*. Commonwealth Agricultural Bureaux, Slough.

NATIONAL ADVISORY COMMITTEE ON NUTRITION EDUCATION (NACNE) (1983). *Proposals for Nutritional Guidelines for Health Education in Britain*. Health Education Council, London.

NATIONAL DAIRY COUNCIL (1988). *From Farm to Doorstep ... A guide to Milk and Dairy Products*. National Dairy Council, London.

PORTER, J.W.G. (1975). *Milk and Dairy Foods*. Oxford University Press, Oxford.

RANCE, P. (1983). *The Great British Cheese Book*. Papermac, London.

ROBINSON, R.K. (Ed.) (1986). *Dairy Microbiology* (2 vols). Applied Science, Barking.

ROYAL COLLEGE OF PHYSICIANS (1976). *Prevention of Coronary Heart Disease*. Report reprinted from Journal of Royal College of Physicians, London.

SCHMIDT, G.H. AND VAN VLECK, L.D. (1974). *Principles of Dairy Science*. Freeman, USA.

SCOTT, R. (1986) *Cheesemaking Practice*. Applied Science, Barking.

WEBB, B.H. *et al.* (1974). *Fundamentals of Dairy Chemistry*, 2nd edition. Avi, USA.

WRIGHT, G. (1987). *Milk and the Consumer*. University of Bradford, Food Policy Research Unit, Bradford.

CHAPTER 6

Carbohydrates

Carbohydrates constitute one of the three main classes of nutrients. They occur in food as sugars and starches, which are a major source of energy in the diet, and as cellulose which is the main component of dietary fibre.

Sugars are produced in plants from carbon dioxide and water. At the same time oxygen is evolved as shown in the equation for the formation of the simple carbohydrate glucose:

$$6CO_2 + 12H_2O \rightarrow C_6H_{12}O_6 + 6O_2 + 6H_2O$$

Glucose

Water appears on both sides of this equation because it has been shown that all the oxygen evolved originates from the water. The oxygen atoms in the glucose and water molecules on the right-hand side of the equation are those which were originally combined with carbon in the carbon dioxide. The equation is a comparatively simple one, but it shows only the starting materials and final products of a series of complex reactions.

The building-up of carbohydrate molecules by plants is accomplished by photosynthesis. Energy is required to transform the carbon dioxide and water into carbohydrates and this is supplied by the action of sunlight on the leaves. Consequently, photosynthesis does not take place in the dark. Animals are unable to synthesize carbohydrates and this is one of the fundamental differences between them and plants. The sugars formed by photosynthesis are converted into the polysaccharides starch and cellulose. Starch is the main energy reserve of most plants and cellulose functions as a structural support for the plant.

The solar energy used in photosynthesis is stored as chemical energy and this may later be drawn upon by the plant, which, by oxidizing the carbohydrate back to carbon dioxide and water, is able to make use of the energy liberated. Alternatively, animals, by eating the plant, may utilize the chemical energy stored in the carbohydrate molecules.

Carbohydrates contain only carbon, hydrogen and oxygen and, except in rare cases, there are always two atoms of hydrogen for every one of oxygen. Accordingly, carbohydrates have the general formula $C_x(H_2O)_y$ where x and y are whole numbers, and it is from this formal representation as hydrates of carbon that the name carbohydrate is derived. There are, of course, no water molecules as such present in the molecule of a carbohydrate.

Many of the carbohydrates known, particularly the simple ones, do not occur naturally but have been obtained by synthesis in the laboratory, Only the naturally occurring carbohydrates are of interest from the point of view of food science and those containing six, or multiples of six, carbon atoms are particularly important. Familiar examples are glucose $C_6H_{12}O_6$, sucrose $C_{12}H_{22}O_{11}$ and starch, the very large molecules of which are represented by the formula $(C_6H_{10}O_5)_n$.

The simpler carbohydrates, called sugars, are crystalline solids which dissolve in water to give sweet solutions. The simplest of these, such as glucose and fructose, are called *monosaccharides* and they are of great importance as the units or building blocks from which the more complex carbohydrates are built. *Disaccharides*, for example, sucrose, maltose and lactose, contain two connected monosaccharide units which may be alike or different. Disaccharides can be hydrolysed by boiling with dilute acids, or by enzymes, to give the monosaccarides from which they are built up. Sucrose, for example, on hydrolysis gives equal parts of glucose and fructose. Like the monosaccharides, disaccharides dissolve in water to give sweet solutions and so are classified as sugars.

MONOSACCHARIDES

D-glucose, dextrose or grape sugar, $C_6H_{12}O_6$

These are all names for the same sugar which is found in grapes (up to seven per cent) and other sweet fruits. Onions (about two per cent) and tomatoes (about one per cent) also contain glucose, but honey, which contains about 31 per cent, is the richest source of the sugar. It plays an essential part in the metabolism of plants and animals and is the main product of photosynthesis in plants. In animals it is produced during digestion of starch and other carbohydrates and it is a normal component of the blood of living animals. Human blood has about 80–120 mg/100 ml and it is the only sugar which plays a significant part in human metabolism.

Glucose is a white solid; like all sugars it is sweet-tasting but it is less sweet than sucrose. Its alternative name – dextrose – is derived from the Latin word 'dextra' meaning 'right'. If a beam of polarized light is passed through a solution of glucose its plane of polarization is rotated to the right. We need not trouble ourselves unduly about the

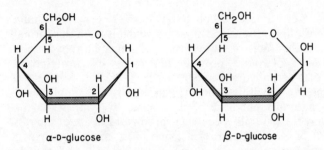

α-D-glucose   β-D-glucose

Figure 6.1 The structure of glucose.

nature of polarized light − it will be sufficient to regard it as light which has been so treated that the wave vibrations all occur in one plane which is called the plane of polarization. Substances such as glucose which rotate the plane of polarization of polarized light in this way are said to be *optically active*, and glucose, which rotates it to the right, is said to be *dextrorotatory*. The effect of optically active substances on polarized light is a consequence of the precise spatial arrangement of groups of atoms in the molecule.

Five of the carbon atoms and one of the oxygen atoms in a glucose molecule are connected together to form a six-membered or hexagonal ring structure. D-glucose exists in two isomeric forms which differ very slightly. Ordinary D-glucose is the α-isomer. Semi-pictorial formula which give an idea of the way in which the atoms are arranged are given in Fig. 6.1: the carbon atoms are numbered to make it easier to refer to them later in this chapter when some important polysaccharides are considered.

The plane of the rings in Fig. 6.1 should be imagined to be projecting forward from the plane of the paper with the thick edge towards the reader and the substituent H, OH and CH₂OH lying above and below the plane of the ring. In fact, the situation is far more complex than this as the ring is not planar and it can fold or buckle to take up various shapes which are referred to as configurations.

D-fructose, laevulose or fruit sugar, $C_6H_{12}O_6$

The sugar fructose is found with glucose in honey, which contains about 35 per cent, and in sweet fruit juices. Fructose is not an important dietary component in its own right but it is produced in the body by hydrolysis whenever food containing sucrose is eaten. It is a laevorotatory sugar which means that when a beam of polarized light is passed through a solution of fructose the plane of polarization is rotated to the left. This is the origin of its alternative name laevulose.

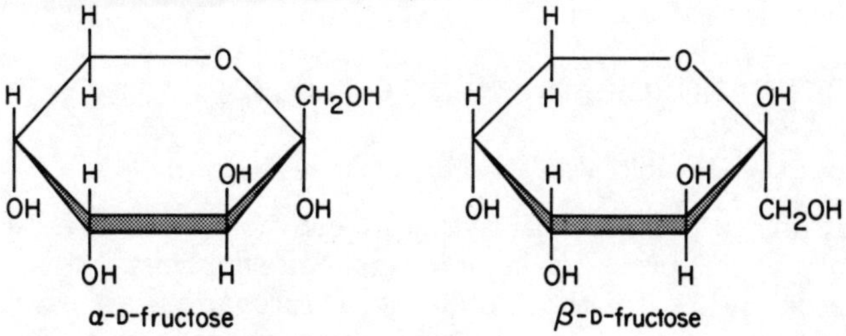

Figure 6.2 The structure of fructose.

Fructose is a C_6 sugar like glucose and it shares the same molecular formula, $C_6H_{12}O_6$. Five of its carbon atoms and an oxygen atom form a six-membered ring with the other groups of atoms disposed about lt as shown in Fig. 6.2.

Fructose and glucose are such similar compounds – both are C_6 sugars produced by plants – that one would expect them to have very similar molecular structures and Figs 6.1 and 6.2 show that this is indeed the case. At first sight the structure of fructose may appear to be identical with that already given for glucose in Fig. 6.l, but close comparison will show that this is not so. The differences between them, though slight, are detected with ease by biological systems such as our bodies. For example, they differ in sweetness and, as will become apparent later, glucose is a key compound in the utilization of carbohydrate foods by the body whereas fructose does not play any part.

Fructose is about twice as sweet as sucrose – ordinary table sugar – but it has the same energy value, about 17 kJ/g. This means that the same sweetening effect is produced by about half the weight of fructose if it is substituted for sucrose. Fructose can thus be used by sweet-toothed slimmers as part of a calorie controlled diet. Fructose can also be used as a sweetener by sufferers from the disease *diabetes mellitus* – sometimes called sugar diabetes. Unlike glucose and sucrose it does not require the hormone insulin for its utilization by the body.

Until fairly recently fructose was not available cheaply but it is now made from commercial glucose (glucose syrup or corn syrup, see p. 116) by treating it with an enzyme which is able to isomerize glucose to fructose.

DISACCHARIDES

Disaccharides are formed by the union of two monosaccharide

molecules, with loss of water:

$$(C_6H_{11}O_5\ )O\vdots H + HO\vdots(C_6H_{11}O_5) \rightarrow (C_6H_{11}O_5)-O-(C_6H_{11}O_5) + H_2O$$

Two monosaccharide molecules   A disaccharide molecule

It is not possible to prepare disaccharides from monosaccharides in the laboratory by this process despite its apparent simplicity. Nature, however, accomplishes it without difficulty. Disaccharides are easily split into their component monosaccharides by enzymes or by boiling with dilute acids. Many disaccharides are known but the most important, and the only ones of interest in food science, are sucrose, maltose and lactose.

Sucrose, cane sugar or beet sugar, $C_{12}H_{22}O_{11}$

Ordinary sugar, whether obtained from sugar-cane or sugar-beet, is substantially pure sucrose. It is a white crystalline solid which dissolves in water to give a dextrorotatory solution. Sucrose is widely distributed in the vegetable kingdom in many fruits, grasses and roots and in the sap of certain trees. It is produced and consumed in far larger quantities than any other sugar. Over two million tons of sucrose are used annually in this country (which is roughly 41 kg per person per year) and of this about one-third comes from home-grown sugar-beet.

On hydrolysis with dilute acids, or the enzyme *sucrase*, sucrose gives equal quantities of glucose and fructose, so the sucrose molecule must contain one glucose unit combined with one fructose unit. The fructose unit in sucrose does not have a six-membered ring but a five-membered ring, as shown in Fig. 6.3.

Sucrose is very readily hydrolysed to glucose and fructose, the latter having the normal six-membered ring structure. Fructose is more strongly laevorotatory than glucose is dextrorotatory, so the mixture

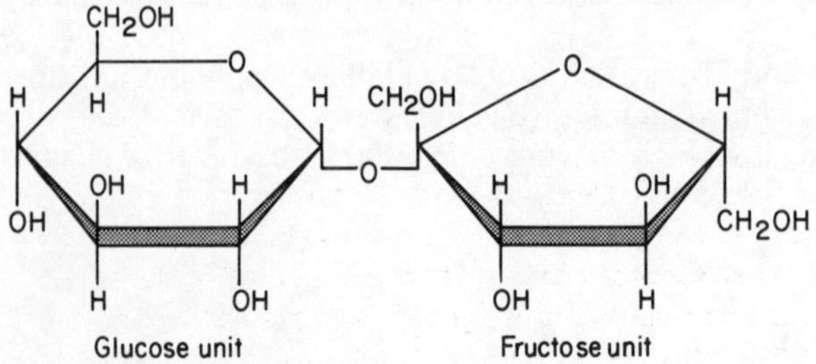

Glucose unit   Fructose unit

Figure 6.3 The structure of sucrose.

of glucose and fructose obtained on hydrolysis is laevorotatory. This change in the sign of rotation is called an *inversion*.

The mixture of glucose and fructose produced by the inversion is called invert-sugar and has been known in the form of honey for many centuries. Bees collect nectar, which is essentially sucrose, from flowers and it is inverted by enzymes during passage through their bodies. Honey is not pure invert-sugar because it contains, in addition to glucose and fructose, some sucrose, about 20 per cent water and small quantities of extracted flavours peculiar to the flower from which it was obtained. When sucrose is used in the preparation of acidic foodstuffs a certain amount of inversion invariably takes place. For instance, if sucrose is used for sweetening fruit drinks it is completely inverted within a few hours. Jams and sweets also contain invert-sugar.

Sucrose is obtained commercially from sugar-cane, which can only be grown in tropical countries, and sugar-beet which can be grown in any temperate climate. Whichever is used, the same product is obtained: there is no difference between the sugar obtained from sugar-beet and that obtained from sugar-cane.

Maltose or malt sugar, $C_{12}H_{22}O_{11}$

The sugar maltose is obtained when starchy materials are hydrolysed by the enzyme *diastase*. Maltose can be further hydrolysed by the ensyme *maltase* or by heating with dilute acid. D-glucose is the only monosaccharide formed during these hydrolyses and this indicates that maltose contains two connected glucose units.

Lactose or milk sugar, $C_{12}H_{22}O_{11}$

Lactose is a white solid which is somewhat gritty in appearance. It occurs in the milk of all animals: cow's milk contains about 4–5 per cent and human milk 6–8 per cent. On hydrolysis of lactose equal quantities of glucose and the monosaccharide galactose are obtained and hence the lactose molecule consists of these two monosaccharides linked together.

SWEETNESS AND SWEETENERS

Why is sugar sweet? A simple question but, as with so many simple questions, the answer is complex. It is even difficult to say with any degree of precision what we mean by sweetness. The term can only be defined in subjective terms and the property is measurable only by tasting and not, as with most other properties, by using an appropriate instrument. Although it is difficult to define sweetness and impossible

Table 6.1   Relative sweetness

Sucrose	1.0	Acesulfame-K	150
Glucose	0.5	Aspartame	200
Fructose	1.7	Saccharin	300
Lactose	0.4	Thaumatin	3000
Sorbitol	0.5	Cyclamate	30
Mannitol	0.7		
Xylitol	1.0		

to measure it in any absolute way it is possible to compare the relative sweetness of different substances. In practice relative sweetness is determined by human beings using that well-tried and exceedingly sensitive organ the human tongue, the tip of which is said to be the area concerned with the sensation of sweetness. By carrying out large numbers of tasting tests with sugars and other substances and adjusting their concentrations until they are of apparently equal sweetness or by finding the lowest concentration at which sweetness can be detected it is possible to draw up a table of relative sweetness, such as Table 6.1.

Most of the substances listed in Table 6.1 are compounds we have not yet encountered, but all of them (except cyclamate) are approved for use as sweeteners in food (see p. 379). Sorbitol, mannitol and xylitol are not sugars but sugar alcohols obtained by reduction of the sugars sorbose, mannose and xylose respectively. Xylitol is produced commercially from birch wood. It has the same energy value as sucrose and hence is of no use as a sugar substitute for slimmers. It is not fermented to acids by bacteria present in the mouth (see p. 127) and hence, unlike sucrose, it is not cariogenic. Xylitol is also used as a sucrose substitute in foods for diabetes sufferers because it does not require insulin for its metabolism.

Assessment of relative sweetness is so subjective that figures such as those given in Table 6.1 should be accepted with reserve and used with caution. Relative sweetness as perceived by an individual may differ from time to time (i.e. a solution which appears to be 'less sweet' on one occasion may appear to be 'more sweet' on another). Moreover, perception of relative sweetness may differ if the sweetness of solid substances rather than the sweetness of solutions of them is compared. It is not surprising, therefore, that the estimations of relative sweetness published by different groups of investigators are far from consistent. For this reason the values given in Table 6.1 should be regarded as representative values.

A great deal of work has been done to see whether there is any link between the sweetness of a compound and its chemical structure. Some progress has been made and it has been suggested by one group of workers that in all sweet substances a 'sweetness-producing' structural feature called a 'saporous unit' could be identified. The saporous unit, which contains two adjacent electronegative atoms, is thought to interact by hydrogen bonding with suitable receptor units in the tongue. Interestingly, proteins which bind sweet-tasting substances in proportion to their sweetness have been isolated from the taste buds of a cow's tongue. Whether a cow's perception of sweetness is the same as that of a human being, however, is purely a matter of speculation.

POLYSACCHARIDES

Polysaccharides are carbohydrates of high molecular weight which differ from the sugars in being non-crystalline, generally insoluble in water and tasteless. A polysaccharide is built up from a large number of connected monosaccharide units which may be alike or different. As in the case of disaccharides, one molecule of water is lost in the union between one monosaccharide molecule and the next. The polysaccharides which are of importance in food chemistry are all built up from monosaccharides containing six carbon atoms and are best formulated $(C_6H_{10}O_5)n$. The value of n varies, but in most cases is quite large. A single cellulose molecule, for example, can contain several thousand connected glucose units. Hydrolysis breaks down a polysaccharide molecule into smaller portions containing various numbers of monosaccharide units and may, if sufficiently drastic, convert the polysaccharide completely to monosaccharide:

$$(C_6H_{10}O_5)n + nH_2O \rightarrow nC_6H_{12}O_6$$

Like the sugars, polysaccharides are built up by plants from carbon dioxide and water and it is probable that the sugars represent an intermediate stage in the photosynthesis of polysaccharides. It has already been mentioned that animals are unable to build up carbohydrates by photosynthesis and for this reason polysaccharides are found predominantly in plants. The polysaccharide *glycogen*, however, is elaborated by man from glucose, and several similar examples of animals syntehsizing polysaccharides from monosaccharides are known. In man, glycogen constitutes a store of carbohydrate; starch performs a similar function in plants and glycogen is sometimes called animal starch. In plants, polysaccharides also serve as skeletal material – a function not paralleled in animals.

Cellulose

Cellulose is the chief structural carbohydrate of plants and, as such, is very widely distributed. All forms of plant life, from the toughest

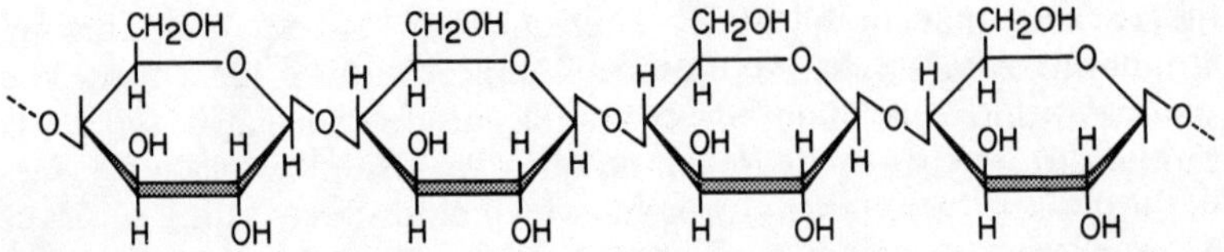

Figure 6.4 Part of a cellulose molecule showing 1–4 connected β-D-glucose units.

tree-trunk to the softest cotton wool, contain cellulose and, indeed, the latter is almost pure cellulose. Whatever its source the constitution of cellulose is the same. Because it is so widely distributed in the vegetable kingdom, cellulose is found, to a greater or lesser extent, in all foods of vegetable origin.

Cellulose can be hydrolysed by heating it with hydrochloric or sulphuric acid. The only monosaccharide obtained in this process is D-glucose and it has been shown that the cellulose molecule consists of a large number of β-D-glucose molecules connected together at carbon atoms 1 and 4 to form very large chain-like molecules as shown in Fig. 6.4.

The number of glucose units connected in this way to form a cellulose molecule varies with the origin of the cellulose, but it is always large and may be as high as 12 000. Bundles of chains lying side-by-side are linked together by hydrogen bonds to give cellulose fibres.

Cellulose is totally insoluble in water and it cannot be digested by man or most other carnivorous animals because the enzymes present in the intestines cannot rupture the β-1-4 links between the glucose units. This is unfortunate because cellulose is the most abundant natural product and if we were able to digest it and obtain nourishment from it an unlimited additional supply of food would be readily available. Although cellulose has no nutritional value it serves a useful purpose as a component of dietary fibre. Much cellulose is removed during the processing of food; for example, the husks of cereal grains, which are mainly cellulose, are usually removed.

It is now generally accepted that dietary fibre is an important part of the diet and removal of large amounts of cellulose during food processing is not looked upon with favour. Cellulose can be digested by horses and by ruminants such as cows. The latter have auxiliary stomachs containing microorganisms which produce enzymes capable of hydrolysing cellulose to glucose.

Starch

Starch is the chief food reserve of all higher plants and is converted, as required, into sugars. It may be stored in the stems, as in the sago palm, or in the tubers, as in potatoes and cassava, from which tapioca is made. Unripe fruits contain appreciable amounts of starch which is converted to glucose as the fruit ripens. It is especially abundant in seeds such as cereal grains and the pulses.

On microscopic examination, starch from various plant sources is found to consist of small particles called granules, the shape and size of which are peculiar to the plant from which they have been obtained. Starch granules are very small and cannot be seen by the naked eye, but they are clearly visible on microscopic examination. They vary in size and the granules of a particular type of starch need not all be of one size. The granules of potato starch, for example, vary in size from 0.0015 cm to 0.01 cm: in other words the largest granules are about seven times as big as the smallest. There are two distinct types of granules in starch obtained from cereal grains such as wheat, barley and rye. Larger granules, which are lenticular in shape (i.e. shaped like lentils or double-convex lenses) and about four millionths of a centimetre in diameter, are accompanied by smaller spherical granules which are only about a quarter the size. The smaller granules outnumber the larger ones by about ten to one but only account for about 30 per cent of the weight of the starch.

Uncooked starchy foods are not easy to digest because the starch granules are contained within the cell walls of the plant which the digestive juices cannot easily penetrate. Cooking softens the cell walls and allows water to enter the starch granules causing them to disintegrate and gelatinize as shown in Fig. 6.5. Overcooking may cause the cell walls to disintegrate completely, producing an unpalatable mush.

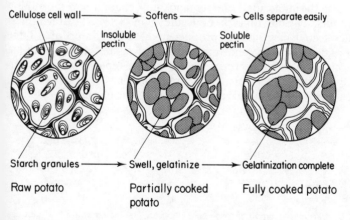

Cellulose cell wall ——→ Softens ——→ Cells separate easily

Insoluble pectin   Soluble pectin

Starch granules ——→ Swell, gelatinize ——→ Gelatinization complete

Raw potato   Partially cooked potato   Fully cooked potato

Figure 6.5 Section of a potato as seen through a microscope.

STRUCTURE OF STARCH   When starch is hydrolysed D-glucose is the only monosaccharide obtained. It was once thought that starch was composed, like cellulose, of strings of connected glucose units, the only difference between the two being that the glucose units in cellulose were the β-isomer and in starch the α-isomer. It is now known, however, that starch is mainly composed of two substances called *amylose* and *amylopectin*. Both of these are polysaccharides and there is usually about three to four times as much amylopectin as amylose.

Amylose is responsible for the blue colour produced when starch reacts with iodine. It can be separated from amylopectin by formation of an insoluble complex with a suitable liquid such as butyl alcohol. The enzyme β-amylase, which is present in cereals, hydrolyses amylose almost completely to maltose. Amylopectin, on the other hand, gives a reddish-brown colour with iodine and only about half of it is converted into maltose by β-amylase, the residue being referred to as a *dextrin*. Glucose syrup (p. 116) contains considerable quantities of dextrins, which are produced by incomplete hydrolysis of starchy materials. When dry starch is heated so-called pyrodextrins are formed. These are brown in colour and soluble in water. Toast and bread-crust both derive some of their brown colour from pyrodextrins.

The molecular weight of amylose varies from about 10 000 to about 50 000 and this corresponds to 70–350 glucose units. The glucose units are connected in a 1–4 manner to form a chain as in cellulose. Maltose is the only disaccharide obtained when amylose is hydrolysed and this shows that the glucose units are the α-isomer and not the β-isomer as in cellulose. The structure of amylose is shown in Fig. 6.6.

The structure of amylopectin is not as simple as that of amylose. To begin with, the molecule is larger and may contain several million glucose units. The amylopectin molecule is, in fact, one of the largest molecules found in a natural product and it is made up of a large number of comparatively short interconnected chains of glucose units. A portion of an amylopectin molecule can be represented diagrammatically as in Fig. 6.7. In this diagram each hexagon represents a glucose unit and AB and DE are the short chains of about 24 glucose

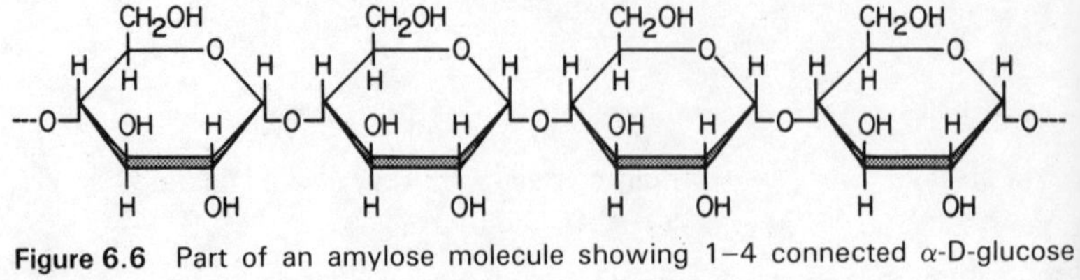

Figure 6.6   Part of an amylose molecule showing 1–4 connected α-D-glucose units.

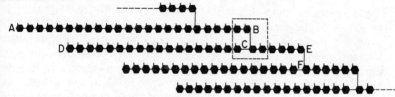

Figure 6.7 Part of an amylopectin molecule.

units connected at BC and EF. The links connecting the chains are from the reducing group at the end of one chain to a primary alcohol group on another, i.e. from C_1 to C_6. The section of Fig. 6.7 enclosed in a broken line is shown in detail in Fig. 6.8.

In a single amylopectin molecule there are large numbers of these chains each containing 20–30 glucose units, depending upon the source of the starch. In Fig. 6.8, only one 'branch' has been shown originating from each chain of glucose units. Most chains have more than one branching point, however, and a most complex three-dimensional bush-like structure can result.

Now that we know something of the structure of amylopectin it is possible to understand why β-amylase can only convert about half of it into maltose. The reason is that amylase splits off pairs of glucose units, in the form of maltose, from the free end of the chains of glucose units in the amylopectin molecule (i.e. the left-hand ends in Fig. 6.7). When the chain has been degraded as far as a branching point the amylase is unable to split off further pairs of glucose units and the product is referred to as a β-limit-dextrin.

In addition to amylose and amylopectin, starch may contain small amounts of non-carbohydrate material such as phosphates and fats.

Figure 6.8 Part of an amylopectin molecule.

The amount and composition of this extraneous material depend upon the source of the starch. Potato starch and wheat starch each contain about 0.5 per cent non-carbohydrate material and phosphates predominate. Wheat starch, on the other hand, contains about 0.75 per cent, made up largely of fats.

GELATINIZATION OF STARCH   If powdered starch is shaken or stirred with cold water it does not dissolve but forms a milky suspension which settles out. When such a suspension is heated, water diffuses into the granules and causes them to swell. This begins at about 60°C and as the temperature increases the granules are progressively disrupted. When the temperature reaches about 85°C a colloidal sol is obtained in which the amylose and amylopectin molecules are completely dispersed in the water. This sequence of events is known as gelatinization. When a starch sol is cooled it congeals into a semi-liquid, semi-solid gel (see p. 57 for a discussion of sols and gels).

Starch resists cold water because the amylose and amylopectin molecules are closely packed in an orderly, almost crystalline, fashion, particularly at the surface of the granules. Hydrogen bonding between adjacent hydrogen and oxygen atoms helps to hold the molecules together and this gives cohesion to the starch particles and increases their resistance to penetration by water molecules. As the temperature increases, however, the molecules of water, amylose and amylopectin acquire additional vibrational energy until eventually the comparatively weak hydrogen bonds between the amylose and amylopectin molecules break down and water molecules are able to penetrate the granules. This starts at about 55–70°C (depending upon the type of starch), which is known as the initial gelatinization temperature.

As gelatinization proceeds the granules swell and a dramatic increase in viscosity occurs. Eventually, the granules lose their separate identity and amylose molecules 'leak out' into the surrounding water to form a sol of somewhat lower viscosity. When the sol cools the thermal energy of the water molecules (and, for that matter, the amylose and amylopectin molecules) decreases, hydrogen bonding is re-established between adjacent molecules and the mixture thickens to become a gel.

Amylose molecules are smaller than amylopectin molecules, but in suspension in water the molecules take up spiral shapes which provide ample opportunity for the formation of hydrogen bonds with adjacent amylose, amylopectin or water molecules. In this way a loose and irregular network is formed in which water molecules are immobilized both by entrapment in network voids and by the weak electrostatic forces of hydrogen bonding. The molecules of amylopectin, though much larger than those of amylose, are relatively more compact and the possibilities of network formation and extensive hydrogen bonding are less pronounced.

The formation of pastes or pasty liquids by heating starchy substances with water or water-containing fluids is of great culinary importance as it is the basis of many dessert, sauce and gravy recipes. Starch-based sauces such as *Bechamel*, *Bordelaise* and *Espagnole* are prepared by heating flour or some other starchy ingredient with butter (or another source of fat) for a few minutes to form a *roux* in which the starch granules are coated with fat. Milk, or a mixture of milk and stock, is then added and heating continued until the starch granules gelatinize. By varying the ingredients and/or their proportions a range of basic sauces, all of which are essentially gelatinized starch, can be prepared.

Several types of starch are used for thickening desserts, sauces and gravies. In Europe and North America wheat starch – in the form of flour – is favoured but potato starch, cornflour and arrowroot are also used. Cornflour (known in North America as cornstarch) is made from maize and is a fairly pure form of starch which gives translucent and glossy sols and gels. Arrowroot as is clear from its name, is a root starch. It is made from the underground stem or rhizome of a West Indian plant with the unmemorable name of Maranta arundinacea.

When a starch-thickened dish is prepared in the kitchen it should be borne in mind that a hot concentrated starch sol becomes a gel when it cools and the near-liquid sol may change into a near-solid gel. This may be acceptable – even desirable – but if it is not the resulting gooey mess will do little to enhance the reputation of the cook. Sauces and desserts based on starch sols and gels should be freshly prepared. Sols may become thin and 'runny' on keeping whereas pastes and gels may 'weep' water and become tough and rubbery. In both cases the change in physical properties is caused by what is known as *retrogradation*, in which the amylose molecules clump together and separate from the sol or gel thus destroying the semi-elastic network on which its properties depend. Starches with a very high amylose content undergo retrogradation less readily than those with a lower amylose content and such starches are commercially available to food manufacturers. Flour-thickened sauces or soups may also become thinner on keeping because of the breakdown of the starch by the enzyme α-amylase in the flour. This can be prevented in manufactured foods (e.g., in 'packet' soups) by the use of enzyme-inactivated flour (see p. 138).

MODIFIED STARCH Starch which has been chemically and/or physically treated so that it will form a gel with cold water or milk is used by food manufacturers under the name of *modified starch*. Amongst other things it is used in the manufacture of a wide variety of 'instant' desserts, mousses, 'toppings' and whips which have been most successfully marketed by the food industry. These products also contain phosphates, such as disodium phosphate, Na_2HPO_4, which, in the presence of calcium ions from milk, form a gel with the modified starch and milk proteins. These concoctions also contain sucrose (the main ingredient), milk solids and a veritable catalogue of additives including permitted gelling agents, emulsifiers, stabilizers, antioxidants, colours and flavourings.

Pectin

Pectin is the name given to a mixture of polysaccharides found in soft fruits and in the cell walls of all plants. Concentrated pectin extracts can be bought for use in jam making (see p. 124). They are made either from citrus fruit residues or from the pulp remaining after the expression of juice from apples. The pectin is extracted by pressure cooking with water or very dilute acid. The extract is filtered, vacuum concentrated, and the pectin precipitated by adding alcohol.

Pectin from various sources differs somewhat in composition and hence it is not possible to specify its structure too precisely. However, all pectins consist essentially of the polysaccharide *methyl pectate* which has long chain-like molecules of several hundred connected units of α-D-galacturonic acid – an acid derived from the monosaccharide galactose. Some of the acid groups have been esterified and converted from free carboxyl ($-COOH$) groups to the methyl ester ($-COOCH_3$). Other sugars (or acids derived from them) may also be present in the chain. This sounds complicated, and so it is, but fortunately the properties of pectin are of more interest to the practising food scientist than its exact chemical composition, and its use as a gelling agent is discussed in connection with jam making in the next chapter.

Pectin is present in unripe fruits and vegetables mainly in the form of its precursor *protopectin*. This is a water-insoluble compound in which most of the carboxyl groups are esterified. Protopectin is responsible for the hard texture of unripe fruits and vegetables: during ripening enzymes present in the plant convert it into pectin.

Glycogen

The polysaccharide glycogen acts as a reserve carbohydrate for man and other animals. Because its function in man parallels that of starch in plants it is sometimes referred to as animal starch. It is a very large molecule consisting of branched chains of α-D-glucose units and its structure is very similar to that of amylopectin but chain branching occurs, on the average, at about every 18–20 glucose units (compared with about 20–30 in amylopectin).

Glycogen is present in man and other animals in the muscles and in the liver; its function is explained in more detail in Chapter 7. There may be several kilograms present in the body of a large animal such as a cow. Despite this, however, glycogen is not a normal constituent of the diet because it is converted into lactic after an animal has been killed.

CARBOHYDRATES IN THE BODY

Foodstuffs rich in carbohydrates are of value primarily as sources of energy. Weight for weight, proteins provide roughly the same amount of energy as carbohydrates, and fats provide twice as much. Of the three, however, carbohydrates are by far the cheapest and the most easily digested and absorbed. Over-indulgence in foods rich in sugar may lead to obesity and to under-consumption of other foods containing essential nutrients. This is equally true for over-consumption of foods containing polysaccharides because these are hydrolysed to monosaccharides before absorption. If more monosaccharide is produced than is required the excess is converted into fat and stored in fat depots until it is required for provision of energy in time of restricted food intake.

In the absence of sufficient carbohydrate the energy requirements of the body can be met by protein and fat, and it is possible to live on a diet containing little carbohydrate. Eskimos, for example, live almost exclusively on protein and fat. In Japan, on the other hand, about 80 per cent of the total energy intake is supplied by carbohydrate. A diet which is low in carbohydrate is not recommended for persons in normal health, however, because carbohydrate 'spares' protein. The body can make best use of protein if carbohydrate is eaten at the same time.

During digestion the carbohydrates contained in food (with the exception of cellulose), are hydrolysed by enzymes to their component monosaccharides. The process starts in the mouth, where saliva, which contains salivary amylase, is intimately mixed with the food and begins to hydrolyse the starch to maltose. The hydrolysis continues in the stomach until the food is acidified by mixture with the gastric juice. The food passes from the stomach to the small intestine, where pancreatic amylase continues the conversion of starch to maltose. Maltase present in the intestinal juices hydrolyses the maltose so formed to glucose. Lactase and sucrase, which are also present, convert lactose and sucrose to glucose, galactose and fructose. The monosaccharides pass from the small intestine into the bloodstream and are carried to the liver (where fructose and galactose are enzymically converted to glucose) and the muscles.

The liver and muscles are able to convert glucose to glycogen which serves as a reserve of carbohydrate for the body. Glycogen is reconverted to glucose as required to meet the energy needs of the muscles and other tissues. The body of a well-nourished man may contain several hundred grams of glycogen about one-quarter of which is stored in his liver. When the muscles and liver can accommodate no more glycogen the surplus glucose is converted by the liver into fat which is stored in fat depots as the body's second line of defence against food shortage. A small amount of glucose circulates in the blood for transport to tissues, drawing upon the liver's carbohydrate stocks or for transport to the liver to be converted to glycogen. After eating a meal the blood may contain up to 0.14 per cent glucose but this figure falls to about 0.08 per cent some two hours or so after eating. This is quite a small amount and is equivalent to only about five grams of glucose.

The oxidation of glucose

The body obtains energy by converting fats, proteins and carbohydrates to glucose and oxidizing that to simpler molecules and ultimately to carbon dioxide and water. The overall process may be likened to the release of energy from fossil fuels by burning them in a power station or boiler. Metabolic oxidation is is infinitely more subtle and impressive, however, and man's most sophisticated fuel burning appliances are primitive compared with nature's delicately balanced and controlled complex of interdependent reactions.

The living cell has to function in an aqueous environment at a constant pH and at a comparatively low and essentially constant temperature. It cannot use heat energy to do work and, bearing in mind the limitations mentioned, it may seem surprising that any energy is made available at all. However, as we shall see below, in fact almost half the energy locked up in the glucose molecule is captured by the body and this is a better conversion rate than that achieved by the most modern fuel-burning power station.

When glucose is completely oxidized to carbon dioxide and water by burning in oxygen a large amount of heat is evolved:

$$C_6H_{12}O_6 + 6O_2 \rightarrow 6CO_2 + 6H_2O + 2820 \text{ kJ}$$

In the body, however, oxidation does not take place in one step but by a complicated and elegant series of nearly 30 reactions, each of which releases only a fraction of the energy which would be made available by complete oxidation of the glucose molecule. Many of the oxidative steps in this sequence of reactions do not involve direct reaction with oxygen at all, but are simple dehydrogenation reactions which can be represented by a general equation:

$$AH_2 \xrightarrow{\ -2[H]\ } A$$

In many of these reactions the two hydrogen atoms removed are transferred to a coenzyme, for example, Coenzyme 1 (nicotinamide-adenine dinucleotide or NAD) or Coenzyme 11 (nicotinamideadenine dinucleotide phosphate or NADP), which act as hydrogen acceptors. The transfer is catalysed by oxidases which are highly specific in the sense that a particular oxidase will only work with one substrate-coenzyme pair.

There are two main stages in the oxidation of glucose by the body. In the first stage it is converted by a series of reactions to pyruvic acid:

$$C_6H_{12}O_6 \xrightarrow{\ -4[H]\ } 2CH_3COCOOH$$ No oxygen required
33 kJ energy made available to the body.

In the second stage the pyruvic acid is oxidized by a further series of reactions to carbon dioxide and water:

$$CH_3COCOOH \xrightarrow{\ +5[O]\ } 3CO_2 + 2H_2O$$ Oxygen required
About 126 kJ energy made available to the body.

The first stage is called *glycolysis* or, since it can take place in the absence of oxygen, *anaerobic glycolysis*. The amount of energy made available by glycolysis is small compared with that released during the second stage. Energy for violent physical exercise – such as running to catch a bus – is required instantly, and as the bloodstream may be unable to supply oxygen sufficiently quickly to permit complete oxidation of glucose, anaerobic glycolysis takes place preferentially. The pyruvic acid produced is reduced to lactic acid and this is carried by the bloodstream to the liver where part of it is oxidized to provide energy for reconversion of the remainder to glycogen and glucose.

The cyclic conversion of glucose to pyruvic acid, lactic acid and back to glucose is known as the Cori cycle after its discoverers, Carl and Gerty Cori. Normally, when energy is not required so quickly, it is obtained by complete oxidation of glucose to carbon dioxide and water via pyruvic acid and about 75 per cent of the cell's energy requirements is provided by the second stage. Conversion of pyruvic acid to carbon dioxide and water occurs by a cyclic process involving citric acid which is often referred to as the citric acid cycle or, after Sir Hans Krebs, who worked out the details of the process, the Krebs cycle. Figure 6.9 summarizes the changes which occur during the digestion absorption and oxidation of carbohydrates by the body.

The biochemical oxidation of glucose is a complex and involved process and it took many years to work out exactly how it occurs. All the separate reactions are enzyme-controlled, ten enzymes being required for the conversion to pyruvic acid and another ten for the second stage of the oxidation to carbon dioxide and water. The whole sequence is of great importance and beauty, but readers may be relieved to learn that we do not intend to give complete details of it here. However, one aspect of the oxidation process, which is common to both the first and second stages, and is indeed a feature of many other biochemical energy transformations, deserves further attention. This is the way in which *adenosine triphosphate* (or ATP) can behave as an energy-bank or energy storehouse for the body. ATP is built up from one molecule of the purine derivative adenine, one molecule of the sugar ribose (which together form the nucleotide *adenosine*) and three molecules of phosphoric acid.

In the body the phosphate groups may be split off successively from an ATP molecule to yield first *adenosine diphosphate* (ADP), then *adenosine monophosphate* (AMP), and finally adenosine itself. The adenosine part of the molecule is unchanged in this series of reactions and if we represent it by Ⓐ we can represent the series of reactions as follows:

$$
\overset{\displaystyle O}{\overset{\displaystyle \|}{\underset{\underset{\textstyle OH}{|}}{\text{Ⓐ}-P}}}-O-\overset{\displaystyle O}{\overset{\displaystyle \|}{\underset{\underset{\textstyle OH}{|}}{P}}}-O-\overset{\displaystyle O}{\overset{\displaystyle \|}{\underset{\underset{\textstyle OH}{|}}{P}}}-OH \xrightarrow{\ H_2O\ }
$$

(ATP)

$$
\overset{\displaystyle O}{\overset{\displaystyle \|}{\underset{\underset{\textstyle OH}{|}}{\text{Ⓐ}-P}}}-O-\overset{\displaystyle O}{\overset{\displaystyle \|}{\underset{\underset{\textstyle OH}{|}}{P}}}-OH \xrightarrow{\ H_2O\ } \overset{\displaystyle O}{\overset{\displaystyle \|}{\underset{\underset{\textstyle OH}{|}}{\text{Ⓐ}-P}}}-OH \xrightarrow{\ H_2O\ } \text{Ⓐ}
$$

(ADP)   (AMP)

+ H_3PO_4 + H_3PO_4 + H_3PO_4

The importance of this series of reactions is the fact that the two terminal phosphate groups are attached by high-energy phosphate bonds (shown in dark print in the equations) and conversion of one mole of ATP to ADP or one mole of ADP to AMP is accompanied by the release of 33 kJ. Conversion of AMP to adenosine, on the other hand, only yields 12.6 kJ. In the reverse reactions the same amounts

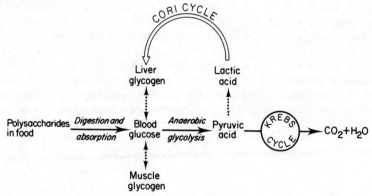

Figure 6.9 Digestion, absorption and oxidation of carbohydrates.

of energy are absorbed:

$$\overbrace{\text{Energy evolved available for synthesis or work}}$$

$$\text{ATP} \underset{\text{33 kJ absorbed}}{\overset{\text{33 kJ evolved}}{\rightleftharpoons}} \text{ADP} \underset{\text{33 kJ absorbed}}{\overset{\text{33 kJ evolved}}{\rightleftharpoons}} \text{AMP} \underset{\text{12.6 kJ absorbed}}{\overset{\text{12.6 kJ evolved}}{\rightleftharpoons}} \text{A}$$

$$\underbrace{\text{Energy absorbed from oxidation of glucose}}$$

The energy-rich phosphate bonds function as energy stockpiles for energy released during oxidation of glucose in the cell. The energy is available for reuse on demand, for muscular contraction or to make possible the synthesis of some other molecule by the body or for any other purpose. When this occurs ATP is converted to ADP or AMP which are then available for reconversion to ATP at a time when surplus energy is available.

During the complete oxidation of one mole of glucose by the body 36 moles of ATP (or their equivalent in related compounds) are formed from ADP and the energy absorbed in this process (and hence available for further use) is $36 \times 33 = 1188$ kJ. This compares with the 2820 kJ evolved when glucose is burned in oxygen and from this we see that the body is able to capture about 43 per cent of the energy of the glucose molecule. The residue is either used up in making other molecules during the oxidation process or appears as heat.

The complicated series of reactions involved in the assimilation and utilization of glucose by the body are controlled by several hormones, the best known of which is *insulin*, which is secreted by the pancreas. In the disease *diabetes mellitus* insufficient insulin is produced by the pancreas; as a result glucose circulates in the blood in abnormally large amounts and is not taken up by the liver or muscles for conversion to glycogen or for oxidation. The body is thus unable to utilize carbohydrate foods and has to resort instead to fats and proteins to make up its energy deficiencies. Unfortunately, increased utilization of fat in this way leads to accumulation of certain poisonous products of fat metabolism in the liver and the bloodstream and this can have serious consequences (see ketosis, p. 85). Diabetes mellitus can be controlled by careful attention to the diet or, in more severe cases, by regular injections of insulin.

CHAPTER 7

Carbohydrate foods

With the exception of lactose, which is present in animal milk, carbohydrates are of plant origin and they are formed by photosynthesis from carbon dioxide and water. A substantial amount of energy is required to transform the simple molecules of carbon dioxide and water to the more complex carbohydrate molecules and this is obtained from sunlight. When carbohydrates are eaten this process is, in effect, put into reverse to produce energy for use by our bodies. The way in which this is done has already been discussed in Chapter 6.

About one-third of the total dietary intake of carbohydrates in Britain consists of sucrose and glucose syrup, about one-twentieth is lactose and the remainder is starch and dietary fibre.

The carbohydrate content of some foods is shown in Table 7.1.

Table 7.1 The carbohydrate content of some foods

Food	Carbohydrate (%)	Food	Carbohydrate (%)
Rice (raw)	86	Bread (white)	45
Cornflakes	85	Bread (wholemeal)	38
Honey	76	Fried chipped potatoes	34
Spaghetti (raw)	74	Dairy ice-cream	20
Weetabix	70	Bananas	19
Jam	69	Boiled potatoes	18
Porridge oats	66	Grapes	16
Milk chocolate	59	Peanut butter	13
Potato drisps	50	Apples	12

SUGARS IN THE DIET

Monosaccharides

Glucose and *fructose* are the only monosaccharides present to any extent in an average diet. They occur in honey in roughly equal amounts and in 'glucose syrup' which is extensively used by food manufacturers for sweetening their products.

HONEY   The nectar of flowers from which bees make honey consists largely of sucrose. During its passage through the bee's honey sac and during further activity by bees in the hive the sucrose is hydrolysed by enzymes into glucose and fructose and the product – honey – consists of about 20 per cent water and about 76 per cent glucose and fructose. The balance is made up of a small amount of unconverted sucrose, some other disaccharides and minor quantities of minerals, acids, vitamins and flavour-producing substances.

The very variable flavour of honey depends essentially on the flowers from which the nectar was collected. Honey contains over 200 different substances but honey lovers are able to distinguish honey derived mainly from clover, for example, from that originating from other flowers.

Although honey is now mainly eaten as a spread like jam it has been used as a sweetener since ancient times and it was the only sweetening agent used until sucrose became available. It was also used for making the alcoholic drink mead, which is a beer made by fermenting honey.

The nutritive and curative properties of honey are often grossly exaggerated. It is really no more than a flavoured, concentrated solution of glucose and fructose. There is no evidence that honey has special nutritional or medicinal properties or that it can function as an aphrodisiac or delay the onset of old age. Like other sugars its function in the diet is solely to act as a source of energy.

GLUCOSE SYRUP   This is a sweet, syrupy mixture of glucose with other sugars and dextrins which is used extensively by the food industry. It is produced by the hydrolysis of starchy materials, such as maize, with dilute acids or by enzymes. The composition of glucose syrup (or *corn syrup* as it is known in North America) depends upon the extent to which the hydrolysis occurs, i.e. upon how far the amylose and amylopectin molecules in the starch are broken down. The degree of hydrolysis can be controlled to give a product which is largely glucose or a more viscous syrup containing substantial amounts of *dextrins* (i.e. partially hydrolysed amylose and amylopectin molecules). The more completely hydrolysed syrups, which contain a larger proportion of sugars, are sweeter than those containing substantial amounts of dextrins and they are widely used in the manufacture of sweet confectionery and cakes.

If fully hydrolysed glucose syrups are treated with the enzyme *glucose isomerase* up to half the glucose may be converted to fructose. The product is sweeter than glucose syrup because fructose is the sweetest of the sugars. It is claimed that, in addition to its greater sweetness, fructose has the effect of enhancing fruit flavours and 'high-fructose syrups' are widely used in soft drinks and jams.

Sucrose, cane sugar or beet sugar

Ordinary sugar, whether produced from sugar-cane, (which can only be grown in tropical countries) or sugar-beet (which can be grown in temperate zones) is almost pure sucrose. There is no difference between the sugar obtained from sugar-beet and that obtained from sugar-cane. Sugar has been produced commercially from sugar-beet for about 150 years. About 30 per cent of the total world production of sugar and 60 per cent of British sugar is produced from sugar-beet.

PRODUCTION OF RAW SUGAR   Sugar-cane is a type of giant grass which resembles bamboo and may grow to a height of 4.5 metres with a diameter of 3–5 cm. The sugar, which amounts to about 15 per cent of the weight of the cane, is found in a soft fibre in the interior of the cane. From three to eight tons of sugar are normally obtained from an acre of sugar-cane. The sugar is extracted from the cane by crushing and spraying with water. The solution obtained contains about 13 per cent sugar and three per cent impurities, the rest being water. This solution is purified and the clear solution obtained is concentrated by evaporation under reduced pressure until a mixture of sugar crystals and mother liquor is obtained. The crystals are separated from the mother liquor by spinning the mixture very quickly in large drums. These drums have perforated sides through which the mother liquor, called molasses, is forced by centrifugal action, leaving the raw sugar behind. Molasses is used for the manufacture of rum and industrial alcohol.

The sugar content of sugar-beet is similar to that of sugar-cane but only about two tons of sugar can be obtained from an acre of sugar-beet. The sugar is extracted from the shredded beet by steeping in hot water; it diffuses through the cell walls into the water, leaving behind most of the non-sugar solids. Batteries of specially designed diffusers are used to ensure that the maximum amount of sugar is removed by the minimum quantity of water. The solution obtained contains about 14 per cent sugar and four per cent impurities, the remainder being water. It is treated to remove impurities and the clear solution obtained is concentrated by evaporation under reduced pressure to give sugar crystals and molasses which are separated by centrifuging as described for cane sugar. The raw sugar obtained is very similar to that produced from cane sugar but may have a less attractive taste and smell.

SUGAR REFINING   Raw sugar, whether obtained from beet or cane, contains about 96 per cent sucrose. It consists essentially of sugar crystals coated with a layer of molasses. The first step in the refining process is to mix the raw sugar with sugar syrup to produce a stiff semi-solid mixture of syrup and crystals. This is centrifuged and the syrup is forced out, leaving the sugar crystals within the centrifuge. The sugar crystals are then washed with water to remove the adhering syrup. A considerable amount of sugar is contained in the syrup and wash liquors from the centrifuge and this is recovered by evaporation of the water under reduced pressure.

The sugar crystals are next dissolved in water and treated with milk of lime and carbon dioxide to remove the bulk of the impurities. The remainder are removed by allowing the solution to percolate through a deep bed of charcoal which removes coloured impurities to produce a colourless liquid known as *fine liquor*.

All that remains to be done is to concentrate the fine liquor by evaporating the water, and crystallize the sucrose to produce uniform crystals of the correct size. The evaporation is carried out at reduced pressure in steam-heated 'vacuum pans' so that the water can be removed at a temperature much below the normal boiling point of the solution; this prevents any discoloration of the sugar.

At the end of the crystallization the sugar crystals are suspended in syrup. Spinning in centrifugal machines removes most of the syrup, the remainder being removed by spraying with hot water and centrifuging. The syrup and washings contain about 40 per cent of the sugar originally present in the fine liquor and most of this is recovered by decolorizing and recrystallizing. This can be repeated several times. Eventually, however, the colour and quality of the sugar recovered from the syrup do not conform to the high standards required. When this happens the syrup is used for the manufacture of Golden Syrup or is concentrated and crystallized to give a soft brown sugar.

Caster sugar is made in the same way as granulated sugar but the crystallization procedure is modified so that much smaller crystals are obtained. This is done by using a larger number of nuclei or by preventing the crystals from growing to granulated sugar size. Caster sugar is also obtained as a by product of the production of granulated sugar and it is separated from it by sieving the dry sugar. The particle size in *icing sugar* is even smaller than in caster sugar and it is made by pulverizing granulated sugar in special mills. *Brown sugar* is obtained by crystallizing the syrup obtained at the end of the refining process. The sugar obtained is slightly sticky because of the presence of a thin coating of mother liquor on each crystal. Genuine *demerara sugar* is a golden-brown, slightly moist sugar produced in this way in Demerara, Guyana. *London demerara* is a substitute made by coating white sugar crystals with molasses. *Muscovado* (or *Barbados sugar*) has a higher molasses content than demerara sugar and a 'treacly' flavour. *Lump sugar* is made by crystallizing the fine liquor (see above) in such a way that crystals of two different sizes are obtained: the larger crystals give the characteristic sparkle of lump sugar and the smaller ones bind the crystals together

FOODSTUFFS MANUFACTURED FROM SUGAR

Sugar, as we have seen, occurs naturally in fruits but it is also present, sometimes in surprisingly large amounts, in many manufactured foods. Who would imagine that bottled tomato sauce and 'brown sauce' would contain as much as 25 per cent sugar? Some mueslis also contain over 25 per cent sugar and substantial amounts are present in many other breakfast cereals. Ice-cream contains about 20 per cent sugar, sweetened yoghurt up to 15 per cent and even baked beans contain about six per cent! In this section, however, we will be concerned only with those substances such as 'sweets' and jam where sugar is the main constituent.

Sugar confectionery, chocolate and jam

The term sugar confectionery is used to describe the large range of confectionery we commonly call 'sweets'. Boiled sweets, toffees and caramels, the filling used as centres for chocolates, marshmallow, nougat, pastilles and gums are all examples of sugar confectionery. This great variety of tooth-decay accelerators have one thing in common – they are all produced by controlled crystallization of sucrose from a supersaturated solution. The differences that exist between them depend upon water content, the extent to which crystallization of sucrose has occurred, and the presence of fat or milk, which enables emulsions to be formed, and flavouring agents.

When a concentrated solution of sucrose – a syrup – is heated, the temperature at which it will boil depends upon the amount of water present in the solution as shown in Fig. 7.1. The water content of a sugar solution can be accurately determined from a knowledge of its boiling point. If boiling is allowed to continue the concentration of the solution increases as water evaporates and the boiling point increases. Soft sweets, such as toffee, are cooked to a lower temperature than hard sweets such as butterscotch or boiled sweets.

BOILED SWEETS These are traditionally made by boiling sugar solutions with acidic substances to produce a certain amount of inversion. Boiling is continued until the temperature exceeds $150°C$, by which time almost all the water has been driven off and, on cooling, the mass solidifies as a glass-like solid – the familiar boiled sweet. At the end of the cooling period only a small amount of water (a few per cent) is present. The glucose produced by the inversion

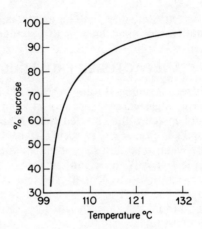

Figure 7.1   The boiling points of sugar solutions.

process prevents the sucrose from crystallizing when the mass is cooled down. It is said to 'cut the grain'. Glucose itself does not crystallize easily from water and in a mixed solution of glucose and sucrose the glucose inhibits the crystallization of the sucrose. The final product is a supersaturated solution of sucrose, with smaller amounts of glucose and fructose, in a very small quantity of water. The amount of invert-sugar produced during the boiling must be carefully controlled because too much will make the product prone to absorb water from the air and become sticky. This is because the fructose in the invert-sugar is hygroscopic. On the other hand, if too little invert-sugar is produced it will be insufficient to prevent crystallization of the sucrose. About 10–15 per cent invert-sugar is the amount required to give a non-sticky non-crystalline product.

Any acidic substance may be added to produce the inversion, and cream of tartar (potassium hydrogen tartrate) is commonly used. The amount of acid used depends upon several factors, including the hardness of the water, but it is quite small and, in the case of cream of tartar, is in the region of 0.15–0.25 per cent of the weight of sugar used. Boiled sweets are made commercially by heating a syrup of sucrose and glucose syrup which prevents crystallization of the sucrose in the same way as invert-sugar. The amount of glucose syrup used can amount to 30–40 per cent but the product is not hygroscopic because it contains no fructose. Glucose syrup is less sweet than sucrose or invert-sugar and so the boiled sweets produced in this way are also less sweet. In addition, the dextrin in the glucose syrup is said to impart a certain toughness to the sweets and can sometimes cause cloudiness.

FONDANT   This is the creamy material used for filling soft-centred chocolates and by biscuit and cake manufacturers for decorative purposes. It consists essentially of minute sugar crystals surrounded by saturated sugar syrup.

Fondant is made by boiling a sugar solution and adding glucose syrup or an inverting agent, as in the case of boiled sweets. No attempt is made to boil off all the water, however, and the mixture is only boiled to 115–120°C compared with 150–165°C in the case of boiled sweets. The syrupy solution obtained is cooled quickly to about 38°C by running onto a rotating water-cooled drum from which it falls into a beater. Here it is agitated violently to induce crystallization which occurs suddenly to produce a very large number of tiny crystals.

FUDGE   This is simply fondant containing added milk solids, fat and chocolate solids which are suspended in the sugar syrup and give it added solidity.

TOFFEE   Fat, milk, sugar and glucose syrup are the main ingredients of toffee. It consists essentially of a dispersion of minute globules of fat in a supersaturated sugar solution. Various grades of sugar are used, depending upon the recipe, from the best quality granulated sugar to raw sugars and treacles which contribute characteristic flavours to the product. As in the case of boiled sweets, sugars other than sucrose must be present to prevent graining. In home-made toffee an acidic inverting agent such as vinegar or citric acid may be used to produce some invert-sugar from sucrose, but in commercial practice it is usual to employ glucose syrup to prevent graining.

The milk used makes an important contribution to the flavour and is added as condensed milk, either full-cream or skimmed. The characteristic colour of toffee is largely due to caramelization of the milk solids during cooking. In addition, the casein of the milk acts as an emulsifying agent. Butter and various vegetable fats are used in the manufacture of toffee, and emulsifying agents, such as glyceryl monostearate or lecithin, may also be incorporated, if insufficient milk solids and butter are present, to aid in the dispersion of the fat and produce a stable emulsion.

In toffee manufacture the ingredients are boiled together until the temperature reaches the required level. The temperature attained in the boiling process largely determines the consistency of the toffee produced because this depends, among other things, on the amount of water in the toffee. Very hard toffee such as butterscotch is heated to 146–154°C, which gives a water content of 3–5 per cent. Ordinary toffees and caramels are heated to 118–132°C when the mixture contains 6–12 per cent water.

Toffee and fudge are made from similar materials and they both contain about ten per cent water. Why then are they so dissimilar in other respects? The short answer is that toffee, like boiled sweets, consists of sugar syrup surrounding other ingredients such as fat and milk whereas in fudge, as in fondant, part of the sugar is present in crystalline form. Toffee is in fact a thick, supersaturated syrup and not a solid at all, and this is why it has such a tough and chewy consistency.

There are innumerable varieties of toffees but in all of them there is a common background 'caramel taste'. This flavour is produced by 'caramelization' or breakdown of some of the sugar molecules which occurs whenever carbohydrates are heated. During toffee making a reaction also occurs between the heated sugars and the proteins present in the milk solids and this produces a brown colour. A *browning reaction* of this type (called the *Maillard reaction* or non-enzymic browning reaction) occurs whenever carbohydrates are heated with proteins (see p. 295).

CHOCOLATE   The essential ingredients of chocolate are cocoa, cocoa-butter and sugar. The cocoa and cocoa-butter are both obtained from cocoa-beans which grow in pods on cacao trees in tropical countries. The pods are egg-shaped, about eight inches long, and about three to four inches in diameter. Each pod contains 20 to 40 beans embedded in a soft, white, starchy pulp. The pods are split open and the beans, with adhering pulp, are scraped out and allowed to ferment for several days. A yeast fungus grows on the pulp and it liquefies owing to fermentation to alcohol. The liquid formed is allowed to drain away from the beans which during the process change in colour from their original light violet to dark brown. After drying in the sun the beans are ready for shipment to the cocoa manufacturers.

In the manufacture of cocoa and chocolate, the beans are first roasted in revolving drums and are then broken into small pieces by passing through special rollers. The husk is removed, leaving behind the small pieces of roasted bean which are known as *nibs*. This roasting process is of great importance because it is at this stage that the characteristic chocolate flavour and aroma develop as a consequence of Maillard-type browning reactions occurring between carbohydrates and proteins present.

The nibs contain about 50 per cent of a fat known as cocoa-butter and during the next operation, in which the nibs are finely ground in mills, the heat generated melts the cocoa-butter to produce a viscous brown liquid as product. When the liquid, which is a dispersion of cocoa in cocoa-butter is cooled, a brown solid known as *cocoa-mass* is obtained. For the production of cocoa a proportion of the cocoa-butter is squeezed out in powerful hydraulic presses. The residue, containing 20–30 per cent cocoa-butter, is finely ground and sold as cocoa powder. It contains about two per cent theobromine and 0.1 per cent caffeine; both these compounds are alkaloids and they are responsible, at least in part, for the bitterness of cocoa and chocolate.

Chocolate is made by mixing cocoa-mass with sugar, cocoa-butter and, for milk chocolate, dried milk or condensed milk. The mixing is carried out in *melangeurs*, in which massive rollers rotate in contact with a heated plate. The mixture then passes to a refining machine where it is pinched between rollers revolving at different speeds. To complete the process the molten chocolate is then 'conched' for a period of up to 24 hours. In the conching process heavy rollers subject the chocolate to severe mechanical treatment and blend all the ingredients into a uniform velvety consistency.

Chocolate for coating purposes, such as is used in covering 'centres' to make individual chocolates or in covering biscuits, contains a larger proportion of cocoa-butter than ordinary block chocolate. The purpose of this is to increase the fluidity of chocolate when warm. Chocolate-coated goods are made by passing the centre or biscuit to be coated through a curtain of molten chocolate in a machine called an *enrober*.

Conversion of liquid chocolate to the familiar solid chocolate bar is not simply a matter of pouring the chocolate into moulds and allowing it to cool. The fats present in cocoa butter can solidify in six different forms, or polymorphs, with different melting points. One of the polymorphs melts at $33.8°C$ and when only this form is present in solid chocolate it will be smooth and glossy and will easily melt in the mouth. To produce as much of this polymorph as possible chocolate is subjected to a special heat treatment process called *tempering*. The molten chocolate is cooled until it just begins to solidify and then it is reheated to just below the melting point of the desirable polymorph. The chocolate is then stirred at this temperature so that a high proportion of the fat will solidify in the preferred polymorphic form when the chocolate is finally moulded or used for coating.

If chocolate is incorrectly tempered, or if it is subjected to a series of temperature changes (e.g., if left in a shop window or a car) it may develop a white coating or 'bloom'. This may look like mould growth but it is in fact a harmless coating of fat crystals.

Molten chocolate is such a capricious substance that it is difficult to handle in the kitchen where precise temperature measurement and control is not easy. For this reason chocolate-flavoured cake coating substitutes are available which contain vegetable fat in place of cocoa-butter. Vegetable fat solidifies in only one form and hence the problems caused by the polymorphism of cocoa-butter fats do not arise.

Chocolate is a nutritious food and a small bar (100 g) of milk chocolate provides about 9 g of protein and 220 mg of calcium – approximately one-eighth and one-half respectively of the RDA of these nutrients for a moderately active man. It would also supply about one-sixth of his energy requirements and 10–15 per cent of the RDA of iron, thiamine and riboflavin.

JAM   Jam is made by boiling fruit with sugar solutions and is essentially a gel or semi-solid mass containing pulped or whole fruit. A gel is really a very viscous solution, or dispersion, which possesses some of the attributes, such as elasticity, of a solid. The gel is formed from the sugar, the acids present in the fruit and the polysaccharide pectin (see p. 108).

Jam contains about 67 per cent dissolved sugar and it inhibits the growth of moulds and yeasts because the water activity (see p. 336) is too low to permit their multiplication.

The quantity of pectin and acid in fruit for jam making is of great importance because gel formation only occurs when the concentrations of sugar and pectin and the pH of the mixture lie within certain limits. Some fruits, such as currants, damsons, gooseberries, lemons and bitter oranges, are rich in both acid and pectin and can easily be made into jam. Others, such as strawberries, blackberries, raspberries and cherries contain little pectin and some must be added before jam can be made successfully from them. A simple way of doing this is to add another fruit rich in pectin and acid, for example, apple or a concentrated pectin preparation (see p. 108). As the percentage of pectin in a jam increases so does the firmness of the gel produced on cooling. A satisfactory gel is obtained with about one per cent pectin, although for a given pH and sugar content the firmness of the gel is influenced by the 'quality' of the pectin as well as by the quantity present.

As we have already seen, pectin consists of large numbers of simpler molecules connected together to form a long thread-like molecule. The length of a pectin molecule depends upon its source. During gel formation the long molecules link loosely together to form a three-dimensional network which gives the gel its stability. If the pectin molecules are too short the gel may lack strength and be runny or soft.

Anyone who has tried to make jam at home knows that the jam obtained from ripe fruit is not as good as that from fruit which is almost ripe. This is because pectin will not form a satisfactory gel until the pH is lowered to about 3.5 and unripe fruit is usually more acidic than ripe fruit. To decrease the pH of a jam mixture an acidic fruit juice, such as lemon juice, may be added or a small amount of citric acid, malic acid or cream of tartar. A low pH during the cooking period may cause inversion of too much sucrose, and also hydrolyse the pectin to some extent. As both these changes are detrimental the pH adjustment is often carried out at the end of the cooking period.

When fruit is boiled with sugar a certain amount of inversion occurs (see p. 99) and this is of the utmost importance because invert-sugar prevents the crystallization of the sugar in the jam when it is kept. Too much invert-sugar, however, is detrimental because it reduces the strength of the gel and may cause the jam to set to a honey-like mass on keeping.

Another reason why fruit used in jam making should not be over-ripe is that pectic substances, particularly *protopectin*, from which pectin is formed during jam making, are present in maximum amount just before the fruit is ripe. If jam factories could make jam only when supplies of fresh fruit were available they would be able to work for only a very short period each year. Moreover, the great amount of fruit harvested during these periods would be far too large to be made quickly into jam by the existing number of jam factories. To overcome this difficulty large amounts of preserved fruits are used. The method of preserving the fruit is very simple; it is kept submerged in a weak solution of sulphur dioxide in water. Fruit can be preserved in this way either before or after cooking. In practice, strawberries, raspberries and blackberries are preserved raw whereas plums, currants and gooseberries are usually preserved as a cooked pulp because sulphur dioxide tends to toughen the skins of the uncooked fruit. The fruit is bleached by the sulphur dioxide during the preserving but the colour returns during the cooking. Almost all the sulphur dioxide is driven off during cooking and the finished jam must not contain more than 100 parts per million.

When jam is made at home the fruit is first cooked until tender. Pectin is extracted during this process and the length of the cooking period depends upon the fruit being used. Cooking is complete when a sample of the fruit forms a coherent pectin clot when allowed to stand for a few minutes with several times its volume of methylated spirit. The clot should be strong enough to withstand pouring from one vessel to another without breaking. When a satisfactory pectin clot is obtained the sugar should be added and the mixture boiled as rapidly as possible. The duration of the boiling period will depend on the fruit used and the size and shape of the pan; it varies from about five to about 20 minutes. A shallow pan, in which a large surface area of the jam is exposed for evaporation, is best because this permits rapid evaporation. The boiling period is usually complete when the temperature of the boiling mixtures reaches $104°C$. The exact temperature at which boiling should be stopped depends upon the acidity and pectin content of the fruit, however, both of which influence the setting properties of the jam.

Jam making is carried out commercially in a much more scientific way than this. To begin with, the recipe is adjusted to give the correct amounts of sugar and pectin with the particular fruit being used and pH is also carefully controlled. Pre-cooked pulp is often used and the boiling is usually carried out in open pans holding about 180 litres of jam. The boiling time is very short and seldom exceeds ten minutes. This short boiling time preserves the gel-forming properties of the pectin and also keeps the amount of invert-sugar formed to between 25 per cent and 40 per cent. The time of boiling and hence the amount of invert-sugar formed can be controlled by altering the amount of water used. When the temperature reaches 104°C the end of the boiling period is imminent. The boiling point is really an indication of the concentration of the solution being boiled, and this can be more accurately determined by measuring the refractive index with a refractometer. From a knowledge of the refractive index the concentration of soluble solids in the jam can be calculated. To the soluble solids are added sugar and pectin together with sugars, acids and other solids extracted from the fruit. The soluble solid content varies from jam to jam but is usually about 70 per cent.

SUGAR IN THE DIET

During the present century world consumption of sugar has tripled. In Britain annual consumption per person is now almost 40 kg, roughly half of which is used domestically, the remainder being contained in a wide range of manufactured food. Not everyone eats the same amount of sugar, of course, and the average consumption is vastly exceeded by some people and not approached by others. In fact, the variation in individual consumption is probably greater for sugar than for any other dietary item except alcohol. Sugar is a fairly cheap commodity and there are no obvious differences in sugar intake between rich and poor in Britain. The amount consumed increases with age up to about 16 years (below this age perhaps shortage of cash does play a part!) and then it tapers off. Males seem to be more sweet-toothed than females at all ages.

Why is so much sugar consumed? The main reason is probably the attraction that sweet-tasting foods have for many people – especially, of course, children. It has even been suggested that sugar can be mildly addictive. Regardless of how strong-willed one is, however, it is practically impossible to forgo all sugar because of its presence – often unsuspected – in a wide range of foods. In many cases the reason for its presence is its sweetness but it also has other properties which commend it to the food manufacturer.

1. It can act as a preservative when it is present in food in high concentrations by making water unavailable to microorganisms.

2. Because it has such an affinity for water very concentrated solutions can be made. These syrupy solutions are smooth to the tongue and improve the 'mouth feel' of foods.
3. It can be used with pectin and acids to form semi-solid gels such as jam.

Not only can sugar do all these things but it is also cheap – hence its presence in so many foods and its prominence in our diet.

Sugar and health

Suspicion has fallen on sugar as the causative agent of a number of diseases which are prevalent in developed countries where sugar consumption is high. Coronary heart disease, diabetes, gallstones, kidney stones, certain types of cancer and even behavioural abnormalities have all been speculatively linked to excessive sugar consumption. A great deal of research work has been carried out to see whether such links exist. The consensus of informed opinion is that there is no convincing evidence that sugar causes, or is a contributing cause of, any of these so-called modern diseases.

SUGAR AND DENTAL DECAY   It is now generally accepted that the eating of sugar-rich foods is one of the main causes of tooth decay. Sucrose is the main offender because so much of it is eaten, but glucose, fructose and lactose are equally potent as tooth-rotting agents. Tooth decay (or dental caries, as it is more correctly known) can also be caused by starch because this is partly converted to glucose in the mouth by the action of salivary amylase.

Sugars themselves do no attack the teeth but they are converted to acid by streptococcal bacteria in the mouth and the acids formed attack and erode the hard enamel surface of the teeth. This occurs rapidly after sugary food is eaten but the tooth surface has some ability to repair slight erosion and if enough time elapses between attacks no permanent harm is done. If successive erosions occur too rapidly, however – through constantly eating sweets for example – a cavity will form and the tooth will be permanently damaged.

Research has shown that fluorides help to increase the resistance of tooth enamel to acid attack and fluoridation of drinking water and the use of fluoride-containing toothpaste are beneficial to dental health.

SUGAR AND OBESITY   It is widely believed that people become overweight or obese if they eat too much sugar or sweet food. Weight gain can only occur if the energy content of a person's diet exceeds his or her energy expenditure. Putting this in a more down to earth way we could say – what everyone knows – that if you eat too much you get fat. Excess energy derived from sugar is no different from that derived from other foodstuffs in this respect. Nevertheless, if an already adequate diet is 'topped up' by eating sweets, biscuits and cakes the result is inevitable – and the sweet 'extras' usually get the blame. Apart from its energy content, sugar – whether white or brown – contributes absolutely nothing to the diet. No harm would be done by excluding it completely if that were possible.

A diet rich in sugars, or any other carbohydrate for that matter, which only just provides the daily energy requirement may be deficient in one or more nutrients and this is the main objection to such a diet. The risk is not great, however, because most diets in Britain provide a more than adequate supply of all nutrients.

At present about 20 per cent of the energy content of the average British diet is provided by sugar and most nutritionists think this is too much. The National Advisory Committee on Nutrition Education (NACNE) recommend that the average annual consumption of sugar should be reduced from about 40 kg to about 20 kg per head. The committee which advises the Government on the Medical Aspects of Food Policy (COMA) more cautiously recommend that sugar consumption should not be further increased.

LACTOSE INTOLERANCE   The disaccharide lactose is present in animal (including human) milk. It is normally hydrolysed to its component monosaccharides in the small intestine by the enzyme *lactase*. The monosaccharides glucose and galactose are formed and these are readily absorbed. About 95 per cent of Western Europeans secrete lactase throughout their life but about three-quarters of the population of Africa, India, Eastern Europe and the Middle and Far East develop lactose intolerance between the ages of 15 and 25 when lactase production ceases. They are unable to digest lactose, or are able to cope with only small amounts of it. The undigested lactose is converted to lactic acid by bacteria in the large intestine and this causes flatulence, discomfort and diarrhoea.

Most people who suffer from lactose intolerance are able to take small amounts of milk without too much of a problem. Most cheeses contain little lactose and hence they too can be digested without difficulty. The same is true of yoghurt – a frequent ingredient in Indian recipes – because the lactose which was originally present in the milk from which it was made will have been largely converted to lactic acid.

Man is the only animal to consume milk in any quantity after infancy and it may well be that lactose intolerance is the natural state of affairs. It is possible that an ability to digest lactose has persisted into adulthood only in those countries where milk has become a normal part of the adult diet.

CEREALS

The major part of the carbohydrate intake and about 30 per cent of the energy content of an average British diet is provided by foods of cereal origin. Cereals are cultivated grasses and the grain which we use as a food source is a seed which is really intended as a rich store of nutrients for the grass that would grow from it. The most important cereals from our point of view are wheat, oats, maize and rice. Rye and barley are also cereals but they are not of great importance as food sources in Britain. Civilized societies in all parts of the world depend upon cereals for nourishment because they produce the maximum yield of food from a given area of ground. The cereal grains are obtained by threshing the harvested 'grass' to separate the grain from the chaff, which surrounds each grain, and the stalk which supports the ear.

Cereals are principally sources of carbohydrate but they also contain substantial amounts of protein (from about six per cent in rice to about 12 per cent in oats and Canadian wheat), and because large quantities of cereal products are eaten this may constitute quite a large proportion of the total protein intake. Fats are also found in cereals (from about 1.5 per cent in wheat to about five per cent in oats). Cereal grains contain substantial amounts of vitamins of the B group though, as will become evident, the quantity of these vitamins present in foods manufactured from cereals depends largely upon the degree to which the several parts of the grain have been separated in milling. The amount of moisture in cereal grains is quite small (from seven per cent in oats to about 12 per cent in wheat) and this largely accounts for their good keeping qualities.

Wheat

Wheat is by far the most important cereal as far as people in Britain are concerned. It was first grown in the Middle East some 10 000 years ago, but in the course of centuries its cultivation has spread and varieties of wheat suitable for cultivation in zones as climatically different as the tropics and the North European areas bordering on the Arctic Circle are now known. Some varieties of wheat, known as winter wheat, are sown in the autumn and harvested in the following August but spring wheat, which is sown and harvested in the same year, is grown in countries such as Canada where the winters are severe. Winter wheat, such as English wheat, usually contains less than ten per cent protein and gives a weak flour and a dough which bakes into small close-textured loaves. Spring wheat (such as Canadian wheat) is richer in proteins (12–14 per cent) and because it has a hard and brittle grain it is described as a hard wheat. Such wheat produces a strong flour from which a strong elastic dough can be made. Strong flours give doughs which produce bold, well-risen loaves and they are very suitable for bread making. English flour and other similar 'weak' or 'soft' flours are more suitable for the manufacture of cakes and biscuits and for household use. Flour used for bread making in this country is usually a blend of the two types.

Durum wheat, which is used for making macaroni and spaghetti, is a particularly hard wheat of high protein content.

THE STRUCTURE OF A WHEAT GRAIN   A grain of wheat is normally about one centimetre long and half a centimetre broad. It is egg-shaped with a deep fissure or crease running along one side and a number of small hairs, called the beard, at one end. The grain is enclosed in an outer covering called *bran* which consists of several distinct layers and constitutes about 15 per cent of the whole wheat. Bran contains a high proportion of B vitamins and about 50 per cent of the mineral elements present in the grain and it consists largely of cellulose which is indigestible by humans. The *germ*, which is situated at the base of the grain, is the actual seed or embryo and constitutes about 20 per cent of the whole grain. It is rich in fats, protein, vitamins of the B group, vitamin E and iron. The combined fatty acids present in the fats are mainly essential fatty acids. A membranous tissue called the scutellum separates the germ and the endosperm; it is exceedingly rich in the vitamin thiamine and contains about 60 per

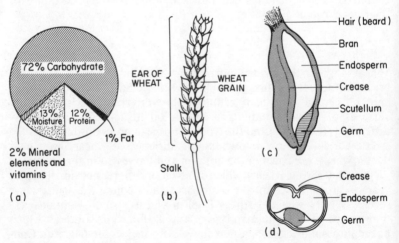

Figure 7.2   (a) The composition of wheat. (b) An ear of wheat. (c) Longitudinal section of a wheat grain. (d) Transverse section of a wheat grain.

cent of all the thiamine present in the grain. The *endosperm* is mainly starch and is intended as a reserve of food for the germ. It is by far the largest component and makes up the major part of the wheat grain. The starch granules are embedded in a matrix of protein and the periphery of the endosperm is composed of a single layer of cells called the aleurone layer. This layer contains a higher proportion of proteins than the endosperm as a whole, but unfortunately it is removed with the bran during the milling of the wheat.

So many varieties of wheat are grown and climatic and other conditions are so variable that it is not possible to give precise figures for the composition of wheat. The figures given in Fig. 7.2 represent average values.

FLOUR MILLING   Wheat grains are almost always reduced to flour before being eaten and this operation is known as flour milling. Archaeological evidence shows that flour was made in hand mills in the neolithic era. In later times windmills or water mills were used in which the wheat was ground between two circular grooved stones, the upper of which revolved while the lower stone remained stationary. The wheat was fed into the centre through a hole in the upper stone and was ground into flour during its passage to the periphery of the stones. In this process *all* the wheat grain was ground up so that the flour produced, called wholemeal flour, contained the germ, bran and scutellum as well as the powdered endosperm. Wholemeal flour is dark in colour and bread produced from it may be rather coarse, depending upon the extent to which the bran particles have been reduced in size.

Modern milling processes differ greatly from the age-old method just described. The milling is carried out using steel rollers in place of revolving flat stones, and the germ, bran and scutellum are removed so that the flour produced consists essentially of powdered endosperm. The process is very complex but in essence it consists of separating the endosperm from the other constituents of the grain, then gradually reducing the size of the endosperm particles by passing them through a series of steel rollers as shown diagrammatically in Fig. 7.3. Before milling, different varieties of wheat may be blended together so that the flour obtained from the blend is best suited to the purpose for which it is intended. After blending, the wheat is passed through a series of ingenious machines which remove stones, weed seeds and other extraneous materials, and may then be washed and brushed to remove adhering dirt.

Imported wheat is often too dry to be milled directly, and it must be 'conditioned' or brought to the optimum moisture content for milling. This can be done by storing the wheat in a moist condition for one or more days. The conditioning can be speeded up by passing the moist wheat through a machine known as a conditioner, in which is it heated to a temperature of 40–50°C for 30–90 minutes. Alternatively, the wheat may be exposed to live steam for a minute or so, followed by rapid cooling in cold water. During the conditioning process the distribution of moisture through the grain becomes more uniform, the bran becoming tougher and the endosperm more friable. This makes easier the separation of endosperm and germ from the bran in the milling operations which follow. English wheat often contains more water than is desirable, and when this is so it must be dried before milling.

The conversion of grain to flour begins with the operation known as breaking, in which the grain, or grist, is passed through four or five pairs of 'break rolls'. These rolls, which are made of steel, are corrugated and rotate at different speeds. The corrugations break the grain apart at the crease and scrape away the endosperm from the bran. After passing through the first break rolls the wheat is sieved through silk or fine wire gauze and separated into a small quantity of flour, termed 'first break flour', small particles of endosperm known as 'middlings' or 'semolina' and coarser particles of bran with adherent endosperm. The bran with its attached endosperm is passed onto the next set of break rolls. Each pair of break rolls is set closer together and has finer corrugations than the one before it and the branny residue from the final sifting is extremely thin with little or no attached endosperm.

The middlings produced at each set of break rolls are graded, or sized, by further sieving in machines known as purifiers. These consist of enclosed reciprocating sieves through which a current of air is blown. The blowing removes particles of endosperm which are still attached to the bran, and these are treated in a separate part of the mill where the endosperm is scraped away from the bran by finely corrugated rolls similar to break rolls.

The graded semolina and middlings are converted to flour by a series of smooth rolls called reduction rolls. There are usually from ten to fifteen sets of reduction rolls and as with the break rolls, the clearance between the rolls decreases from one set of rolls to the next. The reduction rolls crack the semolina particles and gradually produce smaller and smaller fragments without damaging the starch grain themselves. If the endosperm were simply crushed to a fine powder by being passed through a pair of rolls set very closely together the starch granules would be badly damaged and the resulting flour would be of very poor quality. The product from each set of reduction rolls is sieved to remove the flour, and the residue is divided into two parts; the finer of these two fractions is sent to one of the succeeding reduction rolls, and the less fine is sent back down the reduction system. The germ is not friable and so it is flattened rather than

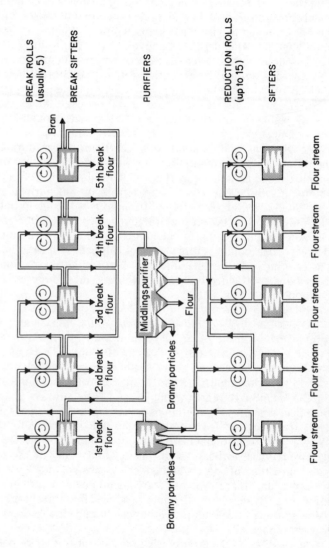

Figure 7.3 Conversion of wheat to flour.

powdered by the reduction rolls, and is easily separated in the sifting operations.

Flour 'streams' are obtained at each of the break rolls and reduction rolls and these differ markedly in composition. They may be mixed in various proportions to give flour suitable for special purposes, or they may all be combined to give what is known as a straight-run flour.

The milling technique described above can be modified to produce more or less flour from a given amount of wheat. The percentage of flour produced is termed the extraction rate of the flour. Wholemeal flour, which contains all the bran, germ, scutellum and endosperm of the wheat grain has an extraction rate of 100 per cent. An extraction rate of 70 per cent, on the other hand, produces flour which is composed almost exclusively of crushed endosperm.

Because vitamins and minerals are lost from wheat during the milling of flour of low extraction rate it is obligatory for millers to add certain nutrients to all flour other than wholemeal flour of 100 per cent extraction rate. Sufficient iron, thiamine, niacin and purified chalk must be added to ensure that 100 g of the flour will contain not less than 1.65 mg of iron, 0.24 mg of thiamine, 1.60 mg of niacin and between 235 mg and 390 mg of calcium carbonate. The iron may be added as ferric ammonium citrate, ferrous sulphate or, rather surprisingly, as very finely divided metallic iron. Flour of any extraction rate may be produced provided that the above nutrients are present in the stated amounts.

SPECIAL FLOURS   *High- and low-protein flours.*   The starch particles present in a sample of ordinary flour vary in size from below 5 μm to about 12 μm in diameter (1 μm = 10^{-6} m). The larger particles, of diameter greater than about 35 μm, consist of starch granules embedded in a protein matrix. Particles of sizes between 17 and 35 μm are, in the main, free starch granules and little protein matrix is present. Below about 17 μm the particles are mainly small starch granules, fragments of free protein and small particles of protein matrix with adherent fragments of starch. If flour is subjected to a further size-reducing process in an impact mill many of the larger particles consisting of starch embedded in protein matrix can be broken down and more separate starch granules and protein fragments released. By separating the flour produced into three fractions it is possible to obtain two high-protein flours and one low-protein flour as shown in Table 7.2.

Because the sizes of the flour particles are so small it is not possible to separate the fractions by conventional sieving operations, but this can be done by using a piece of equipment known as an air classifier. In the classifier flour particles in air are subjected to powerful centrifugal forces by being made to follow a spiral path at high speeds

Table 7.2 Protein contents and yields from an English wheat flour (Parent flour = 9.5% protein content)

Particle size range	Yield (%)	Protein content (%)
35–120 μm	43	11.5
17–35 μm	45	6.0
<17 μm	12	15.6

and at the same time, of course, each particle is experiencing frictional drag. The centrifugal force on each particle is proportional to its weight, whereas frictional resistance varies with particle size and not with weight. Because of this difference the coarser particles migrate to the outside of the spiral path and finer particles accumulate in the centre. The fraction of largest particle size is noticeably coarser, because of the absence of fine particles, than normal flour and it has been used for biscuit making and for self-raising flour. The intermediate low-protein fraction is valuable as a high-ratio cake flour (see p. 67) because of its fine and even particle size.

Agglomerated flour.   When ordinary flour is added to water the particles float on the surface and tend to stick together to form lumps which are difficult to disperse. An 'instant' flour, which is easily wetted by water and can be dispersed in water without difficulty, is manufactured by allowing flour to fall through jets of steam. The outer surfaces of the flour particles are wetted and if they are then allowed to fall through cold jets of air the particles stick together, or agglomerate. The moisture content of the clusters is adjusted by passing the flour through heated chambers and oversized particles are reduced to a uniform particle size. As well as being easily dispersible in water, agglomerated flour, or instant flour, is free-flowing and dust-free because it consists of small clusters of flour particles above 100 μm in diameter. When agglomerated flour is added to water the clusters are penetrated by water as a result of capillary action and the wetted particles sink. Agglomerated flour is especially useful in soup and gravy powders or as a thickening agent, but it is likely to find many other applications in convenience foods.

Self-raising flour.   When bread is made it is aerated by carbon dioxide produced by fermentation, i.e. by the action of yeast on sugars present in the dough. Yeast can be successfully employed as a raising agent only with high-protein (i.e. 'strong') flours which give a strong elastic dough. Weak doughs, such as those used in pastry, are not fermented with yeast but a chemical raising agent (see p. 301) is used instead. Self-raising flour contains an acidic raising agent and sodium bicarbonate incorporated into it in suitable amounts so that carbon dioxide is formed during cooking and this aerates the dough.

Enzyme-inactivated flour is discussed on page 138.

FLOUR IMPROVERS   It was formerly the practice to store flour for several weeks after milling before it was used for bread making. During this period its bread making characteristics improved, as a result of the partial oxidation of the proteins which form gluten during dough making. The gluten obtained from 'aged' flour is stronger and more elastic than that obtained from freshly milled flour. The ageing period can be dispensed with if the flour is treated with a minute quantity of one of a number of oxidizing agents which are called flour improvers, and it is common practice for all flour (except wholemeal flour) to be treated in this way. The substances used as flour improvers, the proportions in which they may be used, and their serial numbers (see p. 337), are listed in Table 7.3.

The first four of the improvers listed increase the whiteness of the flour by bleaching the *carotene* and *xanthophyll*, which are always present and give the flour a slight yellow tinge, and they are sometimes referred to as bleaching agents rather than flour improvers. Unlike the other flour improvers listed in Table 7.3 ascorbic acid is a reducing agent. The actual flour improver, however, is not ascorbic acid itself but dehydroascorbic acid. This is an oxidizing agent and it is formed in the bread mix by enzymic reduction of ascorbic acid.

The way in which improvers perform their functions is not fully understood, though many explanations have been put forward. It is likely, however, that the improvers produce some cross-linking between adjacent protein molecules by the formation of disulphide

Table 7.3   Flour improvers

Serial number and name		Formula	Maximum permitted level (parts per million parts of flour) and restriction on use (if any)
925	Chlorine	Cl_2	2500 (for cakes only)
926	Chlorine dioxide (Dyox)	ClO_2	30
E220	Sulphur dioxide	SO_2	200 (for biscuits or pastry only)
—	Benzoyl peroxide	$(C_6H_5CO)_2O_2$	50
924	Potassium bromate	$KBrO_3$	50
E300	Ascorbic acid	$C_6H_8O_6$	200
927	Azodicarbonamide	$N_2(CONH_2)_2$	45

Note: None of these compounds may be used with wholemeal flour

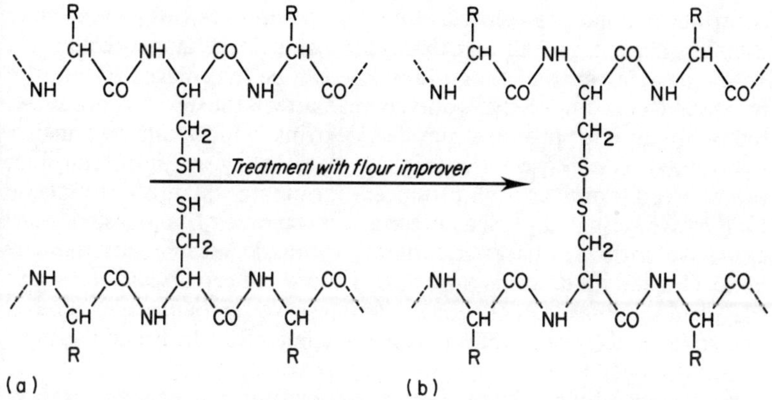

Figure 7.4 Action of flour improvers. (a) Two protein molecules with adjacent sulphydryl groups. (b) One larger protein molecule formed by cross-linking.

links ($-S-S-$) from neighbouring sulphydryl ($-SH$) groups as shown in Fig. 7.4.

The sulphydryl groups belong to the amino acid cysteine which forms part of some protein molecules. The increase in molecular weight and molecular complexity produced by such a process of cross-linking would be expected to produce a corresponding increase in the strength and elasticity of the gluten formed on treatment with water. This, of course, is the characteristic effect produced by flour improvers.

THE ENZYMES OF FLOUR Wheat flour contains α- and β-amylases which are capable of hydrolysing the amylose and amylopectin of starch. The hydrolysis does not occur to any significant extent in dry flour but begins immediately a dough is made. β-amylase attacks the damaged starch grains inevitably present as a result of milling and hydrolyses some of the amylose to maltose. It can also hydrolyse amylopectin and produce maltose by splitting off pairs of glucose units from the ends of the chains. With amylopectin only the 'free' ends of the chains can be attacked and hydrolysis ceases before the glucose unit which connects two chains together is reached. The links between the chains cannot be hydrolysed by β-amylase and a high molecular weight dextrin, which is not susceptible to further attack by β-amylase, is ultimately obtained. α-Amylase, which is present in flour made from sprouting wheat, attacks amylopectin in an entirely different way: it splits the links between the chains to produce low molecular weight dextrins. These dextrins differ in structure, as well as in size, from those produced by β-amylase. They consist, like amylose, of strings of connected glucose units whereas the dextrins produced by β-amylose are branched and three-dimensional like amylopectin.

The dextrins produced by α-amylase can be hydrolysed to maltose by β-amylase. Conversely, α-amylase can attack the dextrins produced by β-amylase to produce simpler dextrins which can be further hydrolysed to maltose by α-amylase. Clearly, by acting in conjunction, α- and β-amylase can produce a far greater quantity of maltose than either acting alone. The presence of too much α-amylase in flour can have a disastrous effect when it is made into bread: the low molecular weight dextrins present cause marked crumb-stickiness and may lead to collapse of the loaf. α-amylase is active up to about 60°C and so its action may continue for some time after the bread has been put in the oven.

In practice, flour is more likely to be deficient in α-amylase because it is usually milled soon after being harvested. If the α-amylase content is low the amount of fermentable sugars present in the dough may be so low that insufficient carbon dioxide is produced during bread making. Consequently, the bread produced will be close-grained and lacking in volume. This can be prevented by *adding* α-amylase to the flour. Dough made from such flour rises strongly during fermentation and produces softer bread of greater volume.

A high α-amylase content may be undesirable in flour intended for use in gravies, thickening agents or powdered soups. If such a gravy or soup is kept hot the α-amylase may break down the starch to such an extent that it is no longer able to act as a thickening agent. *Enzyme-inactivated flour* can be produced for special purposes such as this, and it is made by heating wheat with steam to 100°C. The wheat is subsequently dried and converted into flour in the usual way. Enzyme-inactivated flour is not suitable for bread making because the elastic properties of the gluten are destroyed by the steaming process.

Flour also contains peptidases (i.e. protein-splitting enzymes) and it is believed that these play a part in the 'ripening' of the dough. Peptidases also make available much α- and β-amylase which is combined with protein and would otherwise be unavailable. Lipases and lipoxidases also occur in flour and act upon the fats present. Lipases catalyse the hydrolysis of fats to glycerol and fatty acids whereas lipoxidases catalyse oxidation. Flour which has been stored for a long period may have a tallowy smell and flavour owing to the presence of oxidation products of fats. Phytase is another important enzyme present in flour. It splits up phytic acid and phytates (see p. 147) and is active during the fermentation and early stages of baking.

Bread making

A type of bread, known as unleavened bread, can be made by mixing flour and water and then baking it. This is the forerunner of modern bread, but the product is hard and unattractive to most palates and the resemblance to bread, as we know it, is slight.

Bread is traditionally made from flour, water, salt and yeast. It has a honeycomb structure and may be regarded as a solid foam with a multitude of pockets of carbon dioxide distributed uniformly throughout its bulk. Sugars naturally present in flour, and the maltose made available by the action of amylases, are hydrolysed to glucose and this is fermented by zymase present in the yeast. Alcohol and carbon dioxide are formed and the latter aerates the dough.

$$(C_6H_{10}O_5)_n \xrightarrow[\text{flour}]{\text{amylases in}} C_{12}H_{22}O_{11} \xrightarrow[\text{yeast}]{\text{maltase in}}$$

$$\text{Maltose}$$

$$C_6H_{12}O_6 \xrightarrow[\text{yeast}]{\text{zymase in}} C_2H_5OH + CO_2$$

Most of the alcohol formed during fermentation is driven off during baking and many thousands of gallons of alcohol enter the atmosphere daily from bakeries.

Small amounts of carboxylic acids are produced during the fermentation period as well as carbon dioxide and alcohol. The acids formed lower the pH of the dough and this affects the colloidal state of the gluten (see below) and assists in ripening the dough. The carbon dioxide retained by the dough also lowers its pH and so, in addition to its leavening function, it has a beneficial effect on the gluten structure. Some protein breakdown occurs during the fermentation period owing to the presence of proteolytic enzymes. During this period the yeast cells multiply and the yeast contributes substantially, together with the fermentation products, to the flavour of the loaf.

Two of the proteins present in flour – *gliadin* and *glutenin* – become hydrated and form an elastic complex called *gluten* when flour is kneaded with water. It is the presence of this elastic gluten which makes the manufacture of bread possible, because it forms an interconnected network which contains the carbon dioxide within the loaf and prevents its escape. The gluten is uniformly distributed throughout the dough and the carbon dioxide becomes trapped as small pockets of gas. As gas production continues the gluten strands are stretched and it is thought that bonds between adjacent protein molecules are broken and re-formed to produce an elastic, gas-retaining, three-dimensional network. A ripe dough, that is one which is ready for baking, is springy and elastic; it can be fairly easily stretched out and shows a capacity to recover its former shape. An under-ripe dough is extensible, i.e. it can be stretched, but lacks elasticity. If fermentation is allowed to continue unhindered the dough becomes over-ripe and, as dough in this condition cannot be stretched far without breaking, its power to retain carbon dioxide is lost.

Salt is used in dough making and its most important function is to improve the flavour of the bread which has a flat insipid taste without it. The weight of salt added is about two per cent of the weight of flour used, although this figure varies slightly, more salt being used in the north of England than in the south. The amount of water needed to make a dough varies with the quality of the flour, but it is roughly half the weight of the flour used. Its temperature must be adjusted so that the fermented dough has a temperature of 24–27°C. The quantity of yeast needed depends upon the time and temperature of fermentation but is usually about 0.3–1.0 per cent of the weight of the flour used. When bread is baked the carbon dioxide expands, the starch gelatinizes, and the gluten coagulates to produce a more or less rigid loaf. The changes which occur during baking are considered in more detail on page 142.

DOUGH MAKING   In the traditional method of bread making known as the *long-fermentation process* the warm ingredients are thoroughly mixed to form a dough which is allowed to ferment in bulk. The dough is covered to prevent the formation of a skin and it is allowed to ferment for a period of one hour or longer, depending upon the quantity of yeast used and the temperature of the dough. Yeast functions best at a temperature of 26°C and the dough should preferably be kept at or about this temperature. The dough is then thoroughly kneaded, or 'knocked back', to expel some of the carbon dioxide and tighten up the dough. This has the effect of bringing yeast cells into contact with a new environment and assists further fermentation. The dough is then covered again and allowed to ferment for a further period – usually about two and a half hours – but, as before, this depends upon the amount of yeast used and the ambient temperature. At the end of this time it is divided into pieces of the required weight which are shaped into balls.

A good deal of carbon dioxide is lost during the dividing and moulding process and the dough is given a further period of fermentation, referred to as first proof or intermediate proof, to allow more carbon dioxide to be formed. During this period the gluten fibres recover after the rather harsh treatment they have received during dividing and shaping. When the first proving period of 10–15 minutes is over the dough is moulded to its final shape and after placing in baking tins, or on baking sheets, is given a final proof of about 45 minutes at a somewhat higher temperature (usually 32–35°C), during which it becomes fully inflated with carbon dioxide and assumes its final shape.

In a modern bakery the mixing, knocking back, dividing and moulding to shape are carried out mechanically and the proving periods are often spent on conveyor belts passing through temperature-controlled chambers. Because of the fairly long fermentation period, however, a considerable weight of dough is being dealt with at any one time in a large bakery and this has to be manhandled from place to place and, of course, it occupies considerable space. In modern bread making methods the fermentation period is replaced by a short period of intensive mixing (see below) and this overcomes one of the main drawbacks of the traditional bread making process.

Most bread in Britain is now made by a short-fermentation process which is known as the *Chorleywood bread process* because it was developed by the Flour Milling and Baking Research Association at Chorleywood in Hertfordshire. In this method of bread making a normal recipe is used except that about twice as much yeast is required and flour improvers are added to the bread mix. The long period of fermentation of the traditional bread making method is replaced by a short period of high speed mixing in the presence of about 75 ppm (based on weight of flour used) of a mixture of ascorbic acid and potassium bromate. Azodicarbonamide, fast-acting flour improver, may also be used in addition to the other two compounds.

Dough made by the Chorleywood bread process is called *mechanically developed dough*. The ingredients for the dough are fed into a powerful mixing machine which is equipped with a meter to indicate how much energy is expended during the development period. The mixing and development are completed in as short a time as possible – usually about five minutes – and when the appropriate amount of energy has been expended the machine switches off. The mixing machine rapidly stretches the gluten and this replaces the stretching brought about more slowly by gas evolution in the long-fermentation process. It is believed that bonds between adjacent protein molecules are ruptured by the severe mechanical treatment and that this is followed by a rapid repositioning of the protein molecules. At this point mixing must stop and bonds are rapidly re-formed to give the required elastic network structure. If mixing is continued an inferior product is obtained.

When the mixing process has been completed the dough is divided into pieces of the correct weight for the loaves required and these are shaped into balls and given a first proof period of 6–10 minutes before being moulded to shape and being placed in baking tins. After a second proving period of about 50–60 minutes the bread is ready for baking in the normal way.

Yeast is necessary in mechanically developed dough making even though one of its jobs – the stretching and reorientation of gluten fibres – is done mechanically. The yeast is still required for producing

Table 7.4 Advantages of the Chorleywood bread process

Time	Less than two hours including baking – a saving of about 60 per cent.
Premises	Dough-room areas are reduced by 75 per cent. Temperature and humidity control not required.
Materials	Flour of low protein content may be used. Low fermentation losses.
Product	Variability of product reduced. Bread has a lower staling rate.

carbon dioxide, which aerates the bread in the usual way, and for flavouring.

Almost any flour can be used in the Chorleywood bread process and good bread can be obtained from weak flour. This means that English flour can be used, whereas for the normal bulk fermentation methods the use of strong flour is necessary. The reason for this is that a certain amount of protein is lost during the long-fermentation period and this does not occur to any extent with mechanically developed doughs.

Control of temperature is less important in the Chorleywood process than in long-fermentation methods and good results can be obtained throughout the range $27-32°C$. Because of the large amount of energy expended during development of the dough its temperature may increase by $11-24°C$ during the process.

About 80 per cent of the bread made in Britain, including wholemeal bread, is made by the Chorleywood bread process. The advantages of the Chorleywood bread process compared with the long-fermentation process are summarized in Table 7.4.

The special mixing machinery required for the Chorleywood process is usually found in large-scale bakeries, or 'plant bakeries' as they are known. Bakers who do not have the specialized equipment can eliminate, or considerably reduce, the time taken for fermentation by using ordinary low-speed mixers and adding about 50 ppm (based on the weight of flour used) of the naturally occurring amino acid L-cysteine to the ascorbic acid/potassium bromate mixture used in the Chorleywood process. This process, called *activated dough development* is used for about ten per cent of the bread made in Britain.

CHANGES DURING BAKING Bread is baked at a temperature of about $232°C$ for a period of $30-50$ minutes, depending upon the type of bread and the size of the loaf. During baking, the dough first rises rapidly because the pockets of carbon dioxide in the loaf expand as the temperature increases. At first there may also be some slight increase in the activity of the yeast, resulting in increased production of gas, but this diminishes as the temperature increases, until at a temperature of about $54°C$ fermentation ceases. As the temperature increases, the water present causes the starch granules to swell and gelatinize, and during this period the starch probably abstracts some water from the gluten. Hot gluten is soft and devoid of its characteristic elasticity, and gelatinized starch now supports the structure of the loaf. The gluten begins to coagulate at about $74°C$ and the coagulation continues slowly to the end of the baking period. The temperature of the interior of the loaf never exceeds the boiling point of water, in spite of the high temperature of the oven. Water and much of the carbon dioxide and alcohol formed during the fermentation escape during baking. Considerable dextrin formation occurs at the outside of the loaf as a result of the action of heat and steam on the starch; the sugars formed are converted to caramel which imparts an attractive brown colour to the crust. Maillard reactions (see p. 295), which occur between the carbohydrates and proteins present, also contribute to the brown colour of the crust.

BREAD QUALITY   A loaf of bread has certain characteristics by which its quality is judged. The dough should rise to produce an upstanding loaf, the interior of which should be uniform in porosity and firm and elastic to the touch. The crust should be golden-brown in colour and should be crisp and brittle rather than tough, otherwise it would be difficult to cut and chew. Bread produced by the traditional long-fermentation process may be less consistent in quality than that produced from mechanically developed dough. A dough which has been insufficiently fermented will have a starch gel which is too stiff to permit expansion during baking and a small dense loaf will result. In an over-fermented dough, on the other hand, extensive starch breakdown will have occurred and the starch gel will be weak. The dough will be unable to stand the increased internal gas pressure which occurs during baking. Individual gas bubbles will coalesce to form large pockets and gas escape at the surface will prevent the loaf from rising properly. In addition, such a loaf will contain larger quantities of dextrins than are found in a properly fermented loaf, and this will cause the interior of the loaf to be darker in colour.

STALING   When bread is kept and becomes stale its crust becomes soft and leathery and it loses its appealing flavour. At the same time the interior of the loaf loses flavour and becomes less elastic through crumb-staling.

Crust-staling is largely due to the diffusion of water from the interior of the loaf. It occurs more rapidly with wrapped bread because the moisture is unable to escape.

Crumb-staling is not due to drying out but is caused by retrogradation (see p. 107) of amylopectin in the gelatinized starch which makes up the bulk of the loaf. This takes place as the bread ages and the amylopectin molecules become more regularly arranged. In fresh bread the amylopectin molecules and the chains of glucose units of which they are built up are arranged in a completely haphazard and random manner in partially swollen starch granules. In stale bread, on the other hand, many amylopectin molecules may be arranged in a cluster and the chains of glucose units lie parallel to each other, almost as if they had been combed into position. The presence and absence of order in stale and fresh bread respectively has been confirmed by X-ray studies.

During the change from partially gelatinized starch to the more-or-less crystalline form, water is released and is absorbed by the crust, which is why it loses its crispness and becomes leathery. If the bread is left uncovered the crust will, in time, become hard again owing to diffusion of moisture into the air but the resulting dry crust bears little resemblance to a crisp fresh crust.

Staling can be prevented by drying, because bread containing less than about 16 per cent moisture does not stale. Surprisingly, excess moisture can also prevent staling and wet bread remains fresh for long periods. Neither of these methods is of any practical importance, however. Temperature can play an important part and bread kept under controlled moisture conditions remains fresh above $60°C$ and below $-10°C$. At temperatures between these extremes bread becomes stale with a maximum rate of staling at about $0°C$ and this has been put to practical use in the preservation of bread and sponge-cakes (which normally stale in the same way as bread) by deep-freezing. Wrapped bread stays fresh longer than unwrapped, although crust staleness may develop more quickly because of moisture retention, which prolongs the period for which the crumb remains soft.

Another method of delaying staling is the incorporation into bread of emulsifying agents, although it is not yet clear how these substances delay the onset of staling. The use of emulsifying agents in bread is controlled by the Bread and Flour Regulations 1984, and only the substances specified in the Regulations may be used as emulsifiers.

TYPES OF BREAD   So many varieties of bread are made that it is sometimes difficult to be sure exactly what type of bread one is buying or eating. Certain types are defined in the Bread and Flour Regulations 1984, and these are listed in Table 7.5. It is only in these cases that the name given to a particular type of bread has any real significance.

Bread as a food

Like all other foods of cereal origin, bread is eaten mainly as a cheap

Table 7.5 Types of bread

Wholemeal bread	Made from flour obtained by milling whole wheat grains including the bran and germ and no other cereal. May contain ascorbic acid if made by the Chorleywood bread process. Addition of L-cysteine or any of the flour improvers listed in Table 7.3 is prohibited. May contain any other permitted additional ingredients or additives (see white bread below).
Brown bread	Must contain at least 0.6 per cent crude fibre (calculated on the dry matter) and flour other than wholemeal flour. May contain any other permitted additional ingredients or additives.
Wheatgerm bread	Must contain at least 10 per cent processed wheat germ (calculated on the dry matter). May contain any other permitted additinal ingredients or additives.
White bread	Defined by exception as bread whch is not wholemeal, brown or wheatgerm. May contain a wide range of additional ingredients including milk and milk products, liquid or dried eggs, wheat germ, rice flour, oat grain (or oatmeal) soya bean flour (in limited quantitites), salt, vinegar, oils and fats, malt extract, malt flour, sugars, wheat gluten, various seeds, wheat, malted wheat, rye or barley and, in limited quantities, starch. May contain one or more of 46 permitted additives.
Soda bread	Contains sodium hydrogen carbonate (sodium bicarbonate $NaHCO_3$) as an ingredient.
Wheatmeal bread	Use of this name is prohibited by the Bread and Flour Regulations to avoid confusion with wholemeal bread.

source of energy. It contains about 40–45 per cent available carbohydrate and has an energy value of 900–1000 kJ/100 g. Because considerable amounts of bread are eaten its other constituents also contribute substantially to the daily intake of nutrients. It contains 8–9 per cent protein and significant amounts of minerals and vitamins.

The nutrients in wheat grains are not present in the same proportions in all parts of the grain, and so a change in the extraction rate (see p. 134) produces a change in the composition of the flour produced. In particular, unmodified flour (i.e. flour to which nutrients have not been added) of low extraction rate contains smaller amounts of the B vitamins (thiamine, riboflavin and niacin) and iron than flour of higher extraction rate. This is because these nutrients are mainly concentrated in the bran, germ and scutellum, all of which are removed in the production of low extraction rate flour. To compensate for the loss supplements of thiamine, niacin and iron are added to all flour produced in Britain other than wholemeal flour (see p. 134).

Table 7.6   Composition of 100 g of bread

Nutrient	Extraction rate of flour used		
	72% (White bread)	85% (Brown bread)	100% (Wholemeal bread)
Protein	7.8 g	8.9 g	8.8 g
Fat	1.7 g	2.2 g	2.7 g
Sugars	1.8 g	1.8 g	2.1 g
Starch	43 g	39 g	36 g
Dietary fibre	2.7 g	5.1 g	8.5 g
Phytic acid	4 mg	202 mg	360 mg
Calcium	100 mg	100 mg	23 mg
Iron	1.7 mg	2.5 mg	2.5 mg
Thiamine	0.18 mg	0.24 mg	0.26 mg
Riboflavin	0.03 mg	0.06 mg	0.06 mg
Niacin	1.76 mg	1.86 mg	1.7 mg
Energy	990 kJ	950 kJ	920 kJ

Calcium carbonate (chalk) is also added to all non-wholemeal flour as a means of enhancing the calcium content of the diet.

The composition of bread made from flour of different extraction rates is shown in Table 7.6. Because bread is somewhat variable in composition the figures given should not be regarded as constants but rather as representative values.

The amount of bread eaten varies greatly between individuals but surveys show clearly that average consumption is decreasing. Despite this, however, the amount of wholemeal bread being eaten is increasing. Total bread consumption per head is greatest in households with low incomes and four or more children. Families in this category also eat the smallest proportion of wholemeal bread.

Surveys show that the average bread consumption in Britain is about 125 g per day – about four 'thickish' slices or five thin slices. This provides about 1250 kJ of energy which is about one-tenth of the estimated daily requirement for a moderately active man (see Appendix I). It also makes a substantial contribution to the RDA of iron, calcium, thiamine, niacin and protein. Bread also contains valuable amounts of dietary fibre and about one-third of the dietary fibre intake in Britain is provided by bread and cereal products.

Bread is an important source of protein, but unfortunately the protein is of low quality and it contains only about three per cent of the essential amino acid lysine (see Table 8.7, p. 173). This is not of great importance, however, because most people eat more protein than they need and it is unlikely that there will be an overall deficiency of lysine. In addition, bread is not normally eaten on its own and it is more than likely that the other foods eaten at the same time will compensate for the lysine shortfall.

The calcium content of the notional daily average bread consumption of 125 g is about 30 mg for wholemeal bread and 125 mg for other bread. The reason for the big difference is, of course, the enrichment of non-wholemeal flour with calcium carbonate. Calcium in food reacts with the phytic acid present in bread, especially wholemeal bread (see Table 7.6), to form insoluble calcium phytate. At one time it was believed that the body was unable to absorb calcium from calcium phytate. It is now known, however, that if phytic acid forms a regular part of the diet the body is able to adapt to its presence and counteract its bad effect on calcium absorption. The effect of phytic acid on calcium absorption is also nullified by proteins present in the diet and the consensus of expert medical opinion is that phytic acid in bread made from high extraction rate flours does not prevent adequate absorption of calcium or other minerals.

All flour in Britain, except wholemeal flour, has iron added to it to compensate for that removed during milling. Despite this, wholemeal bread is considerably richer in iron than bread of lower extraction rate. It also contains more phytic acid but, just as with calcium, it is now believed that its presence has little or no effect on the absorption of iron. Nevertheless, several studies have shown that iron is poorly absorbed from bread by the body. Although about ten per cent of the average iron intake is provided by bread and other flour products, most of it passes through the body. Thus the difference in iron content between wholemeal bread and other types of bread is probably of little nutritional significance.

Other wheat products

BISCUITS   Biscuits are made from flour with the addition of other ingredients such as salt, fat, sugar and flavouring agents. Baking powder is sometimes added to make them rise a little and some biscuits, such as cream crackers, are leavened with yeast in much the same way as bread. The dough is rolled to a thin sheet, cut into appropriate shapes and quickly baked at a high temperature. The water content of biscuits is only about three per cent compared with about 39 per cent in bread. The energy value of sweet biscuits may be twice as high as that of bread because of their low water content and the extra sugar and fat they contain.

PASTA   Pasta is the collective name given to a number of wheat-flour products which are cooked by boiling in water rather than baking. *Macaroni*, *spaghetti*, *vermicelli* and *ravioli* are all pastas and they are made from the endosperm of a particularly hard (i.e. protein-rich) variety of wheat known as *durum wheat*. Milling of durum wheat produces *semolina* which consists of hard endosperm particles. Semolina is much grittier than normal wheatflour because the endosperm particles are considerably larger. Semolina and water can be made into a stiff dough from which the various pasta shapes – ribbons, tubes, spirals or sheets, to name but a few – are made.

When pasta is cooked in boiling water it absorbs up to three times its weight of water and becomes soft but, providing it is not over-cooked, it does not disintegrate and form a paste as a normal flour dough would. Because of its high water content large quantities of cooked pasta are required to provide a sustaining meal and this is why pasta is usually accompanied by other more nutritious foods such as cheese or a meat-rich sauce.

Other cereals

Oats, rye, barley, maize and rice are all important cereals but their contribution to the British diet is much less than that of wheat.

OATS   Oats are richer in fats and mineral elements than other cereals and their protein content is also high. Flour made from oats is not suitable for making bread, however, because the proteins it contains do not form an elastic complex like the gluten of wheat when mixed with water.

Oats are prepared for human consumption by cleaning, then drying and storing for a period before removing the closely adherent husk. The product, known as *groats*, can be ground to produce oatmeal or rolled into flakes after being partially cooked by steam. The rolling ruptures the cell walls and flattens the grains and this makes subsequent cooking easier.

Oatmeal contains roughly 11 per cent protein, 66 per cent carbohydrate, 9 per cent fat and 6.5 per cent dietary fibre: it has an energy value of about 1580 kJ/100 g. It must be remembered, however, that porridge made with water contains only about one-eighth of its weight of oatmeal.

RYE   Rye can be grown in areas where the climate is too severe for wheat. The nutrients in rye are present in roughly the same amounts as in wheat. It is much less valued than wheat, however, because the proteins it contains do not give a strong gluten during dough making and, while large quantities of bread are made from rye flour, the product differs markedly from bread as we know it. Rye flour gives stodgy loaves which are unattractive to those accustomed to wheat bread. However, rye-bread is still a staple article of diet in Northern Europe. It may contain substantial proportions of wheat flour and this improves the appearance of the loaves obtained. Rye-bread is not eaten to a large extent in Britain, but crisp rye-biscuits (e.g., 'Ryvita') made from crushed whole rye grain are popular.

BARLEY Barley is not eaten as a cereal in this country but is grown extensively for the manufacture of malt for brewing and for animal fodder. Bread is never made from barley because its proteins do not form a gluten when mixed with water and an aerated loaf cannot be obtained.

MAIZE Maize is not widely used as a food for human consumption in Britain. In other countries, particularly America and South Africa, it forms an important part of the diet. It cannot be used for making bread because its proteins do not form a gluten. As a provider of energy maize is as efficient as other cereals, but in other respects it is less desirable. Like all cereals its proteins are deficient in the essential amino acid lysine. Tryptophan, another essential amino acid, which can be converted by the body to the vitamin niacin, is also only present

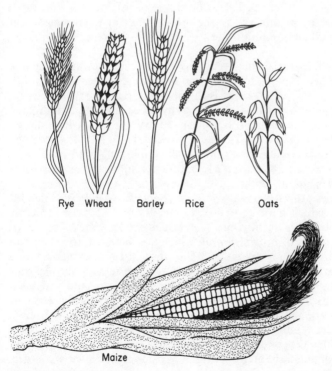

Rye   Wheat   Barley   Rice   Oats

Maize

Figure 7.5 Cereal grains.

in small amounts. Maize contains niacin but it is bound to hemicellulose and is not available to the body when maize is eaten.

The deficiency disease *pellagra* is caused by a shortage of niacin and at one time it was a serious problem in parts of the world where maize was a staple food. It is still a serious health problem in parts of Africa and India but because of improved standards of nutrition it is now rarely seen in Europe or America.

A good deal of the maize crop is now used as cattle fodder or for conversion to glucose syrup or 'corn syrup'. It appears in the British diet mainly as cornflakes, which are toasted, malt-flavoured rolled maize. The nutritional deficiencies of maize are counteracted in cornflakes by means of lavish additions of iron, niacin, thiamine, riboflavin, vitamin B_6, vitamin B_{12}, vitamin D and folic acid. The amounts added are such that an average serving provides about one-quarter of the RDA of the added nutrients for a moderately active man (or one-third for a child).

Apart from cornflakes, maize is represented in British diets by sweetcorn and corn-on-the-cob, which are eaten as vegetables, and cornflour which is used in custard powder, blancmange powder and as a thickening agent (see p. 107). Cornflour consists of little but starch and is made by washing away the protein and fat from maize flour with dilute alkaline solutions.

RICE This is the main cereal of Eastern countries and although not a great deal is eaten in Europe or America it is one of the world's most important food crops. It is the poorest of all cereals in protein, fat and mineral elements and can hardly be called a nutritious food although, like all cereals, it is a cheap source of energy. Rice has an energy value of 1530 kJ/100 g and to provide the 12 MJ required daily by a moderately active man, almost 800 g of rice would have to be eaten. This would weigh over 2 kg after cooking because of the large quantity of water absorbed.

When rice grows the grains are surrounded by a loose, inedible outer husk which must be removed. The grain itself has a structure similar to that of a wheat grain. The starchy endosperm is surrounded by several layers of brownish bran which are removed, together with the germ, by a milling process. Unlike wheat, rice is not made into a flour and after milling the grains are polished to remove the aleurone layer, or silverskin, which surrounds the endosperm and the membranous scutellum which separates the endosperm from the germ.

Many of the nutrients of whole rice are lost when the outer layers and germ are removed. The finished product, polished rice, contains about 85 per cent starch and 7 per cent protein. Some B-group vitamins remain in the endosperm but the amounts present are largely dissolved out and lost if rice is cooked in a large volume of boiling water. At one time many people who ate mainly boiled rice suffered from the deficiency disease beriberi which is caused by a lack of thiamine (see p. 253). This vitamin is required by the body to make use of carbohydrates in the diet and the amount required is related to the quantity of carbohydrate eaten. Thus a diet which is rich in carbohydrates and low in thiamine is particularly likely to cause beriberi.

In India rice is steeped in water after harvesting and it is then steamed and dried before milling. During this process, which amounts to parboiling, much of the vitamin content of the bran, germ and scutellum diffuses into the starchy endosperm and is retained there in the finished product. For this reason beriberi occurred less extensively in India than in other rice-eating parts of the world.

Beriberi is now a relatively uncommon disease, possibly owing to a general improvement in standards of nutrition. An understanding of the way in which it occurs and the availability of cheap synthetic thiamine have also been important factors in its eradication.

Dietary fibre

NATURE OF DIETARY FIBRE   It is somewhat difficult to give a clear explanation of what dietary fibre is or, indeed, to describe its function in the diet. The term dietary fibre is used to describe those parts of the diet which are not broken down by the enzymes of the stomach and small intestine and hence enter the large intestine, or bowel, unchanged. It consists of the skeletal material of plants – the tough structural parts of stems, seeds, husks and leaves.

Dietary fibre is almost entirely composed of polysaccharides, especially cellulose, hemicelluloses, pectin and protopectin. Non-carbohydrate material such as lignin (a constituent of wood) may also be present to a minor extent. Starch is not classed as a dietary fibre because it is efficiently digested and absorbed in the small intestine.

Dietary fibre was formerly known as 'roughage'. The newer name is an improvement, if only because it does not conjure up visions of intestinal abrasion associated with the word 'roughage'. Nevertheless, it is a far from ideal term. To begin with there are other fibrous components of food – notably muscle fibres in proteins – which are *not* classed as dietary fibres because they are digested and absorbed. In addition, not all the compounds which constitute dietary fibre are fibrous in nature. Pectin, for example, can hardly be regarded as fibrous and even the cellulose component of dietary fibre is fibrous only at a microscopic level and bears little resemblance to the cellulose fibres of cotton wool. Foods rich in dietary fibre are not usually noticeably fibrous in character: dietary fibre does not confer coarseness or 'stringiness' and it cannot normally be detected by the tongue.

SOURCES OF DIETARY FIBRE Inhabitants of so-called developed countries eat much less dietary fibre than people in more primitive societies, where food undergoes little or no 'processing' before consumption. The average Briton eats only about 20 g of fibre per day compared with the 70–100 g per day common in some rural African communities. The stems, skins, bran and husks of plants are rich sources of dietary fibre and it is unfortunate that they are, to a large extent, removed during harvesting, in the food factory or in the kitchen, before food is eaten. Cereals and cereal products (especially wholemeal bread), leafy and root vegetables and fruit and nuts all contain dietary fibre. The main sources of dietary fibre in the British diet are shown in Fig. 7.6 and the fibre content of individual foods is given in Table 7.7.

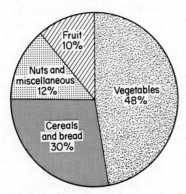

Figure 7.6 Sources of fibre in the British diet.

Table 7.7 Dietary fibre content of common foods

Foodstuff	Fibre g/100 g	Foodstuff	Fibre g/100 g
All-bran	22.47	Macaroni	2.62
Weetabix	9.77	Carrots	2.30
Shredded wheat	9.82	Cream crackers	2.30
Porridge oats	6.49	Rich tea biscuits	2.15
Peanuts	6.00	Apples	1.93
Wholemeal bread	5.79	Brown rice	1.74
Runner beans	4.42	Raisins	1.70
Hazelnuts	4.32	Tomatoes	1.68
Brown bread	4.26	White bread	1.60
Brussels sprouts	3.32	Bananas	1.11
Baked beans	3.20	Lettuce	0.93
Digestive wheatmeal biscuits	2.88	Rice Krispies	0.85
Spaghetti	2.72	Cornflakes	0.64
Cabbage	2.70	White rice	0.50

EFFECTS OF DIETARY FIBRE   At one time it was believed that the presence of 'roughage' in the diet was of value only as a means of easing or preventing constipation. Thus the main selling-point for bran-rich breakfast cereals was their efficiency in promoting regular bowel movements. This is thought to be a consequence of the absorption of water by dietary fibre in the small intestine leading to the formation of softer and larger stools. In addition, dietary fibre speeds the passage of food through the alimentary tract. The 'transit time' for a typical British diet may be as long as 100 hours compared with as little as 35 hours or less if a high-fibre diet is eaten.

The prevention of constipation may not be of earth-shattering importance, except perhaps to a sufferer. Eating enough dietary fibre has other, more important, beneficial effects, however, and there is substantial evidence to show that it helps to prevent many bowel diseases. Appendicitis, diverticular disease (in which distended pockets are formed in the bowel walls) and haemorrhoids are all less likely to occur if a high-fibre diet is eaten.

High-fibre diets have also proved to be of value to sufferers from the disease diabetes mellitus in which the concentration of glucose in the blood exceeds the normal level. Diets with a high-fibre content and, in particular those rich in soluble fibre, are able to slow down the release of glucose to the bloodstream in some people and in this way the symptoms of the disease are minimized. Soluble dietary fibre sounds almost like a contradiction in terms but, as explained before, pectins are regarded as dietary fibres and these, in addition to being non-fibrous, are soluble.

Studies have shown that diets rich in soluble dietary fibre can lower the concentration of cholesterol in some people's blood. This may account for the statistical link between high-fibre diets and a lower incidence of coronary heart disease (see Chapter 4). At present it cannot be said with certainty, however, that a fibre-rich diet is effective in protecting against heart disease. The incidence of bowel cancer is also apparently lower amongst those whose diet is rich in dietary fibre. It has been suggested that this may be connected with the shorter alimentary transit time of a high-fibre diet.

Dietary fibre is unaffected by the enzymes of the digestive system and it passes more-or-less unchanged to the large intestine. Once there, however, it is attacked and broken down by the harmless bacteria which inhabit the bowel and is partly converted to short chain fatty acids, carbon dioxide, hydrogen and methane. The short chain fatty acids – mainly acetic, propionic and butyric (i.e. C_2, C_3 and C_4 acids) are absorbed into the bloodstream and may be used as a source of energy.

The gases formed when bacteria break down dietary fibre in the bowel are not absorbed and, although harmless, they may be a source of embarrassment as many who have followed a high-fibre diet (e.g., the F-plan diet) for weight reduction purposes have found. During bacterial decomposition of dietary fibre the bacteria themselves multiply rapidly. This increases the bulk of the faecal matter and assists in the production of a softer and larger stool.

REQUIREMENTS OF DIETARY FIBRE   It is not known with any degree of certainty how much dietary fibre we need. Dietary fibre is so varied in composition and individual reaction to it so unpredictable that it is impossible to specify a precise figure and the best that can be done is to set a daily target figure. The average British diet provides about 20 g per day and it is recommended (NACNE report, 1983) that this should be increased to 30 g per day.

SUGGESTIONS FOR FURTHER READING

AKROYD, W.R. (1970). *Wheat in Human Nutrition*. FAO (United Nations), Rome.

BIRCH, G.G. (1970). *Glucose Syrups and Related Carbohydrates*, (Symposium Report). Elsevier, Amsterdam.

BIRCH, G.G. AND PARKER, K.J. (Eds) (1983). *Dietary Fibre*. Applied Science, Barking.

BIRCH, G.G. AND SHALLENBERGER, R.S. (1970). *Developments in Food Carbohydrate – I*. Applied Science, Barking.

BRITISH NUTRITION FOUNDATION (1987). *Sugars and Syrups*. BNF, London.

BURKITT, D. (1981). *Don't Forget Fibre in your Diet*, revised edition. Martin Dunitz, London.

CAKEBREAD, S. (1975). *Sugar and Chocolate Confectionery*. Oxford University Press, Oxford.

CANDY, A. (1980). *Biological Functions of Carbohydrates*. Blackie & Son, London.

DEPARTMENT OF HEALTH AND SOCIAL SECURITY (1981). *Nutritional Aspects of Bread and Flour*. HMSO, London.

FAO/WHO (1980). *Carbohydrates in Human Nutrition*. HMSO, London.

HEALTH EDUCATION COUNCIL (1985). *The Scientific Basis of Dental Health Education*. HEC, London.

HOSENEY, R.C. (1986). *Principles of Cereal Science and Technology*. American Association of Cereal Chemists, USA.

KENT, N.L. (1983). *Technology of Cereals*, 3rd edition. Pergamon Press, Oxford.

LEES, R. AND JACKSON, B. (1973). *Sugar Confectionery and Chocolate Manufacture*. Leonard Hill, Glasgow.

MINIFIE, B.W. (1980). *Chocolate, Cocoa and Confectionery: Science and Technology*, 2nd edition. Avi, USA.

NATIONAL ADVISORY COMMITTEE ON NUTRITION EDUCATION (NACNE) (1983). *Proposals for Nutritional Guidelines for Health Ediucation in Britain*. Health Education Council, London.

RAUCH, G.H. (1965). *Jam Manufacture*, 2nd edition. Leonard Hill, Glasgow.

ROYAL COLLEGE OF PHYSICIANS (1976). *Fluoride, Teeth and Health*. Pitman Medical, London.
ROYAL COLLEGE OF PHYSICIANS (1980). *Medical Aspects of Dietary Fibre*. Pitman Medical, London.
TAYLOR, R.J. (1975). *Carbohydrates*. Unilever Booklet, Unilever, London.
WILLIAMS, A. (Ed.) (1975). *Breadmaking, the Modern Revolution*. Hutchinson Benham, London.
YUDKIN, J., EDELMAN, J. AND HOUGH, L. (1971). *Sugar*. Butterworth, Sevenoaks.

CHAPTER 8

Amino acids and proteins

Previous chapters have been concerned with substances containing the elements carbon, hydrogen and oxygen, and it has been shown that these elements combined in the form of fats and carbohydrates are of fundamental importance in food science. In addition to these elements there is a fourth, namely nitrogen, which also plays a vital role in human life and which, when combined in the form of amino acids and proteins, constitutes a third group of compounds which are of basic importance in food and nutrition.

There is a definite relation between elementary nitrogen on the one hand and the complex proteins required by man on the other. The way in which nitrogen is built up by stages into animal protein and then degraded in further stages back again into nitrogen is summed up in the nitrogen cycle (Fig. 8.1).

Elementary nitrogen occurs in almost limitless quantities in the atmosphere, while combined nitrogen is widely distributed in the soil as salts and, in the form of organic compounds, is found in all living matter. Combined nitrogen forms an essential part of the structure of the body, which requires a continuous supply of nitrogen in a suitable form. Unfortunately, the body is unable to synthesize its nitrogen compounds starting from elementary nitrogen; indeed, it is unable to perform such syntheses even when it is provided with a supply of inorganic nitrogen compounds. This means that man must be supplied with nitrogen which has already been converted into a suitable organic form. This produces the paradoxical situation that although nitrogen is abundant in its elementary form, nitrogen compounds which can be utilized by man are scarce. The explanation of this is to be found in the character of the element, which is noted for its inertness. This lack of reactivity makes it difficult to convert the element into its compounds, a process known as *fixation*.

Fixation of nitrogen into ammonia and subsequently into soluble ammonium salts is carried out commercially, but the amounts of fixed nitrogen thus produced are infinitesimal compared with the amounts required by living things. Fortunately, nitrogen fixation carried out with difficulty by chemists is performed with ease in nature, aided by microorganisms such as *Rhizobium*, which enable leguminous plants (e.g., peas and beans) to synthesize protein from nitrogen. Other green

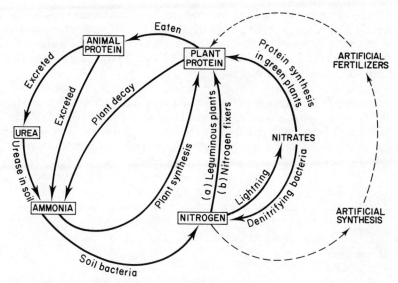

Figure 8.1 The nitrogen cycle.

plants synthesize protein from nitrates present in the soil. Synthesis of protein is opposed by destructive processes which break down protein by stages into nitrogen, thus completing the cycle which is summarized in Fig. 8.1.

AMINO ACIDS

The structure of amino acids is relatively simple and every amino acid contains an amino group $-NH_2$ and a carboxyl group $-COOH$. Amino acids of interest in nutrition have the amino and carboxyl groups attached to the same carbon atom, and to understand their structure it is easiest to start with *acetic acid*. If a hydrogen atom of acetic acid on the carbon next to the $-COOH$ group is replaced by an $-NH_2$ group the simplest possible amino acid, known as amino acetic acid or *glycine* is formed.

CH_3COOH

Acetic acid

CH_2NH_2COOH

Glycine

$CHRNH_2COOH$

General formula

All amino acids encountered in food have a structure which can be expressed by the general formula $CHRNH_2COOH$, where the nature of R can vary considerably, as can be appreciated from Table 8.1 which gives the nature of R for the important amino acids obtained from food and body proteins.

The carbon atom to which the $-NH_2$ group is attached is known as the α-carbon and, except in the instance of glycine, this carbon atom has four *different* atoms or groups of atoms attached to it. Such structures may exist in two different spatial arrangements that are the mirror image of each other. In nature only one of these forms, the L-form, exists.

Table 8.1   Structure of amino acids, $H_2NCHRCOOH$

Name (E = essential)	Abbreviation	R	Isoelectric point
Neutral			
Glycine	Gly	$H-$	6.0
Alanine	Ala	CH_3	6.0
Valine (E)	Val	$(CH_3)_2CH-$	6.0
Leucine (E)	Leu	$(CH_3)_2CHCH_2-$	6.0
Isoleucine (E)	Ile	$CH_3CH_2CH(CH_3)-$	6.0
Norleucine	Nor	$CH_3(CH_2)_3-$	6.1
Phenylalanine (E)	Phe	$C_6H_5CH_2-$	5.5
Tyrosine	Tyr	$C_6H_4(OH)CH_2-$	5.7
Serine	Ser	$HOCH_2-$	5.7
Threonine (E)	Thr	$CH_3CH(OH)-$	5.6
Cysteine	CySH	$HSCH_2-$	5.1
Cystine	CySSCy	$HOOCCH(NH_2)CH_2S_2CH_2-$	4.8
Methionine (E)	Met	$CH_3SCH_2CH_2-$	5.7
Tryptophan (E)	Trp		5.9
Basic			
Ornithine	Orn	$H_2N(CH_2)_3-$	9.7
Arginine	Arg	$\begin{array}{c} NH_2 \\ | \\ HN{=}CNH(CH_2)_3- \end{array}$	10.8
Lysine (E)	Lys	$H_2N(CH_2)_4-$	10.0
Histidine	His		7.6
Acidic			
Aspartic acid	Asp	$HOOCCH_2-$	2.8
Glutamic acid	Glu	$HOOC(CH_2)_2-$	3.2

CLASSIFICATION OF AMINO ACIDS   More than 20 amino acids have been obtained from food and body proteins. They may be classified as being either *neutral, basic* or *acidic*, as shown in Table 8.1. Neutral amino acids are those, like glycine, which contain one amino and one carboxyl group, basic amino acids contain one carboxyl but more than one basic group, while acidic amino acids contain one amino and two carboxyl groups. Those amino acids marked by the letter E are known as *essential amino acids*, meaning that they cannot be manufactured by the body and so must be supplied by the diet. Eight amino acids are known to be essential for adults, and there is growing evidence that a ninth, *histidine*, may also be essential for adults. It has been established that histidine is an essential part of the diet for infants and children when growth is rapid, and it now seems likely that it remains essential even when growth has slowed down or stopped.

PROPERTIES OF AMINO ACIDS   Amino acids are white crystalline substances which are soluble to some extent in water but which are mostly insoluble in organic solvents. The amino group, as its name suggests, is related to ammonia and like ammonia it has basic characteristics while the carboxyl group is acidic. The combination of an amino group and a carboxyl group in the same molecule results in it being able to act as an acid or a base; such a substance is said to be *amphoteric*.

The formulae in Table 8.1 show the arrangement of covalent bonds in amino acids but they do not show the ionic character which amino acids display in solution. In solution amino acids may be more correctly represented as follows:

$$\overset{+}{N}H_3-CH-\overset{-}{C}OO$$
$$|$$
$$R$$

This formula shows the ionic character of an amino acid and that it contains both a positive and a negative group. Amino acids are weak electrolytes and they ionize according to the pH of the system. We can represent this ionization as:

$$\overset{+}{N}H_3-CH-COOH \underset{\text{acid}}{\overset{+H^+}{\longleftarrow}} \overset{+}{N}H_3-CH-\overset{-}{C}OO \underset{\text{alkali}}{\overset{-H^+}{\longrightarrow}} NH_2-CH-\overset{-}{C}OO$$
$$|\qquad\qquad\qquad\qquad |\qquad\qquad\qquad\qquad |$$
$$R\qquad\qquad\qquad\qquad R\qquad\qquad\qquad\qquad R$$

Positive ion   Zwitterion   Negative ion

Thus, if acid is added to a neutral solution of an amino acid a positive ion is formed, whereas if alkali is added a negative ion is formed.

Therefore an amino acid may be neutral, or positively or negatively charged according to the pH of the system.

When an amino acid is neutral, that is when the positive and negative charges are equal, it is said to be at its *isoelectric point* and it is called a *zwitterion* or dipolar ion. Such zwitterions are effective buffers because of their capacity to combine with both acids and bases, thus preventing the change of pH that would otherwise occur. The buffering action of amino acids is very important, particularly in living cells where the cell can only function provided that the pH is maintained within a narrow range.

The isoelectric points of a number of amino acids are shown in Table 8.1. The isoelectric point is important because at this pH many properties have either a maximum or minimum value; for example, electrical conductivity, solubility and viscosity are all a minimum.

IMPORTANCE OF INDIVIDUAL AMINO ACIDS   *Arginine* is a basic amino acid, formed in the liver and kidney and is involved in making urea in the liver. *Lysine* is an essential amino acid and, like arginine, basic in its reactions. It is used for producing *carnitine* in the body, a substance that transports fatty acids within cells. Cereals are deficient in lysine and this is an important factor in planning diets that are adequate in protein (see p. 174).

Cysteine, *cystine* and *methionine* are all amino acids that contain sulphur and they constitute the main source of sulphur in the diet. The body can make cysteine from methionine, the latter being an essential amino acid. Cystine is one of the main amino acids of insulin, and is formed from cysteine in the body.

Glutamic acid is acidic in nature; it is not an essential amino acid though in the body it plays an important part in the metabolism of ammonia. It is of particular interest because its salt, *monosodium glutamate*, is used as a flavouring agent in food. (see p. 380).

Glutamate occurs both in foods and in the body either free or as part of proteins. For example, proteins of foods such as milk, cheese and meat are rich in glutamate while some vegetables, notably mushrooms, tomatoes and peas, have high levels of free glutamate. The body contains glutamate both free and as part of the protein; about one-fifth of body protein is glutamate.

Histidine is a basic amino acid on account of the *imidazole* ring in its structure (shown in Table 8.1). The body's capacity to make it is limited so that during periods of rapid growth, as in infancy and childhood, additional amounts must be supplied by the diet. There is growing evidence that histidine remains essential into adult life. In the body histidine is converted into *histamine* in a process called *decarboxylation*, in which amino acids are converted into related compounds called *amines* by the removal of the acid group. Histamine dilates blood capillaries and stimulates production of acid in the stomach.

Phenylalanine and *tyrosine* are neutral amino acids with a similar structure (see Table 8.1). The body cannot make phenylalanine for itself but can convert it into tyrosine, so that phenylalanine but not tyrosine is essential in the diet. They both contain a benzene ring in their structures which enables them to provide the body with raw materials from which to make the hormones *adrenaline* and *thyroxine*.

Tryptophan occurs as part of the proteins casein (in milk) and fibrin (in blood). It is an essential amino acid concerned with the synthesis of haemoglobin and plasma proteins. It is also of interest because the body can convert it into the vitamin *nicotinic acid*. However, only small amounts of nicotinic acid can be made in this way as it is not an efficient process, 6 g tryptophan producing only 0.1 g nicotinic acid.

Peptides

When two amino acid molecules combine, the acid group in one molecule reacts with the basic group in the other with the elimination of water:

$$\overset{+}{N}H_3\underset{\underset{R^1}{|}}{C}HCOO^- + \overset{+}{N}H_3\underset{\underset{R^1}{|}}{C}HCOO^- \rightarrow \overset{+}{N}H_3\underset{\underset{R^1}{|}}{C}HCONH\overset{\overset{R^2}{|}}{C}HCOO^- + H_2O$$

The product formed is a *dipeptide* and contains the group $-CONH-$ which is the peptide linkage. Substances of relatively small molecular weight containing this group are called *peptides*.

A dipeptide still contains an amino and a carboxyl group and may react with another amino acid molecule to form a *tripeptide*. Theoretically this procedure may be repeated again and again with the formation of *polypeptides*. In practice, amino acids do not react together in this way, but dipeptides, tripeptides and polypeptides may be synthesized indirectly.

PROTEINS

Protein molecules consist of chains of hundreds or even thousands of amino units joined together rather like beads on a string; they are the most complex substances known to man. This fact can be simply illustrated by comparing the molecular weights and formulae of proteins with those of other types of substance. A simple monosaccharide such as glucose has a molecular weight of 180 and a formula of $C_6H_{12}O_6$, whereas a simple protein has a molecular weight reckoned in thousands and a correspondingly complex formula. The protein *lactoglobulin*, for instance, has a molecular weight of about 42 000 and a formula approximating to $C_{1864} H_{3012}O_{576}N_{468}S_{21}$. Large protein molecules are much bigger than this and have molecular weights of several million.

STRUCTURE OF PROTEINS: PRIMARY AND SECONDARY STRUCTURE
The way in which the complex structure of proteins has been worked out constitutes one of the major advances in biochemistry in recent years. The problem is one of awe-inspiring difficulty, but a notable step forward was made when, in 1951, Sanger determined the nature of the protein *insulin*. Insulin is a relatively small, simple protein built up from only 51 amino acid units (Fig. 8.3) whereas large proteins may contain over 500 amino acid units. Persistent research has enabled the structure of even complex proteins to be worked out.

In addition to the elements carbon, hydrogen, oxygen and nitrogen, proteins often contain sulphur and sometimes phosphorus. On hydrolysis, proteins break down to polypeptides and eventually to amino acids, a single protein producing up to about 20 different amino acids. It is clear, therefore, that amino acids are the building units of which proteins are composed – but how are these units joined together to form a protein molecule? The answer is that they are joined together by *peptide linkages*. For example, although protein molecules contain few free amino or carboxyl groups, on hydrolysis about equal numbers of these groups are produced, as would be expected if peptide links are being broken.

Protein molecules are composed of large numbers of up to 20 different amino acids joined together by peptide linkages. X-ray analysis gives some indication of how the peptide chains are arranged in protein molecules. Such chains have a zigzag structure with the R-groups protruding alternately in opposite directions as shown in Fig. 8.2. The first major problem in working out protein structure is that of determining the sequence of amino acid units R_1, R_2, R_3 constituting the chain.

When it is considered that there are hundreds, and sometimes thousands, of amino acid units in a single protein molecule, and that there are over 20 *different* amino acids available to choose from, it is clear that the number of different protein molecules which can be constructed is almost limitless. It has been calculated that for a medium-sized protein containing 288 amino acid units and 12 *different* amino acids there are 10^{300} different possibilities. The magnitude of this number may perhaps be appreciated from the fact that there are less than 10^{10} people living on the earth!

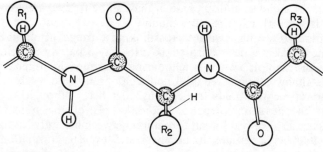

Figure 8.2 The zigzag structure of a polypeptide chain showing its three-dimensional structure.

The amino acid sequence of many proteins is now known, and that of insulin, containing 51 amino acid units, is shown in Fig. 8.3.

The determination of the sequence of amino acids in polypeptide chains reveals what is known as the *primary* protein structure, but this is only the beginning of the problem of working out the complete structure of a protein. In a protein molecule polypeptide chains are linked together in a number of different ways giving rise to molecules of definite shape; this constitutes the *secondary* protein structure. Many of the R- groups in polypeptide chains contain reactive groups (see Table 8.1) which couple with reactive groups in adjacent chains so joining the chains together by cross-linking.

The most important R-group involved in cross-linking is that of *cysteine* which contains the SH-group. When two cysteine units in different polypeptide chains are adjacent, a disulphide bridge —S—S— may be formed between them by oxidation of the SH-groups, thus joining the chains together. Figure 8.3 shows how two such chains are joined together at two different points in insulin. It

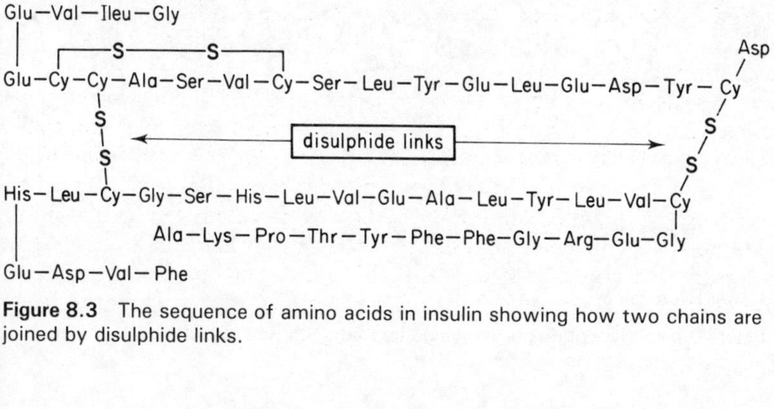

```
Glu—Val—Ileu—Gly
  |       ┌─S────S─┐                                                        Asp
Glu—Cy—Cy—Ala—Ser—Val—Cy—Ser—Leu—Tyr—Glu—Leu—Glu—Asp—Tyr—Cy
         S                                                        S
         |              ←──── disulphide links ────→              /
         S                                                        S
His—Leu—Cy—Gly—Ser—His—Leu—Val—Glu—Ala—Leu—Tyr—Leu—Val—Cy
  |            Ala—Lys—Pro—Thr—Tyr—Phe—Phe—Gly—Arg—Glu—Gly
Glu—Asp—Val—Phe
```

Figure 8.3 The sequence of amino acids in insulin showing how two chains are joined by disulphide links.

also shows how an internal disulphide bridge can be formed between cysteine units occurring in the same chain.

In addition to strong covalent cross-linking through disulphide bridges other weaker types of cross-link also play a part in protein structure. For example, when neighbouring R-groups in different polypeptide chains contain free $\overset{+}{N}H_3$ and $CO\overset{-}{O}$ groups the resulting electrostatic attraction holds the chains together, although the strength of the attraction depends upon the pH of the system. Cross-links are also formed by salt formation between basic groups in one chain and acidic groups in another, and by ester formation between hydroxyl groups in one chain (e.g., threonine and serine) and groups such as phosphate in another. Cross-links may also be produced by the formation of *hydrogen bonds*, although such links formed between hydrogen in one chain and, for example, oxygen in a neighbouring chain are much weaker than true chemical bonds. Figure 8.4 illustrates how such bonds are formed between adjacent polypeptide chains.

CLASSIFICATION OF PROTEINS: TERTIARY STRUCTURE   Animal proteins can be classified according to their molecular shape as either *fibrous* or *globular*. Plant proteins are more difficult to classify, but

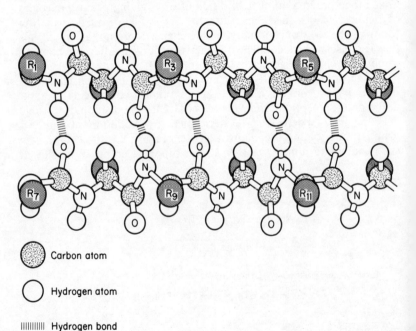

◉ Carbon atom

◯ Hydrogen atom

||||||||| Hydrogen bond

Figure 8.4 Polypeptide chains joined by hydrogen bonds in a protein molecule.

generally speaking they can be divided into *glutenins* or *prolamines*. The nature of these different types is summarized in Table 8.2.

Fibrous proteins, which are simpler than globular proteins, are made up of individual zigzag polypeptide chains which are held together by cross-links to form elongated or fibrous molecules with a fairly stable but elastic structure. They are characterized by being rather insoluble substances.

The fibrous proteins *keratin* and *collagen* have been much studied and will serve to illustrate typical features of this class of protein. Keratin is the main protein of hair. In its natural form, known as α-keratin, a hair or wool fibre consists of many polypeptide chains with the form of an α-helix (Fig. 8.5). These chains are held together by hydrogen bonds and also disulphide bridges provided by the sulphur-containing amino acid cystine. The chains are embedded in an insoluble protein matrix. When α-keratin is subjected to moist heat and stretching the hydrogen bonds break, the α-helix structure disintegrates and a permanently stretched, inelastic form, known as β-keratin is formed. This is the basis of the 'permanent' waving of hair.

Collagen is the most abundant protein in the body. It occurs mainly in skin, cartilage and bone and is the body's major structural protein. The main amino acids in collagen are glycine, proline and hydroxy-proline, and these prevent the formation of an α-helix structure. Instead the polypeptide chains wrap round each other in threes to form a triple helix (Fig. 8.6) which has a rope-like structure. The

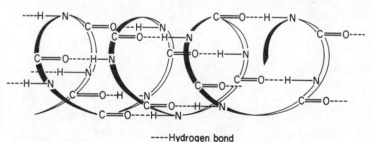

----Hydrogen bond

Figure 8.5 An α-helix showing how the links in the coil may be held together by hydrogen bonds.

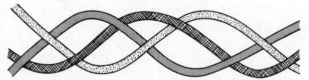

Figure 8.6 Collagen, showing its triple helix structure.

strands of the 'rope' are held together by hydrogen bonds and the whole structure has great tensile strength as well as being insoluble in water.

Globular proteins are more complex than fibrous proteins because the α-helix chain is folded in various ways to form molecules with an irregular but bulky shape. The particular way in which folding takes place depends upon the points in adjacent coils at which disulphide and other cross-links are formed. One of the complexities of determining the structure of globular proteins is that there is no general pattern of folding and so the exact nature of folding – called the *tertiary* structure – needs to be determined for each protein individually. Figure 8.7 gives an impression of how the α-helix is folded in the *myoglobin* molecule, about three-quarters of the chain being in the form of an α-helix. It can be seen that the structure is complex and non-symmetrical, although the overall shape is approximately spherical. Globular proteins are very important in the body as they include all proteins found within body cells and many food proteins.

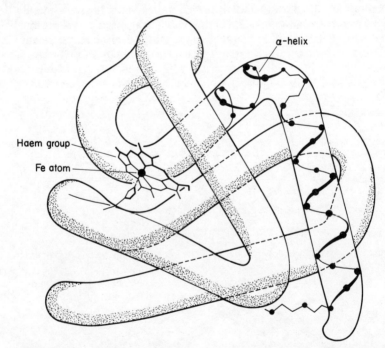

Figure 8.7   An impression of a molecule of *myoglobin* showing the helical nature of a section of the chain and the way in which the chain is folded (after R. E. Dickerson).

The structure of plant proteins is less well understood than those of animals, but they can conveniently be divided into two categories. *Glutelins* are characterized by their insolubility in neutral solutions and their solubility in acids and alkalis. *Prolamines*, on the other hand, are insoluble in water but soluble in alcohol. Examples of both classes are shown in Table 8.2. It is worth re-emphasizing the importance of the presence of both *glutenin* and *gliadin* in wheat. In combination these constitute *gluten*, the significance of which in bread making was discussed on page 139.

SIMPLE AND CONJUGATED PROTEINS   The proteins encountered so far consist entirely of combined amino acids. They are distinguished from *conjugated* proteins whose molecules contain combined amino acids and also a non-protein component, called a *prosthetic* group. The main types of conjugated protein are summarized in Table 8.3.

The nucleic acids *ribonucleic acid* (RNA) and *deoxyribonucleic acid* (DNA) are of particular interest and importance because they play an essential part in the synthesis of all body proteins, including enzymes. Almost all human cells contain *nucleoproteins* in which nucleic acids are bound to different proteins. RNA occurs mostly outside the nucleus while DNA occurs inside. DNA carries the genetic code which determines the nature of the proteins that must be made to maintain hereditary character. DNA passes this information to RNA which controls the synthesis of these proteins. DNA and RNA do not need to be supplied by the diet as the body can make all it requires. Nevertheless, certain foods do contain them, particularly fish roe, liver and kidneys.

Table 8.2 Simple classification of proteins

Type	Solubility/function	Examples/source
Animal		
Fibrous	Insoluble, elastic proteins forming structural part of tissues	Keratin (hair) Collagen (connective tissue) Elastin (tendons, arteries) Myosin (muscle)
Globular	Relatively soluble. Part of fluids of all body cells. Many food proteins	Enzymes, protein hormones Albumins, globulins (blood) Casein (milk), albumin (egg-white)
Plant		
Glutelins	Insoluble in neutral solutions. Soluble in acids and alkalis	Glutenin (wheat) Hordenin (barley) Oryzenin (rice)
Prolamines	Insoluble in water Soluble in alcohol	Gliadin (wheat) Zein (maize)

Table 8.3 Classification of conjugated proteins

Type	Prosthetic group	Examples
Nucleoproteins	Nucleic acid	DNA combined with protamines RNA combined with ribosomes
Lipoproteins	Lipid	Distinguished by density e.g., LDL (low density) see Table 5.11
Chromoproteins	Coloured group containing a metal	Haemoglobin (blood)
Glycoproteins	Carbohydrate	Some enzymes and hormones Mucins (egg-white)
Phosphoproteins	Phosphate	Casein (milk)

Properties of proteins

The molecular weight of an average protein is about 60 000, and they are without characteristic melting points. Some have been obtained in crystalline form. The solubility of proteins in different solvents may show wide variations, especially in water, salt solutions and alcohol, and this diversity is used for classifying them. Although we talk about the solubility of proteins they do not give true solutions but, because of their relatively large size, they form colloidal dispersions or sols. The colloidal character of proteins is important in many foods as we have already seen in connection with the colloidal food systems, such as milk, butter and ice-cream, discussed in Chapters 4 and 5.

The properties of proteins are similar in many ways to those of the amino acids from which they are constructed. For example, they contain free amino and carboxyl groups at the ends of the polypeptide chains, and as these carry a positive and negative charge, proteins form zwitterions; consequently they are amphoteric and they act as buffers. The net charge on protein molecules varies with pH and is zero at the isoelectric point. The isoelectric point is important in considering the behaviour of food proteins because at this pH many properties are either a maximum or minimum; some values are shown in Table 8.4.

Table 8.4 Molecular weights and isoelectric points of some food proteins

Protein	Source	Molecular weight	Isoelectric point
Casein	Milk	34 000	4.6
β-Lactoglobulin	Milk	35 000	5.1
Ovalbumin	Eggs	44 000	4.6
Gliadin	Wheat	27 000	6.5
Gluten	Wheat	39 000	7.0
Gelatin	Bones	Variable	4.9
Myosin	Meat	850 000	5.4

The properties of fibrous proteins are distinct from those of globular proteins. The former are relatively insoluble, being resistant to acids and alkalis, and they are unaffected by moderate heating, whereas the latter are soluble and are affected by acids, alkalis and heating. Globular proteins are very sensitive to chemical and physical conditions on account of the weak cross-links that hold the folded α-helix chains in position. A small change of pH or a small rise in temperature is sufficient to disrupt such cross-links and cause the chains to unfold, a process which is known as *denaturation*. When proteins are denatured their properties are completely altered; biological activity is destroyed, solubility decreased and viscosity increased. Moreover, the change is irreversible.

Denaturation may be brought about by controlling pH and occurs most readily at the isoelectric point when proteins are least stable. For example, casein, which is the main protein of milk, has an isoelectric point of 4.6, and this explains why, when milk turns sour and the pH drops, it is quick to curdle. It also explains why milk is coagulated or clotted in the stomach so easily by rennin, for the pH is low and not far removed from the isoelectric point of casein. The denaturation of milk proteins is discussed in more detail on page 73.

Many proteins are denatured by heat. For instance, if egg white is heated, coagulation begins at about 60°C when the protein *ovalbumin* starts to separate out as a solid. As the temperature is raised, coagulation continues until the whole mass is completely solid. It is clear that coagulation occurs in the cooking of protein foods. For example, the proteins in lean meat coagulate on heating. Coagulation starts at about 60°C, as with albumin, and if cooking temperatures are kept somewhat below 100°C, coagulation is slow and the coagulated protein is not too hard. In this state protein is most digestible. However, if a temperature of 100°C or over is used, as in boiling and roasting, coagulation is more rapid and the denatured protein forms a rather hard solid mass.

Partial coagulation of proteins may be brought about by beating them into a foam. For example, when egg-white is beaten the foam which is formed is stabilized by the partial coagulation of the ovalbumin. If such a foam is heated it becomes rigid due to further coagulation of the ovalbumin. Such foaming occurs most readily at the isoelectric point when the ovalbumin is least stable, and it may be promoted by addition of an acidic substance which lowers the pH to a value near the isoelectric point.

Proteins are precipitated out of solution by certain salts such as ammonium sulphate, sodium chloride, mercuric chloride and lead acetate. Thus, the presence of mineral salts affects the denaturation of proteins by heat treatment. The fact that lead acetate causes precipitation of ovalbumin in egg-white explains why white of egg is used in cases of lead poisoning, for by reaction with egg-white a soluble lead salt is rapidly converted into an insoluble compound which is not assimilated.

TESTS FOR PROTEINS   In order to detect the presence of nitrogen (and sulphur) in a substance *Lassaigne's test* may be used. The substance to be tested is heated with sodium which reacts with nitrogen and sulphur, forming sodium cyanide and sodium sulphide respectively. Sulphide ions give a purple colour on addition of sodium nitroprusside solution and cyanide ions give a blue precipitate or colour with a mixture of ferrous and ferric ions.

Lassaigne's test detects the presence of nitrogen in a substance, but further tests are required to determine whether the compound containing nitrogen is a protein. Proteins may be detected by means of colour tests. For example, in the *Biuret test* proteins give a characteristic purple colour when they are heated in the presence of strong alkali and copper sulphate solution. A similar colour is given by any substance containing more than one $-CONH-$ grouping, so that the test is not specific for proteins but is also positive for all polypeptides.

Most proteins give a positive result in the *Xanthoproteic* reaction in which the suspected protein is heated with concentrated nitric acid. In the presence of protein the solution turns yellow and, on the addition of alkali, orange. *Millon's* reaction may be used to detect proteins which on hydrolysis yield *tyrosine*. The only common protein which does not give a positive result in this test is gelatin. The substance is warmed with Millon's reagent (which contains mercurous and mercuric nitrates in nitric acid), and if a protein is present a white precipitate turning red is obtained.

GELS AND GELATIN   Gels are remarkable colloidal systems in which large volumes of liquid are immobilized by small amounts of solid material, the liquid constituting the dispersion medium and the solid the disperse phase. Gels are of considerable importance in food preparation on account of the rigidity which such systems possess. For example, gels formed by polysaccharides – such as pectin gels in jam and starch gels in cooked starchy foods – have been discussed in Chapter 7. Proteins also form gels and the gel formed by *gelatin* is of particular importance in food preparation.

Gelatin is made from the protein *collagen* which occurs in the skin and bone of various animals. The molecules of collagen are made up of three chains which are intertwined to form a triple helix and held in position by covalent cross-links and other weaker links (see Fig. 8.6). When collagen is treated with hot water and acid or alkali all the weak cross-links and, according to the harshness of the treatment, a proportion of the covalent cross-links, are broken down. The resulting product, commercial *gelatine*, is soluble and contains a range of protein molecules and about 12 per cent water and one per cent mineral salts. It is produced both in granular form and in thin sheets, the former being the more common.

When cold water is added to gelatin it swells owing to the absorption of water. This is because the different protein molecules that constitute gelatin are in the form of zigzag polypeptide chains that are weakly linked together to form a three-dimensional network. Water becomes entangled and immobilized in this network in much the same way as water is held by a sponge. Additional water is bound to the gelatin by hydrogen bonding. If hydrated gelatin is heated with water above 35°C it liquefies and forms a sol. On cooling, the sol 'sets' and becomes solid, a process which is known as *gelation*. As little as one per cent gelatin is sufficient to produce such a gel. The gel formed is semi-rigid, although it loses its rigidity on heating or shaking; the solution thus formed is not coagulated by heat.

The setting powers of gelatin are utilized in the preparation of food. Commercial table jellies are made from syrups of glucose and sucrose. Gelatin is added to the hot syrup and, after it has dissolved, acid (usually citric), flavouring and colouring are added and the mixture is cooled until it sets. Gelatin is also responsible for the setting of stews and of broth prepared by boiling bones. If a gelatin sol is cooled until it is viscous but not firmly set, it can be beaten into a foam. A foam is most easily formed at the isoelectric point of gelatin (see Table 8.4), when the gelatin particles adhere to each other most strongly. During beating, air is incorporated into the mixture and because the gelatin at this stage has a certain elasticity it is able to stretch and surround the air bubbles without breaking. Whipped cream and flavouring may be added to such foams in making gelatin desserts.

Gelatin is also used as a stabilizing agent for emulsions, and its use in this connection was discussed with reference to ice-cream on page 63.

Protein quality

The quality of protein food may be judged by its protein content, the number and amounts of essential amino acids it contains and the degree to which its protein is digested and absorbed by the body. The highest quality protein foods are those which provide all the essential amino acids in the proportions needed by man (Table 8.5).

PROTEIN QUALITY: BIOLOGICAL EVALUATION The *biological value* (BV) of a protein food is usually measured by feeding the protein under test to young rats as the only source of nitrogen and at a level below that required for maintaining nitrogen balance (see

Table 8.5 Essential amino acid content of high quality animal proteins (mg/g protein)

Amino acid	Eggs	Milk	Beef	Suggested pattern for adults[*]
Histidine	22	27	34	16
Isoleucine	54	47	48	13
Leucine	86	95	81	19
Lysine	70	78	89	16
Methionine and cystine	57	33	40	17
Phenylalanine and tyrosine	93	102	80	19
Threonine	47	44	46	11
Tryptophan	17	14	12	9
Valine	66	64	50	5

[*] FAO/WHO report 1985, assuming a safe level of protein intake for adults of 0.7 g per kg body weight

p. 175). The biological value is calculated by measuring the nitrogen intake and the amounts of nitrogen lost in the urine and the faeces. The nitrogen lost in urine measures the nitrogen that has been absorbed and used by the body while the nitrogen lost in the faeces measures the nitrogen that has not been absorbed.

$$BV = \frac{\text{Nitrogen intake} - (\text{Nitrogen lost in urine} + \text{faeces})}{\text{Nitrogen intake} - \text{Nitrogen lost in faeces}} \times 100$$

$$= \frac{\text{Retained nitrogen}}{\text{Absorbed nitrogen}} \times 100$$

BV is defined as the percentage of absorbed protein that is retained in the body, i.e. that which is converted into body protein.

The BV value of a protein takes no account of the digestibility of the protein, and if this is taken into account we measure the *net protein utilization* (NPU). NPU is defined as the percentage of protein eaten that is retained in the body.

$$NPU = \frac{\text{Retained nitrogen}}{\text{Nitrogen intake}} \times 100 = BV \times \text{digestibility}$$

As most proteins have a digestibility exceeding 90 per cent (see Table 8.8), BV and NPU values do not differ greatly as is demonstrated by Table 8.6. As biological evaluation of protein quality is usually done on rats there is no guarantee that the results for BV and NPU values are accurate for humans. In practice it is believed that the results are reasonably reliable, though they tend to underestimate protein quality for man, especially adults. The NPU value of a good mixed diet is about 70.

PROTEIN QUALITY: CHEMICAL EVALUATION   The protein quality of a food can be evaluated in chemical terms by measuring its amino acid content and comparing it with that of a reference protein. Whole egg protein is usually taken as the reference protein and given a score of 100. Chemical score values are found to match up fairly well with biological values, as shown in Table 8.6.

LIMITING AMINO ACIDS   An inspection of Table 8.5 shows that eggs, milk and beef are high-quality proteins because they contain all the essential amino acids and in sufficient amounts to meet the needs of an adult. However, in other protein foods one or more amino acids may be present in amounts that are below human requirements. The amino acid which is furthest below the human requirement is known as the *limiting amino acid*. The limiting amino acid in a variety of foods is shown in Table 8.7.

Table 8.6   Protein quality of some foods comparing BV, NPU and chemical values

Food	BV	NPU	Chemical score
Whole egg	98	94	100
Milk	77	71	95
Soya flour	70	65	74
Wheat	49	48	53
Maize	36	31	49
Rice	67	63	67
Gelatin	0	0	0

Table 8.7   Limiting amino acids in animal and plant foods

Animal source	Limiting amino acid	Plant source	Limiting amino acid
Milk	None	Wheat	Lysine
Eggs	None	Maize	Tryptophan
Beef	None	Legumes	Methionine
Cheese	Methionine	Soya	Methionine
Gelatin	Tryptophan	Nuts	Methionine

COMPLEMENTARY PROTEINS   Where a particular protein is deficient in an essential amino acid, the disadvantage may be overcome simply by eating more of the protein in question. However, in some circumstances, such as in times of food shortage or with infants, this may not be possible.

Normal diets contain a mixture of proteins and the fact that any one protein is of high or low biological value is of no great significance. The important dietary requirement is that the total protein intake should supply all the essential amino acids in suitable proportions for our needs. Thus, although a certain protein may be of low quality because it lacks a particular amino acid, if it is eaten together with a second protein which lacks a *different* essential amino acid, the mixture is of high biological value. Such amino acids are said to *complement* each other. This principle is illustrated by a mixture of gelatin and bread. The limiting amino acid of wheat is lysine whereas that of gelatin is tryptophan. As gelatin is relatively rich in lysine the two complement each other. Examples of other complementary proteins are fish and rice and maize and beans.

Provided that a diet contains a mixture of different proteins it is unlikely to be deficient in any essential amino acid. Even diets which are low in protein quality, such as some vegetarian diets of under-developed countries, usually fulfil adult requirements for essential amino acids. It is only in the case of children, with their relatively greater need for protein to sustain growth, that such diets lack sufficient essential amino acids.

The general conclusion reached is that provided diets contain a variety of protein sources the total mixture will have high biological value. Only where 70 per cent or more of dietary proteins comes from one staple food, which is very low in protein (such as cassava or plantains) need there be concern that protein quality will fall below an acceptable level.

DIGESTIBILITY OF PROTEINS   While a great deal of research has been devoted to protein quality, little consideration has been given to digestibility. As it is now believed that few diets provide insufficient amounts of essential amino acids, except for infants and pre-school children, more attention needs to be given to digestibility, especially of diets in developing countries.

It is evident from Table 8.8 that while animal proteins are highly digestible, plant proteins may have a much lower rate of digestibility. This factor needs to be taken into account when considering protein requirements (see below). The digestibility of protein foods is related to their content of dietary fibre which increases excretion of nitrogen in faeces and hence reduces the apparent digestibility.

Protein requirement

In considering how much protein should be supplied to the body in the diet, account must be taken of its nature; consequently it is much more difficult to estimate the optimum dietary intake of protein than

Table 8.8 Relative digestibility of different foods (FAO/WHO, 1985)

Food	Digestibility
Reference: egg, milk, cheese, meat, fish	100
Wheat, refined	101
Peanut butter	100
Peas, mature	93
Rice, polished	93
Oatmeal	90
Wheat, whole	90
Maize	89
Beans	82

that of carbohydrate or fat. Whereas a certain amount of protein of high nutritional value may satisfy the body's protein requirements, a much larger amount of low-quality protein will be needed. It is, therefore, only possible to calculate the minimum protein intake necessary for health on the assumption that this amount of protein furnishes the body with the minimum of essential amino acids that it requires.

Protein requirement for an individual is defined as the lowest level of dietary protein intake that will balance the losses of nitrogen from the body in persons maintaining energy balance at modest levels of physical activity. In children and pregnant or lactating women the protein requirement is taken to include the needs associated with the deposition of tissues or the secretion of milk at rates consistent with good health.

When such a definition is generalized to cover a class of people the protein requirement is expressed as a *safe level of intake*, previously called the *recommended intake*. It is expressed as the amount of protein that will meet or exceed the requirements of nearly all the individuals in the group.

In practical terms protein requirement is measured when a person is in *nitrogen balance*, i.e. when the nitrogen intake from diet is equal to the nitrogen output in the urine, faeces and skin. The requirement is simply the minimum amount of protein needed to maintain nitrogen balance. In seeking to measure such a protein requirement two reference points have been established, namely those of a young child and a young male adult. For other groups estimates have been made based on the reference points.

FOOD AND AGRICULTURE ORGANIZATION (FAO) STANDARDS   The earliest attempt to determine protein requirement was made by the League of Nations (Technical Committee on Nutrition, 1936) when an

Table 8.9 Estimate of daily amino acid requirements in mg/kg body weight (based on FAO/WHO figures, 1985)

Amino acid	Infants 3–4 months	Children 2 years old	Boys 10–12 years	Adults
Histidine	28	?	?	8–12
Isoleucine	70	31	28	10
Leucine	161	73	44	14
Lysine	103	64	44	12
Methionine (+ cystine)	58	27	22	13
Phenylalanine (+ tyrosine)	125	69	22	14
Threonine	87	37	28	7
Tryptophan	17	12	3	3
Valine	93	38	25	10
Totals:	742	351	216	91–95

intake of 1 g protein/kg body weight was recommended for adults. Such a value had the merit of simplicity, but since then a series of recommendations has been made by the FAO/WHO which has gradually reduced the recommended amount. It is now considered (FAO/WHO, 1985) that a daily protein intake of 45 g, or about 0.7 g/kg body weight of good quality protein of NPU of at least 70, should maintain nitrogen balance in most adults.

For those with higher protein requirements additional amounts of protein are recommended. For example, pregnant women require an additional 6 g protein per day, while lactating women require an additional 11–16 g per day, depending upon the stage of lactation.

The FAO has also estimated daily requirements of individual amino acids for different groups and these are summarized in Table 8.9, which shows the considerable effect that growth has on amino acid needs.

BRITISH STANDARD   In Britain the recommended daily amounts (RDA) of protein are based on a different premise to the FAO/WHO figures. It is considered that a diet which provides less than ten per cent of total food energy as protein would be likely to be unpalatable to most people in Britain. In addition such a diet might be deficient in other nutrients such as easily absorbable iron, vitamin B_{12}, riboflavin, nicotinic acid and zinc, which are often associated with protein. The RDA of protein given in Appendix I are therefore set at a level that provides ten per cent of total food energy on the assumption that dietary protein has an NPU of 75.

The consequence of relating recommended protein allowance to energy expenditure is that the British recommended values are somewhat higher than the FAO/WHO values. For example, for moderately active young men between 18 and 34 years old the British recommendation is 63 g protein per day compared with the FAO/WHO value of 45 g. While the FAO/WHO figure based on nitrogen balance experiments undoubtedly provides the minimum protein needed by the body, many nutritionists believe that it is wise to have a rather higher intake to provide for any particular additional protein needs such as might be required during illness. The British figure while being empirical may nevertheless constitute a safer recommendation.

HUMAN ADAPTABILITY   Although a great deal of effort has gone into establishing minimum protein requirements, it is known that adults can adapt to a wide range of protein intake. It is considered that the health of most adults is unaffected by protein intakes in the range 45–150 g protein per day. The reason why humans can adapt so well to different intakes of protein depends upon the action of the liver which has the ability to transform amino acids into urea, which can then be excreted from the body. When protein intake is high the enzyme *argininosuccinase* is active and is responsible for the formation of urea with consequent loss of protein from the body in urine. When protein intake is low enzymes responsible for using amino acids for protein synthesis are active, so that in these conditions little protein is lost from the body.

It is worth noting that the body's adaptability with regard to protein intake is in sharp contrast to its adaptability with regard to energy intake. In the latter instance the energy requirement is exactly determined by energy expenditure. If the intake exceeds or falls below energy output the body is unable to adapt and responds by storing energy in the form of fat or wasting away respectively.

The total amount of protein in the body can also vary to a considerable degree without serious consequences for health. The body of a well fed adult contains about 11 kg protein but up to 3 kg can be lost without significant detrimental consequences.

HIGH PROTEIN INTAKE   Is a high intake of protein desirable or harmful? In attempting to answer this question the health of groups of people with both high and low levels of protein intake has been studied. The warriors of the Masai tribe of Central Africa, for example, appear to be healthy and of good physique and have a high-protein intake of up to 300 g per day. On the other hand, many people in developing countries have a low protein intake of around 45 g per day and provided that their diet is adequate in other respects they too are healthy.

These findings confirm that the body can adapt to a wide range of protein intake values. It is believed that a high-protein intake throughout life does no harm; on the other hand it is also believed that a high protein intake is in no way necessary.

The notion that people with high energy needs, such as athletes, benefit from a high protein diet, based perhaps on a high consumption of beef-steak, is without foundation. Protein intake for athletes, like that for the general public, should provide about ten per cent of energy requirements.

PROTEIN-ENERGY MALNUTRITION (PEM)   We have already established that adults rarely suffer from protein deficiency even when they have to subsist on poor diets consisting mainly of vegetables. However, protein deficiency is common amongst children in underdeveloped countries. Alternatively, children in such countries may simply lack sufficient food; that is they may exist on diets that are lacking in both energy *and* protein (and other nutrients). In between these two extremes of lack of protein and lack of food lie a variety of diets which are lacking in varying combinations of protein and energy (and other nutrients). The whole range of such diets give rise to what has become known as *protein-energy malnutrition (PEM)*.

PEM constitutes the greatest health problem in underdeveloped countries today. In such countries PEM is a major cause of death and as many as half the children do not reach the age of five.

Protein deficiency in children produces the disease known as *kwashiorkor* which arises when, after a period of breast feeding, children are weaned onto a diet in which the staple food is either cassava or green bananas (matoke) and therefore deficient in protein. Kwashiorkor causes the body to swell and creates patches of pigmentation in both hair and skin; it also produces apathy. It is also associated with other deficiency diseases such as *pellagra* due to deficiency of nicotinic acid.

Lack of food in children produces semi-starvation, a condition which is known as *marasmus*. While it is caused by lack of food it is made worse by susceptibility to repeated infections caused by poor hygiene. Marasmus produces shrunken, dehydrated children with wasted muscles; it is often accompanied by diarrhoea.

PEM is the result of poverty and ignorance. Tradition also plays a part resulting in the father of a family being given any meat or other protein food which is available, the rest of the family having to make do with what other food remains.

PROTEINS IN THE BODY

DIGESTION   During digestion proteins are broken down into amino acids. Their degradation is brought about progressively by peptidases as explained in Chapter 2. Peptidases are hydrolysing enzymes which operate by catalysing the hydrolysis of the peptide links in the protein molecule, so breaking down the protein into smaller units.

In the stomach the gastric glands secrete pepsinogen which, at the low pH of the gastric juice, becomes activated forming the enzyme pepsin. The action of pepsin is extremely specific; it catalyses the hydrolysis of only those peptide links which are joined to particular groupings. Moreover, it is an endopeptidase and acts only on inner peptide links of polypeptide chains. Figure 8.8 shows part of a protein molecule, and the only point at which pepsin could act is indicated. As a result of the action of pepsin, proteins are broken down into smaller peptone units. The enzyme rennin is also present in the gastric juice and brings about the coagulation of the casein of milk.

In the small intestine, endopeptidases such as trypsin and chymotrypsin continue the hydrolysis of proteins and complete their breakdown into peptones. The action of these enzymes is just as specific as that of pepsin, each enzyme attacking only a certain type of link. The points at which they could attack a typical protein fragment are illustrated in Fig. 8.8. Peptones are further broken down by a group of exopeptidases, called erepsin, which are present in the intestinal juice. These enzymes catalyse the hydrolysis of peptones into dipeptides which are broken down into amino acids by a series of

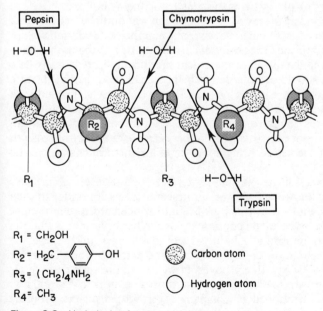

$R_1 = CH_2OH$

$R_2 = H_2C$ —⟨ ⟩— OH

$R_3 = (CH_2)_4NH_2$

$R_4 = CH_3$

Figure 8.8 Hydrolysis of a protein fragment.

dipeptidases. Proteins are thus completely hydrolysed to amino acids before passing from the small intestine into the blood. The amino acids are, however, rapidly removed from the blood by all the cells of the body, but particularly by the liver.

ENZYMES   The supreme importance of enzymes in the body and their protein nature have already been emphasized in Chapter 2. Now that the structure of proteins has been considered it is possible to gain a clearer insight into the nature and mechanism of enzyme activity. It must be admitted, however, that in spite of the success of current research we still have a great deal to learn about the way enzymes act, although this is hardly surprising considering the complexity of protein structure. Moreover, the results of recent research show that few generalizations may be made in this field and that each enzyme needs to be investigated individually.

Enzymes are globular proteins which, as we have already seen, are the most complex proteins known and consist of folded α-helix chains, the method of folding being irregular and dependent on the nature of the cross-links formed between adjacent helical coils. Such structures are very sensitive to both chemical and physical conditions on account of the ease with which the cross-links are broken. This explains why enzymes are so sensitive to changes of temperature and pH. Moreover, the zwitterion structure of enzymes enables them to resist changes of pH in living systems by acting as buffers.

The lock and key theory of enzyme action was outlined in Chapter 2 and this theory has the merit of extreme simplicity. Modern developments have confirmed the essential correctness of this theory but have shown that the details are a good deal more complicated. It appears that the lock should be regarded as being flexible and that a substrate must not only fit the enzyme lock exactly but that it must also induce a change in the enzyme structure, thus causing a reorientation of the enzyme groups involved in catalysis. Thus, the specificity of an enzyme is due not only to a good fit between lock and key but also to the ability of the key to bring about certain structural changes in the lock.

The structures of the enzymes lysozyme, ribonuclease and carboxy-peptidase are known and show that these molecules are folded in such a way as to contain a jaw-like groove into which substrate molecules fit. In the case of carboxypeptidase it has further been shown that the substrate protein brings about structural changes in the enzyme molecule which are essential to its catalytic activity. Not all enzyme molecules possess jaw-like grooves, however, and in such cases (e.g., chymotrypsin) some other mechanism must operate.

As already mentioned in Chapter 2, certain substances promote enzyme activity and three main types have been identified, namely *coenzymes, cofactors* (or activators) and *prosthetic groups.* Coenzymes are large organic molecules such as several members of the B group of vitamins. The coenzyme is not firmly bound to the enzyme, but may become attached to it during enzyme activity only to be released later so that, like the enzyme itself, it may be reused.

Cofactors are usually ions, either metallic or non-metallic, which become temporarily attached to an enzyme during enzyme activity but which are later released. The action of chloride ions as a cofactor for the enzyme salivary amylase has already been mentioned (p. 20).

Prosthetic groups are non-protein groups which are permanently bound to enzymes to form conjugated proteins (see Table 8.3). The enzyme *catalase* present in vegetables (see p. 345) is an example in which the prosthetic group contains a ferrous ion and is known as *haem.*

METABOLISM   The present picture of protein metabolism is quite different from earlier theories which assumed that the proteins of living tissues were stable and only required replacement after a long period of service. It is now known that far from being stable, body proteins are in a constant state of flux and are continually being degraded and resynthesized in a similar manner to fat molecules. On the one hand there are the complete body proteins, and on the other a reservoir of amino acids derived partly from food and partly from degraded body proteins. During life the proteins and amino acids are in dynamic equilibrium with each other. Such an equilibrium involves the continual hydrolysis of the peptide links of protein molecules and the continual resynthesis of proteins from amino acids.

The synthesis of proteins takes place throughout all the cells and tissues of the body. The amino acids that are required for this synthesis are taken from the liver, both essential and non-essential amino acids being involved. If non-essential amino acids are not available from the pool of amino acids in the liver they may be made in body cells by a process of *transamination* in which an *amino* group is *trans*ferred to a substance which does not contain nitrogen thereby converting it into an amino acid. This process is controlled by enzymes known as *transaminases* and by the coenzyme *pyridoxal-5-phosphate* (PLP) derived from vitamin B_6. Such syntheses are extremely rapid, the amino acids being correctly assembled into proteins through the action of DNA and RNA.

PROTEIN AS A SOURCE OF ENERGY   Although the primary function of food proteins is the provision of amino acids for the production and maintenance of body proteins (including enzymes), they are ultimately broken down to urea with the liberation of energy. The secondary role of proteins is, therefore, as a supplementary energy source.

The first step in the breakdown of amino acids is normally *deamination*, in which the amino group is removed from the amino acid as ammonia. The ammonia formed in deamination is poisonous and is rapidly converted into urea, which is then excreted by the kidneys in urine.

Deamination, which requires the presence of the B vitamins nicotinic acid and riboflavin in the form of dinucleotides, results in the formation of an organic acid which may be either *glucogenic* or *ketogenic*. Glucogenic acids are broken down into glucose while ketogenic acids are converted into fatty acids. Both these substances may be broken down to produce energy as described previously, illustrating how carbohydrates, fats and deaminated amino acids all follow closely related metabolic pathways. If energy is not required by the body the non-nitrogenous residues of deamination are converted into fat.

CHAPTER 9

Protein foods

There is no such thing as a pure protein food, i.e. one containing 100 per cent protein. As can be appreciated from Table 9.1 even those foods which are highest in protein do not contain more than 45 per cent, and even this value is exceptionally high.

Table 9.1 The protein content of some animal and plant foods

Animal foods	Protein (%)	Plant foods	Protein (%)
Cheese, Cheddar	26	Soya flour, low fat	45
Bacon, lean	20	Soya flour, full fat	37
Beef, lean	20	Peanuts	24
Cod	17	Bread, wholemeal	9
Herring	17	Bread, white	8
Eggs	12	Rice	7
Beef, fat	8	Peas, fresh	6
Milk	3	Potatoes, old	2
Cheese, cream	3	Bananas	1
Butter	< 1	Apples	< 1
		Tapioca	< 1

In assessing the protein value of food it is often useful to know the proportion of the energy in the foods provided by protein rather than the protein content alone. Most nutritionally adequate diets provide 10–15 per cent of the energy in the form of protein. Thus Table 9.2 makes it easy to assess whether foods are poor, adequate or good as protein sources in relation to their energy content.

It will be remembered from the previous chapter that it is diets based on cassava and plantains which give rise to kwashiorkor in children. This is because when these foods form the main energy source they supply insufficient protein to maintain growth and health.

Some foods which are valuable as sources of protein, such as milk and cheese, have already been considered in Chapter 5. In this chapter other foods, particularly valued as sources of protein, namely meat, fish, soya and eggs, will be described.

Table 9.2 Protein content of foods expressed in terms of their contribution to the energy provided by each food

Value of food as protein source	Per cent of total energy from protein
Poor	
Cassava	3
Sweet potatoes, plantains	4
Adequate	
Potatoes	8
Rice	8
Wheat flour	13
Good	
Peanuts	19
Milk	22
Beans and peas	26
Beef, lean	38
Soya beans	45

MEAT

Lean meat is the flesh or muscular tissue of animals. Its composition is different from that of the internal organs, such as kidneys and liver, which are referred to collectively as *offal*. The composition of flesh of different animals shows considerable variation and the composition of even a single type of meat, such as beef, varies according to breed, type of feeding and the part of the animal from which the meat has come.

Muscle tissue consists of about three-quarters water and one-quarter protein together with a small variable amount of fat, one per cent mineral elements and some vitamins. Structurally the tissue is composed of microscopic fibres, each of which is made up of cells. The main constituent of the cells is water in which the proteins and other nutrients are either dissolved or suspended. Numbers of fibres are held together by connective tissues to form a bundle. A quantity of such bundles is enveloped by a tough sheath of connective tissue which forms a tendon joining the muscle to the bone structure (Fig. 9.1).

The cells of the muscle fibres contain a number of proteins, the most important being *myosin* (7 per cent) and *actin* (2.5 per cent). Myosin is a relatively large elastic protein which can exist in stretched and unstretched forms and which is classified as a fibrous protein (Table 8.2, p. 167). Actin has two forms; a small globular form of molecular weight about 70 000 and a fibrous form in which a series of globular units are arranged in a double chain. The cells also contain

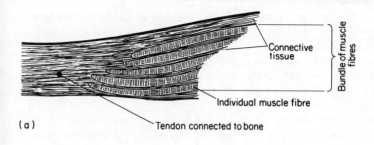

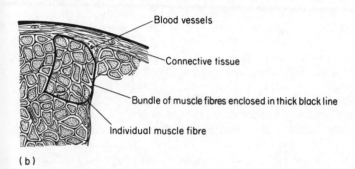

Figure 9.1 Meat. Diagram of muscle showing fibres and connective tissue. (a) Longitudinal section. (b) Transverse section.

ATP (adenosine triphosphate) which, as we saw in Chapter 6, provides the energy used when muscle fibres contract. After death ATP is broken down and in its absence myosin and actin combine to form rigid chains of *actomyosin*. In this state, known as *rigor mortis*, meat is rigid and tough and it is therefore not consumed until, after a period of storage known as *conditioning*, the stiffness has diminished and tenderness and flavour have improved.

The changes that occur during conditioning are complex, as might be expected from the protein nature of meat; moreover, they are affected by a number of variables of which temperature, pH and length of storage are all important. Although the changes that constitute conditioning are complicated and still not completely understood it is a very important process resulting in the conversion of the muscular tissue of animals into the 'meat' of our diet.

After an animal's death the glycogen present in its muscular tissue is broken down by stages to lactic acid with consequent fall in pH. The final pH is usually about 5.5, which is close to the isoelectric point of the main muscle proteins (isoelectric point of myosin is 5.4). As we have already noted proteins are least stable and most readily de-natured at the isoelectric point, and during conditioning proteins of muscular tissue are denatured, though the proteins of connective tissue are not. Denaturation is followed by some breakdown of denatured proteins resulting in the formation of peptides and amino acids and an increase in tenderness. A further important change results from the fact that the solubility of proteins is a minimum at the isoelectric point and hence during conditioning some water is lost from meat.

During conditioning the colour of meat changes from reddish to brown and this is associated with the conversion of the main muscle pigment *myoglobin* into *metmyoglobin*. The myoglobin molecule consists of a protein part, consisting of folded chains of α-helices as shown in Fig. 8.7 together with a non-protein coloured *haem* group containing an iron atom in the ferrous state. During conditioning myoglobin is denatured and oxidized, ferrous iron being converted into ferric iron, and the resulting metmyoglobin is brown in colour.

The *connective tissue* of meat which surrounds the bundles of muscle fibres is mainly *collagen*, while the walls of the muscle fibres are mainly *elastin*. Both these proteins are classified as fibrous proteins (see Table 8.2), collagen containing inelastic polypeptide chains joined by cross-links and elastin having an unstretched α-helix form. Collagen and elastin, being insoluble and tough, are difficult to digest. However, when meat is cooked in the presence of moisture the collagen is converted into gelatin, which is soluble in water. This makes the digestion of connective tissue much easier, and enables the digestive juices to come into intimate contact with the myosin of the muscle fibres. The greater the age of the animal and the more active its life, the greater is the amount of connective tissue and the thickness of the walls of the fibres. Thus, meat of old animals is more difficult to digest than that of young ones and muscular tissue of active animals is more difficult to digest than that of inactive ones.

Embedded in the connective tissue is a variable amount of *invisible fat*. This makes the digestion of meat more difficult because it coats the muscle fibres with a thin oily film which resists the action of the digestive juices. In addition to the invisible fat there is a much larger amount of visible fat which is stored in the fat depots of the animal's body. Such fat is mainly found under the skin and around internal organs and is, therefore, not a part of lean meat.

The fat of meat is of some nutritional interest because it is predominantly saturated in character as is evident from Table 9.3. Chicken fat is notable in having a greater proportion of PUFA than other types of meat fat. Consumers are increasingly showing a preference for lean meat and for types of meat or meat products low in saturated fatty acids and high in PUFA.

Apart from protein and fat, meat contains small quantities of mineral elements and vitamins, but is notable for its lack of carbo-

Table 9.3   Percentage of different types of fatty acids in meat and poultry fat

	Saturated	Monounsaturated	Polyunsaturated
Beef	44	50	4
Lamb	52	40	5
Pork	43	47	8
Chicken	35	48	16

hydrate, in which respect it resembles eggs. Meat is a good source of iron although the amount present is small except in certain organs, such as kidney and liver, which are relatively rich sources. Meat is a good source of zinc, but a poor source of calcium. Meat is a useful source of the B group of vitamins, notably niacin. The amount of thiamine in meat is not large except in the case of pork, which contains about 0.6 mg per 100 g of meat. Riboflavin occurs in useful quantities, especially in internal organs such as the kidneys. Lean meat contains very little vitamin A, and practically no vitamin D or ascorbic acid. Table 9.4 summarizes the differences in nutrient content between beef, lamb, pork and chicken and highlights the considerable differences in the nutrient content of lean and fat meat.

The *flavour* of meat is one of its main attractions and is largely due to the presence of a variety of substances known as *meat extractives* that are soluble in water. These include some produced during muscular activity in the living animal, such as lactic acid and some derived from ATP; also substances that are the result of protein metabolism, such as amino acids (e.g., glutamic acid) and urea. Meat extractives aid digestion by stimulating the secretion of saliva and gastric juice.

Although much is known about the flavour of meat it is a complex matter and is incompletely understood.

Table 9.4   Composition of 100 g edible portions of meat and poultry

	Beef		Lamb		Pork		Chicken
	lean	fat	lean	fat	lean	fat	
Energy (kJ)	517	2625	679	2762	615	2757	508
Protein (g)	20	9	21	6	21	7	21
Fat (g)	5	67	9	72	7	71	4
Calcium (mg)	7	10	7	7	8	7	10
Iron (mg)	2	1	2	1	1	1	1
Niacin (mg)	5	–	6	–	6	–	8
Riboflavin (mg)	0.2	–	0.3	–	0.3	–	0.2

Dash indicates values are difficult to obtain; small amounts may be present

Cooking of meat

The main reasons for cooking meat are as follows.

1. To improve its *texture* by making it tender and digestible.
2. To improve its *flavour*, especially by the development and retention of extractives.
3. To improve its *colour* by making it more pleasing to the eye and palate.
4. To improve its *safety* by destroying bacteria,

TENDERNESS   As already mentioned, tenderness of meat depends on the age of the animal, the amount of activity in its life and correct conditioning after slaughter. Muscle tissue which has been involved in much activity develops longer and thicker muscle fibres and more connective tissue to hold them together than tissue which has been rarely used. This explains why, in the same animal, the neck and the leg will always be tougher than fillet or rump.

In addition to improving the tenderness of meat by cooking, it can be made more tender in a number of other ways.

1. The most effective way of tenderizing meat is to inject an animal with a *proteolytic enzyme* before slaughter. This reduces the time required for conditioning by 1–2 days. Care must be taken not to over-tenderize the meat or an unpleasant flavour and mushy texture are produced. Proteolytic enzymes may also be added to meat after slaughter to help break down muscle fibre and connective tissue. Juice from the papaya fruit, for example, contains the proteolytic enzyme *papain*, and may be added to meat before cooking. Its enzyme activity is most effective at low cooking temperatures. The enzymes *bromelin*, obtained from pineapples and *ficin*, obtained from figs, may also be used to tenderize meat.
2. Meat, particularly in the form of thin sections such as steaks, can be made more tender by *mechanical breakdown* of muscle fibre and connective tissue brought about by pounding with a heavy object such as a mallet.
3. If *salt* is added to meat it increases the capacity of meat proteins to hold water when the meat is cooked and this promotes tenderness.
4. Meat may be *marinated*, that is soaked, in an acid solution of wine, vinegar and spices for several days before cooking. The meat may then be cooked in the marinade to improve flavour.

CHEMICAL AND PHYSICAL CHANGES   The tenderness of meat considered in the previous section can be further improved by cooking and, as we have already mentioned, this is one of the principal reasons for cooking meat. Cooking improves tenderness by increasing the denaturing of muscle protein initiated during conditioning and by converting collagen to gelatin, this latter process being assisted by water.

Cooking brings about chemical changes including the breakdown of nucleoproteins and similar substances which result in producing the desirable flavour of cooked meat. Cooking also changes the colour of meat. A cut meat surface is bright red in colour because the reddish-coloured *myoglobin* is oxidized in air to *oxymyoglobin*. When meat is allowed to stand, myoglobin is converted into brown-coloured *metmyoglobin*. Cooking produces an attractive brown colouration, partly due to metmyoglobin and also due to non-enzymic browning.

Cooking also produces a number of structural and physical changes in meat. Protein is coagulated and this reduces its hydration resulting in water, together with water-soluble substances dissolved in the water, being lost from the meat. The consequence of this loss of water is that meat shrinks during cooking. Fat is melted during cooking and if very high temperatures are used some fat may become charred.

The way in which tenderness, flavour and colour are affected by cooking, and the way in which individual nutrients are affected, depends upon whether dry or moist heat methods of cooking are employed; also on the cooking time and the quality of the meat being cooked.

DRY HEAT METHODS   Dry heat methods of cooking meat include roasting and grilling. As the temperature rises to $60°C$ some proteins start to coagulate. If the temperature remains low, coagulation is slow and the protein will be in its most digestible form. On the other hand, if high temperatures are used coagulation is more complete and the protein becomes harder and tougher. Heat alone does not increase tenderness very much because, apart from coagulation, the tough elastin remains unchanged and conversion of collagen into soluble form is slow. Thus dry heating methods are best used for cooking meat which contains little connective tissue.

At high temperatures some protein may be rendered unavailable by reaction with carbohydrate (non-enzymic browning) as described later in Chapter 14. One of the amino acids liable to be rendered unavailable in non-enzymic browning is lysine and the loss of this amino acid which occurs during the cooking of meat has been studied under a variety of conditions. For example, it has been found that beef cooked for three hours loses one-fifth of its lysine at a cooking temperature of $120°C$, but that when the temperature is raised by $40°C$, the amount lost increases to one-half.

At a temperature a little above $60°C$ meat starts to shrink because of the contraction of the proteins of the connective tissue. Shrinkage results in some of the 'juice' in the meat being squeezed out and the higher the cooking temperature the greater is the shrinkage and loss of juice. In roasting, for example, shrinkage causes the weight of meat to decrease by about one-third. The juice is mainly water but it also contains mineral salts, extractives and small amounts of water-soluble vitamins.

During roasting in open containers, water reaching the surface of the meat rapidly evaporates, leaving behind the non-volatile material. This causes the brown outer layer of the meat to have a good flavour but also to be dry. Occasional basting of the meat with hot fat can prevent dryness. In closed containers and at lower temperatures the rate of evaporation is much slower and the juice drips from the surface of the meat, so reducing its flavour. However, where such methods are used the juice is usually collected and eaten with the meat as gravy.

Fat near the surface of the meat melts during cooking and most of it passes into the cooking vessel, although a small portion of melted fat penetrates the lean meat. The greater the cooking temperature the greater is the loss into the cooking vessel and if too high a temperature is used some fat may become charred.

Some B vitamins are lost when meat is heated. Thiamine, folic acid, pantothenic acid and pyridoxine are the most sensitive to heat and about 30–50 per cent of these vitamins are lost during dry-heat cooking.

MOIST HEAT METHODS   Moist heat methods of cooking meat include boiling, stewing, simmering and braising. The particular advantage of these methods relates to the relatively low temperature of the cooking medium. In stewing and simmering (and particularly when using slow cookers) the conduction of heat through the muscle tissue to the centre of the meat is a slow process and thus the denaturation of protein to the toughening and shrinking stage is delayed and tenderness is at a maximum.

Tenderness is also improved by the action of water and heat because these bring about the conversion of collagen to soluble gelatin, so allowing the muscle fibres to separate from each other. The other protein of connective tissues, elastin, remains unchanged, so that parts of an animal containing a high proportion of this protein, such as the neck, never become tender no matter how long cooking is continued.

Moist heat methods of cooking involve inevitable loss of water-soluble substances by leaching into the water used for cooking. In such methods the surface of the meat is in contact with the cooking water. When protein shrinks, juice is forced to the surface of the meat and soluble matter dissolves in the water. Thus, mineral salts, extractives and thiamine are lost from the meat, which in consequence has less flavour and less nutritional value than if it had been cooked by a dry heat method. However, as such meat is normally eaten with the liquor in which it has been cooked this is not important from a nutritional point of view.

Meat products

A number of meat products are made by processing those parts of an animal which cannot be sold as carcass meat. Such parts include meat scraps recovered mechanically after most lean meat has been removed by hand, and offal. Most processed meat products are made from minced or ground meat which may be compacted or moulded and to which other ingredients, such as fillers, fat, preservatives and other additives including salt and monosodium glutamate (MSG) may be added.

The main types of processed meat products are sausages, which may be fresh, cooked or dry and other meat products such as burgers, pies, pasties, rolls and balls. The composition in terms of meat content is controlled by the Meat Product and Spreadable Fish Product Regulations 1984. Table 9.5 gives the minimum legal requirement of lean meat for a number of such products, the importance of which is growing with the increase in 'fast food' catering.

SAUSAGES *Fresh sausages* are made from raw meat, fat, a filler (rusk, bread, flour), water and seasoning. The mixture is filled into a casing and the product must be cooked, usually by frying or grilling, before it is eaten. *Cooked sausages* use similar ingredients to fresh sausages except that the fillers (binders) used may be corn flour (liver sausage), rice flour (polony) or oatmeal (black pudding). Additional

Table 9.5 The legal minimum lean meat content of meat products

Food	Lean meat (%)	Comment
Sausages: pork	32	80% of meat must be pork
beef	25	50% of meat must be beef
liver	25	30% of meat must be liver
Beefburgers	52	80% of meat must be beef
Hamburgers, luncheon meat	52	
Corned beef	115(a)	Must contain only corned beef
Chopped meat	58	
Meat spreads	45	70% must be of named meat
Pate: liver or meat	35	
Meat pie, pudding	10–12 ⎫	Refers to total weight of pie.
Sausage rolls, pasties	5–6 ⎬	Higher value relates to cooked
	⎭	products.

(a) On a fresh weight basis as it loses water during processing ingredients give a special character to certain products such as black pudding, in which fresh pigs blood (defibrinated to prevent clotting) is a main ingredient. These products are cooked before sale though some, such as black pudding and frankfurters, are heated through before eating.

Dried sausages are similar to the fresh variety except that they are larger, the meat is cured either before processing or early in the process, and they are dried under controlled conditions. *German salami* is a typical example of a dried sausage. It is made from lean beef and pork with pork fat which are finely minced, cured, mixed with garlic and other seasoning, moistened with Rhine wine, dried in air for two weeks and cold smoked.

OTHER TYPES   Many other different types of meat product are made, some being produced from meat trimmings left over from carcass meat (e.g., burgers) while others are made from offal (e.g., brawn, haggis). *Beefburgers*, the staple item of so many fast food outlets, are made from minced beef trimmings, pork fat, cooked chopped onions, breadcrumbs (or rusks) and seasoning. They are moulded into shape and fried or grilled before serving. *Haggis*, the source of innumerable jokes, is made from sheep's lungs, liver, heart and spleen coarsely chopped and mixed with oatmeal, stock and seasoning, which are put into a sheep's stomach (or synthetic equivalent) before being boiled.

Many processed meat products are prone to spoil and should therefore be stored with care after manufacture. They have been the cause of many cases of food poisoning and for this reason they may be preserved using a permitted preservative. Sausages and burgers, for example, may be preserved with *sulphite* (but it destroys the vitamin thiamine) and cured and pickled meats with *nitrite*.

FISH

The flesh of fish is composed of bundles of short fibres called *myomeres*, which are held together by thin layers of connective tissue composed of collagen. Thus, the protein of fish differs from that of meat in having less connective tissue and no elastin. The absence of tough elastin, and the conversion of collagen into gelatin which occurs during cooking, make the protein of cooked fish easily digestible. Fish contains rather more water and waste matter than meat.

Fish may be divided into two classes: *white fish*, such as haddock, cod, whiting and plaice, which contain very little fat (usually less than two per cent), and *fat fish* such as herring, trout and salmon, which usually contain 10–25 per cent fat.

The flesh of most white fish contains no fat, as it is concentrated in the liver, which is often removed and used as a source of vitamins (e.g., cod-liver oil). A few white fish (e.g., halibut) contain small amounts of 'invisible fat' dispersed in the flesh. Fat fish contain a considerable amount of invisible fat which is of particular interest because it is rich in PUFA. The actual amount varies, being highest just before spawning.

Fish are of value mainly as a rich source of protein, the amount and quality of protein in fish being similar to that in lean meat. Fish contain no carbohydrate but are a good source of phosphorus, though not of calcium unless the bones are eaten. They are not usually a good source of iron, although sardines are an exception. Sea fish are a valuable source of iodine. Fat fish are valuable sources of the fat-soluble vitamins A and D, fish-liver oils being exceptionally good sources of these vitamins. They also contain useful amounts of the B group of vitamins. White fish do not contain vitamins A and D, and usually contain less of the B vitamins than fat fish.

Unlike meat, fish deteriorate rapidly after death because the lack of connective tissue makes the muscle protein more vulnerable to spoilage. Such rapid deterioration and the fact that fish are often caught far from land, has led to most fish being frozen at sea immediately after they have been caught. Consumption of fresh fish has rapidly declined in recent years in Britain, while consumption of frozen fish and frozen fish products, such as fish fingers, has increased.

The varieties of fish consumed have also changed in recent years. In the UK cod, together with haddock, halibut and plaice, have traditionally been the most popular fish. However, restricted fishing in the areas where these fish are found has led to their replacement by others, such as mackerel, coley and whiting. In addition, research is being carried out into newer varieties, of which the blue whiting is the most

Table 9.6 Nutrient composition of 100 g fish

	Cod	Haddock	Plaice	Herring	Mackerel	Salmon
Energy value (kJ)	322	308	386	970	926	757
Protein (g)	17.4	16.8	17.9	16.8	19	18.4
Total fat (g)	0.7	0.6	2.2	18.5(a)	16.3(a)	12
PUFA (g)	0.4	0.3	0.6	3.6	4.4	2.9
Calcium (mg)	16	18	51	33	24	27
Iron (mg)	0.3	0.6	0.3	0.8	1	0.7
Vitamin A (μg)	0	0	0	45(b)	45(b)	0
Thiamine (mg)	0.1	0.1	0.3	0	0.1	0.2
Vitamin D (μg)	0	0	0	22.5	17.5	0

(a) Value varies through the year being highest in July–October
(b) Expressed as retinol equivalents

promising. Fish farming is also a developing industry, both salmon and rainbow trout being produced in this way in Britain.

Cooking of fish

When fish is cooked the changes which take place are similar to those which occur during the cooking of meat. As there is less connective tissue in fish than in meat and no elastin, cooking is not required to make the fish tender but only to render it as palatable and digestible as possible. Fish should be cooked as little and as gently as possible as fish proteins coagulate quickly and easily. If fish is over-cooked the flesh becomes rubbery and dry.

There are less extractives in fish than in meat, and fish should therefore be cooked in such a way that as much flavour as possible is preserved. During cooking, proteins coagulate, collagen is converted into gelatin and some shrinkage occurs. Shrinkage, however, is less than with meat because of the smaller amount of connective tissue. Shrinkage causes water and soluble matter to be squeezed out of the fish and in moist heat methods of cooking, water, extractives and soluble mineral salts are lost. In boiling, for example, over a third of the extractives and soluble salts are lost, so that fish cooked in this way is rather tasteless. Dry heat cooking, on the other hand, causes rapid evaporation of water from the surface of the fish while the non-volatile soluble matter remains behind. Thus, fish cooked in this way has much more flavour than fish which has been boiled or steamed. Fatty fish, such as herring or trout, are best cooked by a dry heat method such as grilling or baking since their high fat content keeps them moist.

Both white and fatty fish may be cooked by frying, and this is particularly effective if the fish are coated in batter before frying because the batter helps to retain the structure of the flesh, even if relatively long cooking times are used.

Fish products

Apart from being preserved by freezing, canning and smoking, fish is also converted into a number of convenience products, such as fish fingers, fish cakes and fish spreads. Fish fingers, for example, are made from blocks of frozen filleted white fish which are cut into fingers, dipped into a batter and breadcrumbs and refrozen.

Of the many fish products available only the composition of fish cakes and fish spreads is controlled by law and the minimum fish content of these is shown in Table 9.7. Fish fingers usually contain 50–70 per cent fish while fish cakes (which also contain potato, herbs and seasoning) must contain at least 35 per cent fish.

Table 9.7   The legal minimum fish content of fish products

Food	Fish content (%)
Fish cakes	35
Fish pastes and spreads	70
Potted fish	95
Potted fish and butter	96 (fish and butter)
Fish paste with one other main ingredient	80

A variety of additives is used in preparing fish products. Permitted additives include colour (yellow in smoked fillet, brown in kippers), flavours and emulsifiers. Antioxidants and preservatives, while not being added to the finished product, may be present in the ingredients used. Frozen, fried fish, for example, contains antioxidant present in the oil used for frying. Other additives which are not controlled by law may also be used. For example, both smoke solutions and polyphosphates may be added as these prevent the loss of water which takes place during traditional curing processes.

EGGS

The hen's egg, which is the only variety to be discussed here, is a most interesting food because it is designed to accommodate a living organism. It contains a sufficient store of nutrients to supply a developing chick embryo with all that it needs during its early stages of growth. It is, therefore, a complete food for a growing chick embryo, and although it is not a complete food for humans, it is, nevertheless, a valuable one. The nutrient content of eggs is shown in Table 9.8.

The egg consists of three main parts – the shell, the white and the yolk – and these are shown in Fig. 9.2. The outer shell forms a hard protective layer composed mainly of calcium carbonate and as it is porous it allows a developing embryo to obtain a supply of oxygen. The colour of the shell may vary from white to brown though,

Table 9.8   The nutrient content of eggs per 100 g

Energy (kJ)	612	Calcium (mg)	52
Protein (g)	12.3	Iron (mg)	2
Fat, total (g)	10.9	Sodium (mg)	140
Fat, saturated (g)	3.4	Vitamin A (μg)	140
Fat, polyunsaturated (g)	1.2	Thiamine (mg)	0.09
Carbohydrate (g)	0	Riboflavin (mg)	0.47
Water (g)	75	Niacin (mg)	3.68
Cholesterol (mg)	450	Vitamin C (mg)	0
		Vitamin D (μg)	1.75
		Vitamin E (mg)	1.6

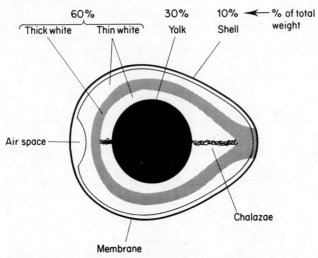

Figure 9.2   The structure of a hen's egg.

contrary to popular opinion, this gives no indication as to the quality of the contents of the egg.

Inside the shell is a viscous colourless liquid called egg-white which in a fresh egg is divided into regions of thick and thin white and which accounts for about 60 per cent of the total egg weight. It is a dilute aqueous sol, being about one-eighth protein and seven-eighths water. The main protein is *ovalbumin*, though smaller quantities of several others are present, including *mucin* which accounts for the viscosity of the liquid. Also present are small quantities of dissolved salts and the vitamin riboflavin.

In the centre of the egg is the yolk, which is a thick yellow or orange oil-in-water emulsion stabilized by lecithin. It is suspended in the white, being held in position by the *chalazae*, and it is a rich source of nutrients, being much more concentrated than egg-white. It is roughly one-third fat, one-half water and one-sixth protein; it also contains a supply of mineral elements and vitamins.

The only type of nutrient not present in an egg is carbohydrate. Its presence might be expected, because it would provide a ready source of energy for the growing chick. However, the size of the egg is limited and fat, which weight for weight has more than twice the energy value of carbohydrate, forms the sole source of energy.

The fat of eggs is concentrated in the yolk which is also a rich source of cholesterol.

The proteins of eggs, particularly those of egg-white, have been intensively studied. Nine proteins have been identified in egg-white

Table 9.9 The principal proteins of egg-white

	Ovalbumin	Conalbumin	Lysozyme	Ovomucin	Ovomucoid
Per cent of total protein	70	9	3	2	13
Type	Albumin	Albumin	Globular	Conjugated	Conjugated
Molecular weight	44 000	74 000	15 000	8 000 000	28 000
Non-protein part	Phosphate Carbohydrate	None	None	Carbohydrate	Carbohydrate
Isoelectric point	4.6–4.8	5.6–6.0	10.5–11.0		3.9–4.5
If coagulated by heat	Yes	Yes	No	No	No

and the nature and properties of the principal ones are summarized in Table 9.9. In addition to those mentioned in the table two globular proteins, identified as G_2 and G_3 and a small amount of a protein called *avidin* are also present. Avidin is of some nutritional importance as it combines with the vitamin *biotin*, rendering it unavailable to the body. However, avidin is inactivated during cooking, so that all the biotin of cooked eggs is available. The presence of the protein *conalbumin* in egg-white prevents the absorption of the iron in eggs.

The main proteins of egg-yolk are the phosphoproteins *lipovitellin* and *lipovitellenin* which comprise about 30 per cent of the total egg-yolk solids. The phosphorus content of these proteins is in the form of phosphoric acid esterified with the hydroxyl groups of hydroxy amino acids. They also contain a lipid part as their names suggest, and this is mainly lecithin.

The proteins of an egg are of high nutritional value and because of the quantity present – about 12 per cent of the edible part – eggs must be considered as a valuable protein food. Moreover, the properties of egg proteins, particularly the ease with which they coagulate, cause eggs to be used in many methods of food preparation.

The nutritional value of an egg may be summed up by saying that it supplies the diet with valuable amounts of iron, phosphorus and protein of high nutritional value and useful amounts of fat, vitamin A and calcium. It also supplies some vitamin D, riboflavin, thiamine and biotin.

Eggs in cooking

When eggs are heated, proteins are coagulated. Over-cooking can produce excessive coagulation of protein and a consequent rubbery texture, particularly of the egg-white. The rate at which egg proteins coagulate depends upon conditions such as pH, salt concentration and temperature. Egg-white coagulates readily into a white solid on heating and at its normal pH of around nine coagulation starts at about 60°C. At higher temperatures the rate of coagulation increases until eventually it is nearly instantaneous. Egg-yolk coagulates less readily than the white and does not coagulate appreciably below 70°C. As coagulation proceeds the viscosity of the yolk increases until eventually it becomes solid.

Cooked eggs form an easily digestible food and, provided they are not over-cooked, there is little loss in the nutritional value of the protein. There is, however, some loss of thiamine and riboflavin though niacin is not destroyed.

Eggs, as we have already noted, have many uses in food preparation. This versatility is mainly due to the properties of egg protein. Eggs may be used as *thickening agents*, in which the coagulation of the egg proteins is used to thicken sauces, custards, soups and lemon cheese. Eggs are also used as *binding agents* when the coagulation of egg proteins gives cohesion to a mixture containing dry ingredients, as in rissoles and croquettes. Eggs are used as *coatings*, as when a mixture of eggs and breadcrumbs is used to coat fish before it is fried. The coagulation of the egg during frying forms a strong coating which holds the fish structure together. Beaten egg is used as a protective covering for fried food because on heating the egg-white hardens quickly; this has the additional benefit of preventing the food being penetrated by the oil during frying.

Egg-white, as we saw in the previous chapter, has the ability to entrap air and form *foams*. Foam formation is brought about by beating the egg-white and is promoted by the addition of acid which lowers the pH to a value near the isoelectric point of ovalbumin. Heating causes further coagulation of ovalbumin and produces a solid foam as when making meringues. Whole eggs when beaten also entrap air and are able to lighten the texture of baked goods such as sponge cakes. They also act as *emulsifiers* and assist the formation of a stable emulsion from a creamed mixture of fat and sugar during cake making. Egg-yolks are used to emulsify oil and vinegar in making mayonnaise.

Whole eggs can be used to improve *texture* and enrich *flavour* especially when making baked goods such as cakes. The yolks also contribute a rich yellow *colour*. The fat of the yolk also exerts a *shortening effect* (see p. 68) in the making of biscuits, shortbread and cakes.

The versatility of eggs in cooking is enhanced by the fact that the white and yolk may be used either separately or together. The making of meringues illustrates the use of egg-white while the making of the filling for a lemon meringue pie illustrates the use of egg-yolk, which is mixed with cornflour and lemon. The making of the exotic-sounding and delicious desert Zabaglione is made by the careful blending together and subsequent very gentle heating of egg yolk, caster sugar and Marsala wine. The gentle thickening produced by the slow coagulation of the egg-yolk contributes the particular smooth, delicate, creamy texture of this dish.

Whole eggs are used to make egg custard when a mixture of beaten eggs (strained to remove the chalazea of the egg) sugar and milk are heated together. The rate and duration of heating can be adjusted to produce either a viscous stirred custard, produced by partial coagulation of the egg proteins, or a solid baked form, produced by more complete coagulation. Slow and even coagulation is essential in making both products if a successful texture is to be achieved. Rapid heating can easily cause curdling.

SOYA

As long as 5000 years ago soya beans were cultivated in China and used as food. It was not until 1804, however, that soya beans arrived in the USA and even then it came about by accident when they were carried as ballast aboard a ship from China. When these beans were cultivated it was for curiosity without any thought of their potential as a food.

It was not until the 1914–18 war that the value of soya beans as a source of oil was recognized and not until the 1939–45 war that shortage of animal protein focused attention on the potential of soya beans as a possible alternative. Today large quantities of soya beans are grown, the USA being the largest producer with Brazil and China as the other two major producers.

Soya beans grow in pods on bushy plants (Fig. 9.3), and in warm temperate climates where they grow best they reach maturity in about four months. Like peas and lentils they are legumes.

As Table 9.10 demonstrates, soya beans have a high food value and they are a valuable source of protein of high biological value. They are also a rich source of oil which may be extracted from the beans. The oil is rich in PUFA, particularly linoleic acid which accounts for about half the fatty acid content. Soya beans are also a useful source of iron, calcium and some B vitamins.

In China soya beans have an amazing variety of uses. They are used as a vegetable and in salads and the seeds are used to produce soya sprouts. The dried bean is roasted and used as a coffee substitute (which has the advantage of being caffeine-free!) and fermented beans form the basis of soy sauces. The bean, or flour derived from it, is used to make a form of 'milk' which has about the same protein content as cow's milk and can be used in a similar way. The milk is used to make a form of cheese or curd known as *Tofu*.

Figure 9.3 The soya bean plant.

Soya beans are often converted into a form of flour which is made by removing the outer seed coat, crushing the beans between rollers to convert them into flakes which are then ground into flour. Soya bean flour is a high protein flour and may be added to wheat flour as a protein supplement. It is an important ingredient in many baby foods and in slimming and 'health' food products. It cannot be used alone for making bread because it is lacking in starch and gluten and its fat content is too high. However, it may be added in small amounts to

Table 9.10   The nutrient content of 100 g soya beans and soya flour

	Cooked beans	Flour	Low fat flour
Energy value (kJ)	648	1871	1488
Protein (g)	13.1	36.8	45.3
Total fat (g)	6.8	23.5	7.2
Polyunsaturated fat (g)	3.8	13.3	4.2
Carbohydrate (g)	9	23.5	28.2
Calcium (mg)	87	210	240
Iron (mg)	3.2	6.9	9.1
Thiamine (mg)	0.4	0.75	0.90
Riboflavin (mg)	0.1	0.31	0.36

wheat flour where it helps to improve colour (the enzymes in soya flour help to bleach the yellowish pigments of wheat flour), improves texture and keeping qualities and imparts an attractive nutty flavour to the baked loaf.

Novel protein foods

For over 20 years, from the 1940s to the 1960s, great significance was attached to a perceived world shortage of protein and consequently to ways of replacing expensive animal protein, which was in short supply, with new forms of protein foods derived from plants. It is now appreciated that the world food problem is one of a lack of food rather than a lack of protein and that most mixed diets provide an adequate supply of protein. The original impetus to develop concentrated protein foods, known as *novel protein foods*, has therefore diminished. Nevertheless, animal protein remains expensive and the conversion of plant foodstuffs into animal protein in the bodies of cattle and other animals is an extremely inefficient and extravagant use of the world's food resources. It is still appropriate, therefore, to consider novel protein foods derived from non-animal sources.

Novel protein foods are in two types:

1. Those produced by processing plant foods.
2. Those produced from sources not previously used as food including (a) plants and (b) micro-organisms.

SOYA PROTEIN   The most widely used novel protein foods are those derived from soya beans, the proteins of which, unlike most vegetable proteins, are of high biological value. Such products are often made to simulate meat and are intended as a cheaper alternative to it. A comparison of the amino acid pattern of soya beans with beef (Table 9.11) shows that the former is low in methionine which is the limiting amino acid, but otherwise they are broadly comparable. Soya bean products are made from soya flour and the conversion of soya flour into products having a meat-like texture is now carried out on a large scale (Fig. 9.4). The simplest way of doing this is to convert the flour into a dough, heat under pressure above $100°C$ and extrude through a nozzle into atmospheric or reduced pressure. The sudden drop in pressure causes the material to expand and achieve the desired texture. The material is cut into pieces and dried. It can either be used in its natural form or flavours and colours may be added to the dough so that the final product, known as *textured vegetable protein* (TVP), has a somewhat meaty colour and texture. TVP is used mainly as a meat extender and is available as chunks or granules for addition to meat products such as mince, stews and pies.

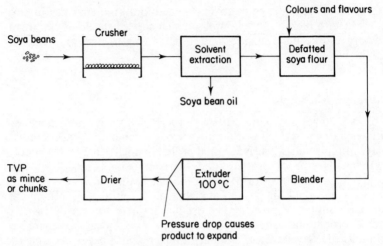

Figure 9.4   How extruded TVP is made from soya beans.

Another type of product, known as *spun vegetable protein* is made by extracting the protein from soya flour, dissolving it in alkali and forcing the resulting solution through the tiny holes of a spinneret to give many fine threads of spun material. The threads of precipitated protein are stretched and twisted into bunches of fibres having a meat-like texture to which additives such as colour, flavour, fat and protein binders may be added. The final product may be frozen or, more usually, dried. Spun products have a more fibrous texture than textured ones, but are more expensive to produce.

Vegetable proteins are regularly used in schools, hospitals and canteens, usually as a meat extender to replace part of the meat in traditional dishes. It is recommended (Food Standards Committee, 1974) that in the UK not more than ten per cent of meat should be replaced by vegetable protein, although in the USA the limit is 30 per cent. Vegetable proteins are also being used as components of simulated meat products such as stews, curries and burger and burger-style dried mixes.

OTHER SOURCES OF NOVEL PROTEIN FOODS   Large amounts of plant materials are grown as a source of vegetable oils. In addition to soya beans, groundnuts, cotton seed and other seeds are used and after the oil has been extracted a protein-rich residue remains. Concentrated protein can be extracted not only from such oil seeds but also from grass and other indigestible, but often abundant, vegetable material and also from fish. At present such processes are only carried out on a small scale.

Table 9.11 Essential amino acid content of novel proteins (mg/g protein) compared with beef

Amino acid	Beef	Soya	Grass	Yeast	Fungi	Bacteria
Isoleucine	53	62	93	45	43	43
Leucine	82	79	130	70	55	68
Lysine	87	53	72	70	51	59
Methionine (+ cystine)	38	16	21	18	10	24
Phenylalanine (+ tyrosine)	75	49	93	44	39	34
Threonine	43	37	67	49	25	46
Tryptophan	12	11	21	14	21	9
Valine	55	53	103	54	60	56

Microorganisms such as yeast, fungi, bacteria and algae are being developed as sources of edible protein usually known as *single cell protein*. The advantages of using unicellular organisms as a protein source are (a) that they can be grown in media which are cheap, such as industrial waste materials, and (b) they grow very rapidly indeed. The media used must contain a suitable source of carbon, which may be a cheap carbohydrate (such as molasses or a waste hydrocarbon produced during oil refining), and a cheap source of nitrogen (such as liquid ammonia or ammonium or nitrate salts). In addition they need a supply of oxygen, water and small quantities of mineral elements, sulphur, phosphorus and possibly vitamins.

From Table 9.11 it can be seen that compared with beef, single cell proteins are low in methionine but otherwise their amino acid patterns are fairly similar. The commercial exploitation of single cell protein is under active development. For example, species of micro-fungi such as *Fusarium graminearum* are being grown on starch waste, yeasts are being grown on petroleum oil and bacteria are being grown on methane and methanol. For the present it is intended that these protein concentrates should be used as animal foodstuffs, but eventually it is hoped that they may be used as new protein foods for humans.

SUGGESTIONS FOR FURTHER READING

BRITISH NUTRITION FOUNDATION (1986). *Nutritional Aspects of Fish*. BNF Briefing Paper No. 10. BNF, London.
DAVIS, P. (Ed.) (1974). *Single Cell Protein*. Academic Press, London.
DEPARTMENT OF HEALTH AND SOCIAL SECURITY (1980). *Foods which Simulate Meat: Nutritional Aspects*. HMSO, London.
FAO/WHO (Reports 1973, 1975, 1978, 1985). *Energy and Protein Requirements*. WHO, Geneva.

FORREST, J.C. *et al.* (1975). *Principles of Meat Science*. Freeman, USA.

GERRARD, F. (1977). *Meat Technology*, 5th edition. Northwood Publications, London.

JONES, A. (1974). *World Protein Resources*. Medical and Technical Publishing Co., Lancaster.

LAWRIE, R.A. (1985). *Meat Science*, 4th edition. Pergamon Press, Oxford.

MINISTRY OF AGRICULTURE, FISHERIES AND FOOD (1974). *Food Standards Committee Report on Novel Protein Foods*. HMSO, London.

STADELMAN, W.J. AND COTTERILL, O.J. (Eds) (1977). *Egg Science and Technology*. Avi, USA.

TAYLOR, R.J. (1969). *The Chemistry of Proteins*, 2nd edition. Unilever Booklet, Unilever, London.

TAYLOR, R.J. (1976). *Plant Protein Foods*. Unilever Booklet, Unilever, London.

WOLF, W.J. AND COWAN, J.C. (1975). *Soya Bean as a Food Source*. Blackwell, Oxford.

CHAPTER 10

Water and beverages

WATER

Without water there could be no life; water is essential to the life of every living thing from the simplest plant and single cell organism to the most complex living system known – the human body. Moreover, while living things may exist for a considerable time without the other essential nutrients, they soon die without water. Living things contain a surprising amount of water, never less than 60 per cent of their total weight and sometimes as much as 95 per cent. About two-thirds of the body is water and all the organs, tissues and fluids of the body contain water as an essential constituent. Only a few parts of the body, such as bones, teeth and hair contain little water.

During life, water is continually being lost from the body, partly in the urine, partly from the surface of the body as sweat and partly as water vapour in the gases expelled in respiration. Small amounts of water are also lost in the faeces. If the body is to function successfully the water that is lost must be replaced, and a balance maintained between intake and output. The main source of water for the body is food and drink, although some is produced when nutrients are oxidized to produce energy. For example, when glucose is oxidized it breaks down to form carbon dioxide and water. A kilogram of glucose produces just over half a litre of water on oxidation. The balance between input and output for a man living a sedentary life in a temperate climate and having a diet which provides 8.8 MJ per day is shown in Table 10.1.

Water is unlike the other essential nutrients in that most of it does not undergo chemical change within the body. Whereas proteins, for example, are broken down into amino acids during digestion, most water passes through the body unchanged. The functions performed by water are mainly due to physical action and depend upon its ability to transport nutrients through the body, to dissolve substances or hold them in colloidal suspension and, above all, to remain liquid over a wide range of temperatures. This latter property enables water to provide a liquid medium in which the thousands of reactions necessary to life can occur.

Table 10.1　Water balance in the body in a temperate climate

Source	Water intake (cm³/day)	Source	Water loss (cm³/day)
Food	1120	Urine	1300
Drink	1180	Lungs	300
Oxidation of		Skin	920
nutrients	280	Faeces	60
Total:	2580	Total:	2580

Although most of the water in the body is involved in physical changes, some is concerned with chemical changes. Some chemical changes, such as the enzymic and hydrolytic breakdown of nutrients during digestion, involve the uptake of water, whereas others, such as the oxidation of absorbed nutrients to provide the body with energy, release water. Apart from the water involved in such chemical reactions in the body, water is also required by all plants in order to synthesize carbohydrate during photosynthesis.

The structure of water and ice

A water molecule contains two atoms of hydrogen covalently linked to an atom of oxygen, the bond angle being about 105° as shown in Fig. 10.1a. The electronegativity of oxygen, that is its attraction for electrons, is greater than that of hydrogen. This results in unequal charge distribution over the molecule, the oxygen atom carrying a partial negative charge ($\delta-$), balanced by partial positive charges ($\delta+$) on the hydrogen atoms. Such molecules are said to be *polar* and they attract each other; in water hydrogen bonds are formed between adjacent polar molecules as shown in Fig. 10.1b. It will be noted that, because of the shape of water molecules, such intermolecular attrac-

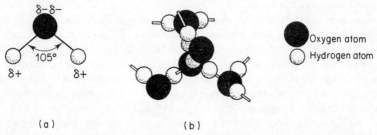

Figure 10.1　(a) A water molecule showing its polar character and bond angle. (b) Hydrogen bonding in water gives rise to its tetrahedral structures.

tion results in water molecules grouping together to form tetrahedral structures in which water molecules are linked to four others.

Hydrogen bonds are weak electrostatic links which are easily broken, and in water such bonds are constantly being formed, broken and re-formed. If water is heated the thermal energy of the molecules is increased, and their resulting increased motion favours the breaking, rather than the formation, of hydrogen bonds. In water vapour no hydrogen bonds are present and the water molecules exist as single units. On the other hand, if liquid water is cooled the loss of thermal energy and the resulting decreased motion of the water molecules favours the formation of hydrogen bonds. In ice hydrogen bonding is so extensive that all the water molecules are linked together by hydrogen bonds so forming a rigid and regular structure. The tetrahedral structure of groups of water molecules is preserved, but the tetrahedra are further linked together to form layers of hexagonal rings which are joined to give a very open structure. The open structure of ice compared to that of water explains why ice has a larger specific volume and smaller density than water.

The physical characteristics of water

Water is the commonest of all liquids and perhaps for this reason it is often considered to be unremarkable. Because it is ubiquitous its presence, like that of air, is taken for granted. Yet it is fortunate that water is so readily available for, in reality, it is a most remarkable liquid, having properties which make it uniquely suited to the endless purposes for which it is used or needed, including the support of life itself.

Water is a colourless, odourless and tasteless liquid which, under normal atmospheric conditions, boils at 100°C and freezes at 0°C. These facts are well known and cause no surprise. Yet in view of its low molecular weight it is surprising that water should be a liquid at all. The reason is that, because of hydrogen bonding, individual molecules group together forming unstable units with an effective 'molecular weight' that is much higher than that of a single molecule.

Although water boils at 100°C at normal atmospheric pressure, its boiling point decreases or increases as the pressure is lowered or raised. At the top of Mount Everest, for example, its boiling point is about 72°C, whereas in a pressure cooker working at maximum pressure (1.05 kg/cm^2), it boils at 120°C. The temperature at which water boils is also affected by the presence of dissolved substances. These increase the boiling point by an amount which is proportional to their molecular concentration. Thus in jam making, when fruit is cooked in water containing dissolved sugar, the boiling point is greater than 100°C. If boiling is continued the concentration of the solution increases and consequently the boiling point rises. Indeed, as mentioned on page 126, the end of the boiling period can be judged from the boiling point, a knowledge of which enables the composition of the mixture to be determined.

The freezing point of water is lowered by the presence of dissolved solids, the lowering being proportional to the molecular concentration of dissolved material. Thus water in plant tissues and foods does not freeze until below $0°C$. This fact must be taken into account when preserving foods by cold storage, where temperatures of $-18°C$ are normally used to ensure that the water in the food is frozen. When water freezes its volume increases due to an increase in hydrogen bonding and the formation of a hexagonal structure which is very open. This causes ice to float on water thereby conserving the heat of the water beneath; this has important consequences in nature as it enables aquatic plant and animal life to continue beneath ice surfaces. Other noteworthy effects of ice formation are the bursting of plant tissues which occurs when the plant sap freezes, and the breakdown of the tissues of frozen food which occurs if large ice crystals are allowed to form.

The specific heat capacity and the specific latent heat of vaporization of water are high compared with their values for other liquids. If liquids where there is no hydrogen bonding are heated all the thermal energy supplied increases the kinetic energy of the molecules and hence the temperature. In water, however, some of the thermal energy is used in breaking hydrogen bonds and therefore more heat energy must be supplied to obtain a given temperature rise than in liquids containing no hydrogen bonds. This is important in the body because the high specific heat capacity of water allows it to act as a heat reservoir, so preventing its temperature from rising quickly when heat is absorbed.

The specific latent heat of vaporization of water is high for a similar reason, namely that energy is required to break hydrogen bonds, making necessary a greater supply of heat to vaporize a given mass of water than would be needed for the same mass of a non-hydrogen bonded liquid. The high specific latent heat value of 2300 kJ/kg means that 2300 kJ are required to vaporize a kilogram of water at a given temperature. Thus, when sweat evaporates from the surface of the body a relatively large amount of heat is absorbed from the skin, which is thereby kept cool.

WATER AS SOLVENT   Water has been called the 'universal solvent' and, although this is not a completely accurate description, it does have unique solvent properties which enable it to dissolve a very large number of substances. It is sometimes called an ionizing or polar solvent because it will dissolve electrovalent substances such as acids

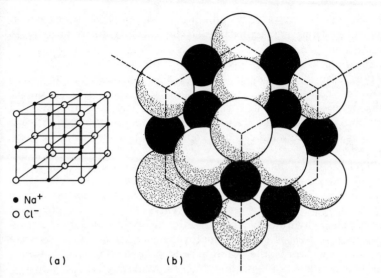

• Na$^+$
○ Cl$^-$

(a)   (b)

Figure 10.2 A sodium chloride lattice.

and salts. It will also dissolve some covalent compounds (e.g., sugar and urea), though others (e.g., fat) do not dissolve. The importance of the solvent action of water cannot be emphasized too strongly; it enables water to dissolve a large number of substances which are essential to plant and animal life. These dissolved substances can then be transported through the organism to areas where they are needed. In the absence of water, or some other solvent, they could not be utilized by living things.

The explanation for the excellent solvent properties of water is to be found in the nature of the molecule itself: to be more specific it depends upon the polar character of the water molecule. To understand this consider the way in which a typical salt, such as *sodium chloride*, forms a solution in water. Sodium chloride exists in the form of *ions*; in it sodium exists as a positively charged particle or ion and chlorine exists as a negatively charged ion.

A crystal of sodium chloride contains many millions of sodium ions and an equal number of chloride ions. The ions are arranged in a three-dimensional geometrical pattern known as a space lattice, as shown in Fig. 10.2a. This represents about one five million million millionth part of a 1 mm cube of sodium chloride. It shows clearly the geometry of the space lattice but suffers from the defect that the sodium and chloride ions appear to be widely separated whereas in fact they are closely packed. In Fig. 10.2b a portion of the lattice has been enlarged and is shown in a more realistic way. In the interior of a sodium chloride crystal each sodium ion is surrounded by six equidistant chloride ions and, similarly, each chloride ion is surrounded by six equidistant sodium ions.

A sodium ion is not associated with a particular chloride ion and in this sense it is wrong to think of a sodium chloride molecule. However, it is convenient and customary to speak of the sodium chloride molecule, and its formula is written NaCl and not Na^+Cl^-. The lines in the diagram do not, of course, represent bonds between the ions but merely indicate the geometrical properties of the space lattice.

When sodium chloride dissolves in water this orderly space lattice breaks down and the ions are separated from each other and become free to move. This process of separating the ions is difficult to achieve because it requires a large amount of energy. Water is able to effect this because it hydrates the ions, and the energy liberated in hydration is sufficiently great to compensate for the energy required to separate them. In other words, the overall process of dissolving a salt in water, that is separating and hydrating the ions, requires little if any energy, and therefore occurs readily. The result is that ions never exist free in solution; they are always hydrated, in which form they are firmly bound to surrounding water molecules.

Water is also a good solvent for substances which, although they are not ionic, contain polar groups and which are able to form hydrogen bonds and other weak electrostatic links with water molecules. Thus low molecular weight alcohols, for example ethanol (C_2H_5OH) – which contain polar hydroxyl groups – are readily soluble in water, although as molecular weight increases and the proportion of non-polar hydrocarbon chain increases, solubility decreases. Molecules which contain a number of hydroxyl groups, such as simple sugars, are very readily soluble in water because the greater number of polar groups that they contain increases attraction between them and water. The ability of water to dissolve covalent substances providing that they contain a reasonable proportion of polar groups is very important in the body. It means that after food has been broken down during digestion into relatively small polar molecules, such as simple sugars and amino acids, it is dissolved by the body fluids, which are mainly water, and transported through the body in solution.

Water supplies

There is no such thing as pure natural water. Rain water, which is the purest form of natural water, contains small amounts of dissolved gases, such as oxygen, carbon dioxide and, due to industrial pollution, it may also contain dissolved oxides of sulphur and nitrogen (when it is known as 'acid rain'). It also contains small quantities of dust.

Other types of natural water, such as spring and river water, in addition to the impurities of rain water, contain dissolved salts and in the latter case further impurities from vegetation and drainage.

An adequate supply of clean, wholesome water is one of the essentials of modern life and water for domestic consumption must be carefully treated before use so that it is not injurious to health. A considerable proportion of the water supply in Britain is obtained from reservoirs which receive their water from moorland catchment areas. The water which is withdrawn from such reservoirs is treated in a number of ways before being supplied to the consumer. The essential stages are *settlement*, *filtration* and *sterilization*. Some water used for drinking is also treated by *fluoridation* as described on page 239. Water which has been treated in these ways is wholesome and suitable for human consumption, although for industrial purposes it may also require to be softened. Since 1985 the quality of water supplies has had to comply with EEC regulations which lay down strict requirements for the bacteriological and chemical quality of water supplies.

Water which needs to be softened is said to be *hard* because it does not easily give a lather with soap. The hardness of water is due to the presence of certain mineral salts – chiefly the sulphates and bicarbonates of calcium and magnesium – and because of this hard water makes good drinking water. For industrial purposes, however, hard water is unsatisfactory because the mineral salts tend to precipitate out and form insoluble deposits in boilers, water pipes and other equipment. Consequently water required for industrial purposes is often treated before use by removing the calcium and magnesium ions by a process of *ion-exchange* in which these ions are exchanged for sodium ions.

STERILIZATION Natural water always contains organic matter and dissolved oxygen and is therefore a natural breeding ground for bacteria. Typhoid fever, cholera and jaundice are caused by the infection of water supplies, and in order that bacteria causing these and other diseases may be eliminated, water which is to be used for human consumption is usually sterilized before use.

Water is normally sterilized by adding 0.5 ppm of chlorine; this small concentration is sufficient to kill all bacteria but not to impart a taste to the water. Water which is to be used for canning or bottling may be sterilized by adding 1 or 2 ppm of ozone. This treatment is more expensive than chlorination but has the advantage that ozone breaks down into oxygen which has no taste.

MINERAL WATERS Mineral waters come from natural springs and sales of bottled mineral water are increasing in the UK. Mineral water contains certain mineral salts such as sodium chloride, sodium carbonate and sodium bicarbonate as well as similar salts of calcium and magnesium. A few mineral waters also contain iron salts or hydrogen sulphide, the latter being responsible for the unpleasant 'bad egg' smell of certain mineral waters. The total of dissolved mineral salts may be as low as 0.1 g/l and is rarely more than 3.5 g/l. Many mineral waters are naturally aerated with carbon dioxide.

Mineral waters, especially the best known such as *Evian*, *Perrier* or *Vichy* waters, have for many years been considered to have special health-giving properties and to be less likely to cause illness than tap water because they are free from bacteria. However, these special claims are quite unfounded and, as already explained, tap water in Britain and most developed countries is quite safe and of high quality.

Mineral waters do allow people to drink water that has a pleasant, sharp taste and that is without the unpleasant taste of chlorine sometimes noticeable in tap water. They may also be preferred by people who object to drinking water that has been treated in various ways, especially by the addition of 'medicine' in the form of fluoride. The drinking of mineral waters in place of alcoholic drinks or drinks containing caffeine can only be of benefit to health.

WATER SUPPLIES AND HEALTH   Although, as already mentioned, water supplies in Britain today are of high bacteriological and chemical quality, there is some concern about the health aspects of soft water.

There is a certain amount of evidence which suggests that the incidence of coronary heart disease (CHD) is linked to soft water supplies. Soft water contains less calcium carbonate, magnesium and sulphate and more sodium than hard water. Soft water may also be deficient in vanadium. In addition, soft water picks up traces of zinc, cadmium, iron, copper and lead from pipes. It seems possible that incidence of CHD may be related to a deficiency of calcium, magnesium and vanadium and/or an excess of copper and cadmium.

Much more evidence needs to be obtained before any link between CHD and soft water supplies can be regarded as established.

NON-ALCOHOLIC BEVERAGES

Although water is essential to man ordinary tap water is not a very popular way in which to consume it. Drinking water in the form of mineral water or with ice added may make it more acceptable to some people, but the great majority of people prefer to take their water in the form of a flavoured beverage – usually a non-alcoholic beverage such as tea or coffee, a fruit juice or a soft drink. Such beverages are more likely to be appreciated for their flavour or stimulating action rather than for their nutritional value, though most fruit juices are a good source of vitamin C.

Tea, coffee and cocoa

TEA Tea leaves, which are the basis of the familiar beverage, come from an evergreen shrub grown extensively in India, China and Ceylon as well as in other countries. The tea shrub is kept small by pruning and only the bud and the two youngest leaves are picked in preparing good quality tea (Fig. 10.3). Fresh tea leaves contain a number of water-soluble constituents including polyphenols, which account for about 30 per cent of the dry weight, amino acids (four per cent), caffeine (four per cent) and traces of sugars. Tea leaves also contain insoluble materials, mainly fibrous material (e.g., cellulose), proteins and pectins and a very small quantity (about 0.01 per cent) of essential oils which contain a large number of volatile components that contribute to flavour and aroma. Tea contains small amounts of the trace elements fluorine and manganese and is a useful source of these elements.

Caffeine is the most important substance present in tea leaves because it acts as a stimulant to the nervous system. The mild stimulating action of caffeine may help to prevent a feeling of fatigue and some people find that it promotes concentration. It is also a weak diuretic, so that drinking tea (or coffee) in place of water stimulates the production of urine.

Tannin is another important component of tea leaves. It is an astringent and contributes a certain bitterness to infusions of tea,

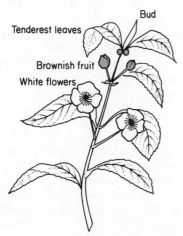

Bud
Tenderest leaves
Brownish fruit
White flowers

Figure 10.3 A branch of the tea shrub showing the tenderest leaves and bud, which are used for making good quality tea.

especially if they have been allowed to brew for a considerable time. Tannin has the property of precipitating proteins in tea and it also interferes with the absorption of iron.

After being picked, tea leaves are dried and in addition to loss of moisture this brings about some chemical changes, including an increase in caffeine content and amino acids. The dried leaves are broken up by passing them between rollers and this process releases enzymes, principally *phenolase*, which are responsible for the so-called 'fermentation' that follows. Ideally 'fermentation' is carried out at 25°C, at which temperature enzyme activity is at a maximum. The leaves are spread out in layers and in the presence of oxygen in the air phenolase catalyses the oxidation of some polyphenols to *o*-quinones which subsequently react to form coloured compounds, particularly the brownish substances generally known as *thearubigins*. The thearubigins contribute most of the colour as well as the astringency, acidity and body to tea infusions. The exact nature of the chemical changes taking place during fermentation are exceedingly complex and remain largely unknown.

After fermentation the leaves are dried further at a temperature which inactivates all enzymes, caramelizes sugars and reduces the moisture content to about three per cent.

The popularity of tea as a beverage depends mainly on the mild stimulating effect produced by caffeine, although a high-grade tea is also prized for its fragrant aroma and delicate flavour. When boiling water is added to tea leaves, the resulting infusion contains a proportion of the soluble constituents of the leaves. An average cup of tea made from 5 g tea contains (apart from any added milk or sugar); 60–280 mg tannin, 50–80 mg caffeine, 0.2–0.5 mg fluorine, 1 mg manganese, 0.02 mg riboflavin and 0.2 mg niacin.

Caffeine is extracted from tea leaves more readily than tannin so that the quality of tea is highest after about five minutes infusion when most of the caffeine (about 80 per cent) but only some of the tannin (about 60 per cent) has been extracted. More lengthy infusion increases the proportion of tannin extracted with consequent increase in bitterness.

COFFEE   The coffee plant, like the tea plant, is an evergreen shrub. It is cultivated in many tropical countries, particularly Brazil. The fruit is like a cherry in appearance and contains two seeds, called 'coffee beans', enclosed in a tough skin or parchment. In order to extract the beans, the outer pulpy part of the fruit is removed and the beans, still surrounded by their skin, are dried in air. The parchment cases are then removed by rolling, so liberating the beans.

The composition of coffee beans is exceedingly complex, over 300 constituents having been identified, and because of this complexity the

Table 10.2 The percentage composition of coffee on a dry matter basis

	Green beans	Roasted beans	Soluble coffee
Protein	13	11	15
Sugars	10	1	7
Starch and dextrins	10	12	5(dextrin only)
Complex polysaccharides	40	46	33
Coffee oil	13	15	<1
Minerals, mainly potassium	4	5	9
Chlorogenic acid	7	5	14
Trigonelline	1	1	4
Phenols	–	2	5
Caffeine (*Arabica*)	1	1.3	4

chemical basis of coffee flavour and aroma remain largely unknown. The main component of green coffee beans are given in Table 10.2. The figures given are average ones and there is considerable variation in composition between different varieties of coffee. For example, the value given for caffeine refers to *Arabica* coffee; *Robusta* coffee, the other main type, contains about twice as much.

Coffee beans are roasted before use; this reduces the moisture content from about 12 per cent to about 4 per cent, converts some of the sugar into caramel and develops flavour and aroma. Although roasting has little effect on caffeine, chlorogenic acid is broken down to caffeic and quinic acids and trigonelline is largely converted to niacin. Roasted coffee beans are a dark brown colour and brittle which makes grinding easy.

When an infusion of coffee is made from ground, roasted coffee beans, about 35 per cent of the constituents of the bean pass into the water. Caffeine is rapidly extracted, especially if very hot water is used; at 95°C about 80 per cent is extracted after two minutes and 90 per cent after ten minutes. A short extraction time of about two minutes is ideal as it favours extraction of caffeine but not of less soluble substances that contribute to bitterness.

A cup of fairly strong coffee contains about 100 mg caffeine together with 100 mg potassium and 1 mg niacin.

French coffee contains added *chicory*, which is the root of the wild endive. The chicory is roasted before use and although it contains no caffeine it does contribute a bitter flavour as well as a brown colour resulting from caramel produced during roasting. The reason for adding chicory to coffee is that it is much cheaper than coffee but legally the mixture must contain at least 51 per cent coffee.

'*Instant*' *coffee* is made by extracting the water-soluble components of roasted beans with hot water and concentrating the resulting liquid extract and converting it to a powder either by spray-drying or by freeze-drying (see Chapter 16). Modern techniques have resulted in the production of 'instant' coffees which retain most of the flavour and aroma constituents of the roasted beans.

Concern about the possible effects of heavy coffee drinking on health has resulted in an increasing demand for decaffeinated coffee. Such coffee is made by removing caffeine with the solvent dichloromethane, which is itself removed by treatment with steam. The resulting coffee must contain less than 0.3 per cent caffeine and less than 10 ppm dichloromethane.

COCOA The processing of cocoa-beans to produce cocoa and chocolate has already been described (p. 122). Cocoa powder, unlike tea leaves and coffee beans, has a considerable nutrient content, being approximately 12 per cent carbohydrate, 20 per cent protein and 22 per cent fat (of which about 13 per cent is saturated fats). It also contains small quantities of vitamins A and B and some mineral elements, notably iron (though it is probable that little of the iron is absorbed). In spite of its relatively high nutritional value, it makes little contribution to the average diet as it is consumed in small quantities.

In addition to small amounts of caffeine cocoa also contains the related alkaloid *theobromine* (dimethyl xanthine) which is a very mild stimulant. An average cup of cocoa contains about 200 mg theobromine and 20 mg caffeine, but these amounts are not sufficiently large to give cocoa a noticeably stimulating effect. Cocoa also contains some tannin.

Drinking chocolate is made from cocoa powder which is treated with alkali to improve solubility and to which sugar and sometimes milk powder, salt and vanilla are added.

Soft drinks

Soft drinks provide water and usually sugar in the form of a pleasantly flavoured drink. Their nutrient value is low apart from the sugar they provide. Some soft drinks contain a little vitamin C derived from the fruit used or added as synthetic vitamin C. Soft drinks include fruit squashes, fruit drinks, cordials, carbonated drinks, cola drinks and ginger beer. Certain soft drinks do not contain sugar. For example, tonic water is simply carbonated water flavoured with quinine to give it a bitter taste; glucose drinks such as Lucozade contain glucose syrup in place of sugar and diabetic and low-calorie drinks contain an artificial sweetener in place of sugar.

In the UK fruit drinks must contain a stipulated minimum fruit and sugar content. For example, undiluted fruit squash must contain at least 25 per cent fruit juice while comminuted fruit drinks (usually

Table 10.3 The nutrient content per 100 g of undiluted fruit juices and soft drinks

	Energy (kJ)	Sugars (g)	Vitamin C (mg)	Sodium (mg)	Potassium (mg)
Pineapple juice	225	13.4	8	1	140
Orange juice (a)	143	8.5	35	4	130
Grapefruit juice (a)	132	7.9	28	3	110
Ribena	976	60.9	210	20	86
Tomato juice	66	3.2	20	230	260
Orange squash	456	28.5	0–60	21	17
Lucozade	288	18.0	3	29	1
Cola	168	10.5	0	8	1
Lemonade	90	5.6	0	7	1

(a) unsweetened

described as made from whole fresh fruit) include both the peel and pulp of the fruit to provide added flavour and must contain at least ten per cent whole fruit when undiluted. Fruit-flavoured drinks contain no fruit and usually contain nothing except water, sugar and synthetic fruit flavour; they are often described as an '-ade' such as lemonade. The nutrient content of a variety of soft drinks is given in Table 10.3.

Fruit juices

Fruit juices are made by extracting the juice from fresh fruit and therefore have a similar nutrient content to whole fruit except that they have lost most of their pectin. Their popularity in Britain is increasing rapidly to the extent that they are now a major source of vitamin C in the average diet. The vitamin C content of fruit juices varies considerably (see Table 10.3) being exceptionally high in Ribena (which is made from blackcurrant juice to which sugar syrup and vitamin C are added), moderately high in citrus fruit juices, somewhat lower in pineapple and tomato juice and very low in apple juice.

Apart from being a pleasant way of receiving useful amounts of vitamin C, fruit juices are of value to people on a low sodium diet as they have a low sodium content but high potassium content (see Table 10.3).

ALCOHOLIC BEVERAGES

Alcoholic beverages are valued on account of their flavour and their stimulating effect and hardly at all as a source of energy; nevertheless it is worth noting that the energy value of wine is about equal to that of milk. There are three main classes of alcoholic beverages: wines, beers and spirits. They are all made from carbohydrate materials and

Table 10.4   Alcoholic beverages

Type	Example	Alcohol content (g/100 ml)	Energy value (kJ/100 ml)
Beers, draught	Bitter	3.1	132
	Mild	2.6	104
Ciders	Dry	3.8	152
	Sweet	3.7	176
Wines			
White	Sweet	10.2	394
Red	Chianti	9.5	284
Fortified	Sweet sherry	15.6	568
Spirits	70% proof	31.7	919
Liqueurs	Cherry brandy	19	1073

the particular ingredient used and the way in which it is processed chiefly determines the character of the drink produced. The main types of alcoholic drink and their energy values are shown in Table 10.4.

MANUFACTURE   Alcoholic beverages are made by a process of *fermentation*. The starting material used depends upon the product required. For example whisky is made from malt or grain, rum from molasses, wine from grapes, beer from malt and cider from apples. If the starting material is essentially starch (e.g., in the form of grain or rice) fermentation occurs in three stages as follows.

Stage 1: Conversion of starch to maltose and dextrins.   This is a hydrolysis which is catalysed by the enzyme *diastase*. Diastase, a mixture of α- and β-amylases (see p. 104) occurs in malt, which is the name given to germinating or sprouting barley. Malt is obtained by steeping barley in water and then removing it from the water and allowing it to stand in warm air for a few days, before it is very slowly and gently dried. During the germination period amylases and also some peptidases are produced and these immediately begin their work of hydrolysing the carbohydrate and protein present in the barley. Their action ceases, however, when the malt is dried.

When ground malt is mixed with a mash of starchy materials in water at 50–60°C and allowed to stand for about an hour the diastase hydrolyses the starch to maltose and dextrins:

$$2(C_6H_{10}O_5)_n + nH_2O \xrightarrow{\text{diastase}} nC_{12}H_{22}O_{11}$$

Starch   Maltose

Stage 2: *Conversion of maltose to glucose.* This is a hydrolysis catalysed by the enzyme *maltase* present in yeast which is added:

$$C_{12}H_{22}O_{11} + H_2O \xrightarrow{\text{maltase}} 2C_6H_{12}O_6$$

Maltose   Glucose

After the yeast has been added the process continues for several days, the temperature being kept at about $30-35°C$.

Stage 3: *Conversion of glucose to alcohol.* *Zymase* (which is the name given to a collection of at least 14 enzymes), also present in yeast, is responsible for the fermentation of glucose into alcohol. Fermentation means *boiling* and the name arose because during the reaction the liquid is agitated by bubbles of carbon dioxide, which produce a frothing or boiling appearance.

$$C_6H_{12}O_6 \xrightarrow{\text{zymase}} 2C_2H_5OH + 2CO_2$$

Glucose

This equation merely represents the start and the finish of the reaction, which is a complex one involving many stages. The result is a solution of alcohol in water, the alcohol amounting to less than 16 per cent of the whole.

It should be noted that only the last of these stages is a true fermentation, as it is only here that a gas is produced. The first two stages are examples of enzymatic hydrolysis although for convenience the whole process is usually referred to as *alcoholic fermentation*.

If the starting material is molasses (as when making rum) instead of a starchy material the initial hydrolysis (stage 1) is not necessary and if the starting material is a monosaccharide (as when making wine) only stage 3 is involved.

Types of alcoholic beverage

WINES   Wines are made from grapes by a process which involves four stages: pressing, fermentation, casking and bottling.

Grape juice contains the sugars glucose (grape sugar) and fructose (fruit sugar), various acids, tannin and nitrogenous materials. Grapes ferment naturally because they contain all the essential ingredients required, namely sugar, water and yeast, the latter being present as a covering or *bloom* on the skin. The bloom contains many varieties of yeast including wine yeasts and wild yeasts. When the juice is extracted from grapes by crushing them in a press, sulphites are added in such concentration that the undesirable wild yeasts are destroyed while the desirable wine yeasts are preserved.

During the fermentation which follows glucose and fructose are converted into alcohol. If the fermentation is continued until almost all the sugars are used up the resulting wine is dry, whereas if it is stopped while some sugars remain it is sweet. As yeasts cannot tolerate an alcohol content greater than 16 per cent, this is the maximum that natural wines can contain and most wines contain 10–12 per cent.

After fermentation wine is transferred to wooden casks or, in modern processes, large containers made of stainless steel. Some wines remain in the cask for several years during which time a slow secondary fermentation takes place because a few live yeast cells remain from the fermentation stage. The wine slowly matures in the cask, although wines which never come into contact with wood and which are matured for relatively short periods are not necessarily of an inferior quality. Finally, the wine is bottled and maturing continues. In particular, very small amounts of esters are formed which contribute to the final *bouquet* or fragrance which is such an attractive feature of a well-matured wine.

Wines are classified by colour as red, white or rosé. The colour of wine is often supposed to depend upon the colour of the grape, although this is not necessarily so, for a black grape may produce a white wine. The colour of the grape is due to pigment just under the skin, a black grape containing a blue-black pigment which turns red in the presence of the acids in the juice. In making a white wine, only the juice of the grape is used, whereas in making a red wine the skins are added during the pressing stage and the alcohol dissolves out the pigment which is carried into the wine. Even the so-called white wines usually have a yellow tinge and may even be brown; this colouration is not due to the grape but is produced while the wine is in the cask, owing to the oxidation of the tannin contained in the wood of the cask. Tannin has an astringent bitter taste and as it is also present in the skin and stalks of grapes, red wines tend to be more bitter than white. Even so, the amount of tannin in a glass of red wine is not likely to be greater than that in a cup of tea.

Some natural wines, such as champagne, *sparkle*: this is due to the secondary fermentation that occurs in the bottle and is brought about by the addition of a pure culture of yeast and a small amount of sugar. The carbon dioxide so produced is stored within the liquid under its own pressure. A bottle may contain five times its own volume of gas. When the cork is drawn the well-known 'pop' of the champagne is heard, due to the sudden release of pressure, when about four-fifths of the gas escapes. Special care is taken at every stage in champagne production to ensure that the product is of the highest possible quality.

A great many different wines are made in many countries of the world, although France remains pre-eminent among wine-producing countries, especially in the making of red wines such as the wines of

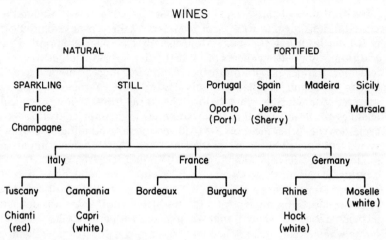

Figure 10.4 Types of wine.

Bordeaux (e.g., St Emilion and Médoc) and Burgundy (e.g., Beaujolais and Mâcon). The German white wines of the Rhine and Moselle and the red Chianti of Tuscany in Italy are also renowned. Figure 10.4 shows the different types of wine that are made with some illustrative examples. Wine is also made in other European countries such as Spain, Greece and Hungary as well as in other parts of the world, notably Australia, USA (California) and North and South Africa.

Fortified Wines. Wines such as port and sherry are said to be *fortified* because extra alcohol is added to give them an alcohol content of about 20 per cent. This has the advantage that they keep well because the alcohol content is sufficiently high to kill microorganisms that spoil natural wines.

Port, which comes from Oporto in Portugal, is made by adding brandy to wine before fermentation is complete. Genuine sherry is made from a special variety of grape which is grown near Jerez in Southern Spain. It is a very dry wine which is fortified after fermentation and it is later blended with sugar if a sweet flavour is required.

Madeira and Marsala are also fortified wines as is Vermouth in which the wine is flavoured with bitter ingredients (and sugar if a sweet wine is desired).

SPIRITS AND LIQUEURS Spirits differ from wines in having been distilled after fermentation; brandy is made by distilling wine, rum by distilling fermented molasses and so on. The special characteristics of these drinks may be associated with a particular ingredient used in the fermentation stage, as with 'Scotch' which is said to owe much to peat fires used in drying the malt. More usually character is imparted by addition of flavouring agents after distillation is complete, as with gin to which juniper berries and other flavouring ingredients are added before a second distillation. Spirits also gain character by being matured for several years before they are drunk. Slow chemical changes occur during this time; some ethyl alcohol is oxidized to acetaldehyde, other alcohols are oxidized to corresponding aldehydes, sweet fruity-smelling esters are formed and these changes impart a mellowness of flavour and fragrance of bouquet to the product.

Whisky, and most spirits on sale in Britain, have an alcohol content of about 30 per cent (see Table 10.4), but are also described as being 70 per cent proof – an ancient term in which 'proof' spirit was defined as being of such a strength that when mixed with gunpowder it would ignite when a light was applied. In Britain proof spirit is now defined as containing 57 per cent alcohol by volume and 50 per cent by weight.

Liqueurs are made by steeping herbs in strong spirits for a week or two and subsequently distilling. The richly flavoured distillate contains essential oils and other flavouring matters from the plants, and to this is added sugar and colouring matter.

Many of the famous liqueurs originated in monasteries and the methods and materials used in preparing them are kept secret. An aura of romance surrounds even those liqueurs with no monastic associations. The recipe of the renowned Scottish liqueur Drambuie, literally 'the drink that satisfies', is said to have been given to an ancestor of the present manufacturer by Prince Charles Edward in 1746. The Prince is said to have divulged the secret recipe as a token of gratitude for the help he received in escaping from the Scottish mainland to the island of Skye. It is made from Scotch whisky, Scottish herbs and honey (presumably from Scottish bees!).

BEER Beer is an alcoholic drink made by the fermentation of malted barley, the essential ingredients used in its manufacture being malted barley, hops, yeast and water. Ale, lager, porter and stout are all beers and they are all made by methods which are similar in principle but which differ in detail. The essential stages of making beer are the same as those described on page 218, except that hops are added to the liquor, known as *wort* which is boiled to extract the flavour of the hops before stage 2 of the process begins. In making typical British beer a top-fermentation technique is used in which the yeast floats on top of the fermenting liquor. In brewing lager, however, a different strain of yeast is used and a bottom-fermentation technique is used in which the yeast works at the bottom of the tank.

Most beers contain 3–7 g alcohol/100 ml. They also contain small amounts of riboflavin and niacin. Mild ales are the weakest, containing only 2.6 g alcohol/100 ml and are inferior in flavour because they are matured for only a few days. Bitter ales have more flavour both because they are matured for longer than mild ales and because more hops are used. Lagers are matured for several months at a low temperature. Stout is characterized by a dark colour and often a relatively high alcohol content; a strong stout contains about 4 g alcohol/100 ml. It is made from special malt mixtures which are heated to a high temperature so as to produce some caramel which imparts a dark brown colour to the liquor. Stout is no more nutritious than any other beer.

Effects of alcohol on the body

Alcohol must be regarded as a foodstuff because in the body it can be broken down to provide energy. In fact it is a more concentrated source of energy than either carbohydrate or protein, and has an available energy value of 29 kJ/g. It is also a drug and affects the central nervous system. These two effects must be considered together when assessing the desirability of alcohol as a source of energy. The nature of the effects of alcohol on the body, varying from mild stimulation when a small amount is consumed to loss of coordination and even death when a large quantity is taken, is indicated in Fig. 10.5. Consumption of a pint of beer produces a maximum level of 0.05 per cent alcohol in the blood.

Unlike most foods, alcohol can be absorbed by the body without prior digestion and this takes place mainly in the small intestine but also through the walls of the stomach. Absorption may take anything from one-half to two hours depending on the concentration of alcohol in the beverage consumed, the amount taken and the nature and amount of food eaten with it or immediately beforehand. An average time for absorption is about an hour.

The fate of alcohol in the body is summarized in Fig. 10.5. After absorption the alcohol is distributed through the body in the bloodstream, and thereafter it is broken down in a series of oxidative steps with liberation of energy. The breakdown process is controlled by a series of enzymes, each step being controlled by its own enzymes. Initial oxidation of alcohol to acetaldehyde is mainly controlled by *alcohol dehydrogenase* and, as its name indicates, this step involves removal of hydrogen. It is followed by further oxidation to acetic acid, the most important enzyme involved in this step being *aldehyde dehydrogenase*. These initial breakdown steps occur in the liver, and the acetic acid produced then becomes part of the general body pool of this substance and is further oxidized, in a complex process, to carbon dioxide and water. Alcohol is oxidized in the body rather

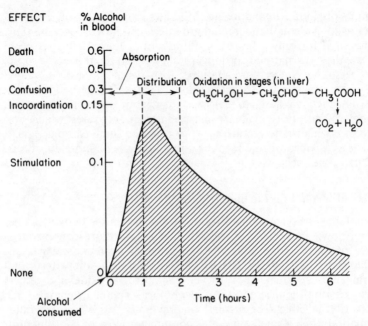

Figure 10.5 The metabolism of alcohol showing the relation between the level of alcohol in blood and its effect on the body (after von Wartburg).

slowly and only about seven grams can be oxidized in an hour. This means that alcohol is removed from the blood at a slow rate and that it can only make a small overall contribution to energy needs.

ALCOHOL INTAKE AND HEALTH　Since 1945 alcohol consumption in the UK has increased rapidly. In a 20 year period between 1962 and 1982 consumption of wine increased by 240 per cent, that of spirits by 95 per cent and that of beer by 22 per cent. The effects of such an increase in alcohol consumption are not fully known but the cost in health and social terms is considerable. The number of deaths caused by alcohol is unknown but it is recorded that the number of deaths caused by cirrhosis of the liver, which is closely linked with alcohol consumption, increased by 60 per cent in the 1970s. It is also recorded that the number of people entering mental hospitals each year as a result of alcoholism or related problems doubled in the 1970s. It is estimated that about three-quarters of a million people in England and Wales are either alcoholics or have severe drink problems.

An excessive intake of alcohol has a damaging effect on health, including cirrhosis of the liver mentioned above. If alcohol replaces fat as the main energy source for the liver, metabolism of fat ceases and lipid accumulates in the liver producing what is known as a 'fatty liver'. Prolonged excessive consumption of alcohol, even when it is part of a nutritionally adequate diet, leads to varying degrees of liver damage ranging from a reversible fatty liver to alcoholic hepatitis and finally irreversible cirrhosis.

Excessive alcohol consumption also contributes to obesity and may lead to malnutrition due to deficiencies of mineral elements and vitamins. While mineral deficiency is relatively uncommon, deficiency of B vitamins – particularly thiamine and folic acid – is relatively common. Such vitamin deficiency may arise in a number of ways, the most likely being that the high energy content of alcohol consumed reduces the intake of other foods some of which would have provided the necessary B vitamins. In addition alcohol may impair absorption of B vitamins and other nutrients.

SUGGESTIONS FOR FURTHER READING

AMERINE, M.A. *et al.* (1980).*Technology of Wine Making*, 4th edition. Avi, USA.

FRANKS, F. (1984). *Water*. Royal Society of Chemistry, London.

HARLER, C.R. (1964). *The Culture and Marketing of Tea*, 3rd edition. Oxford University Press, Oxford.

MATZ, S.A. (1965). *Water in Foods*. Avi, USA.

PRICE, P.V. (1984). *The Penguin Wine Book*. Allen Lane, London.

ROYAL COLLEGE OF PHYSICIANS (1976). *Fluoride, Teeth and Health*. Pitman Medical, London.

URQUHART, M. (1961). *Cocoa*, 2nd edition. Longman Group, Harlow, Essex.

CHAPTER 11

Mineral elements

The elements which occur most abundantly in food are carbon, hydrogen, oxygen and nitrogen. Many other elements may also be present, however, and they are referred to collectively as mineral elements. At least 25 mineral elements occur in foods, sometimes in minute amounts, and may find their way into our bodies. About 16 of these are known to be essential to life and must be present in the diet.

The mineral elements which the body requires in largest amounts are listed in Table 11.1. Others known as *essential trace elements* which, as their name implies, are required by the body in much smaller quantities, are shown in Table 11.4.

Mineral elements are used by the body in a great variety of ways. They may form part of the rigid structure of the body, (i.e. the

Table 11.1 Major mineral elements

Element	Approx adult daily intake	Approx adult body content	Functions in body	Main food sources
Calcium (Ca)	1 g	1000 g	Present in bones and teeth. Necessary for blood clotting, muscle contraction and nerve activity	Milk, cheese, bread and flour (if fortified), cereals, green vegetables
Phosphorus (P)	1.5 g	800 g	Present in bones and teeth. Essential for energy storage and transfer, cell division and reproduction	Milk, cheese, bread and cereals, meat and meat products
Sulphur (S)	0.9 g	150 g	Present in body proteins	Protein-rich foods e.g., meat, fish, eggs, milk, bread and cereals

Table 11.1 *Continued*

Element	Approx adult daily intake	Approx adult body content	Functions in body	Main food sources
Sodium (Na)	4.5 g	100 g	Present in body fluids as Na^+. Essential for maintenance of fluid balance in body and for nerve activity and muscle contraction	Main source is salt (sodium chloride, NaCl) used in food processing, cooking and at the table. Bread, cereal products and meat products are the main sources in processed foods
Chlorine (Cl)	6.0 g	100 g	Present in gastric juice and body fluids as Cl^-	
Potassium (K)	3.3 g	140 g	Present in cell fluids as K^+. Similar role to sodium	Widely distributed in vegetables, meat, milk, fruit and fruit juices
Iron (Fe)	15 g	49 g	Essential component of haemoglobin of blood cells	Meat and offal, bread and flour, potatoes and vegetables
Magnesium (Mg)	0.3 g	20 g	Present in bone and cell fluids. Needed for activity of some enzymes	Milk, bread and other cereal products, potatoes and vegetables
Zinc (Zn)	15 g	2.5 g	Essential for the activity of several enzymes involved in energy changes and protein formation	Meat and meat products, milk and cheese. Bread flour and cereal products

skeleton), or they may be present in the cell fluids or in the body fluids – that is fluids outside the cells. Their main functions and sources, which are shown in Tables 11.1 and 11.4, are discussed in greater detail below.

The figures given in Table 11.1 should be regarded as estimates and not as precisely determined amounts. The quantity of a particular mineral element present in a body will obviously depend on body weight. Whether a person of a given weight is tall, well-built and muscular or short and 'pudding-like' with a comparatively small skeleton will also be relevant. Similarly, the figures for average daily intakes cannot be other than rough averages because of variations in food availability and in individual appetites and tastes.

Average daily intake of minerals must be distinguished from average daily requirement and only part of what is eaten may be absorbed and used by the body. Elements such as sodium, potassium and chlorine, which form soluble salts, are readily absorbed. Others, such as iron, calcium, magnesium and zinc may form insoluble compounds which are less readily absorbed by the body. In such cases an allowance has to be made when estimating dietary requirements to take into account the proportion of the intake which is not absorbed by the body.

MAJOR MINERAL ELEMENTS

Calcium and phosphorus

These two elements account for about 75 per cent of the mineral elements in the body, and both have a number of essential functions to perform. Hence the body must receive a sufficient supply of each of them if it is to remain healthy.

FUNCTIONS OF CALCIUM AND PHOSPHORUS   Almost all the calcium and 85 per cent of the phosphorus in the body are found in the bones and teeth as calcium phosphate, $Ca_3(PO_4)_2$, or more precisely, as the calcium phosphate derivative calcium hydroxyapatite, $Ca_{10}(PO_4)_6(OH)_2$.

The small amount of calcium (about one per cent or 5–10 g) not used for bone or tooth formation is found in the blood and body fluids as calcium ions or in combination with protein. It is involved in muscle contraction (including the maintenance of a regular heartbeat), blood clotting and the activity of several important enzymes. The concentration of calcium in the blood is kept constant by the action of hormones produced by the thyroid and parathyroid glands. Too much or too little calcium in the blood seriously disturbs the function of the muscle fibres and nerve cells.

The phosphorus not present in bones or teeth (about 100 g) is distributed throughout the cells and fluids of the body. Phosphorus is present in the nucleic acids which form part of all cells and are concerned with manufacture of the body's proteins and transmission of hereditary characteristics. Adenosine triphosphate (ATP) plays a key role in the complex processes by which the body obtains energy through the oxidation of nutrients (see p. 112). It also plays an essential part in the metabolism of fats and proteins. Phosphate ions are present in the blood and help to keep its pH value constant at about 7.4.

CALCIUM AND PHOSPHORUS REQUIREMENTS   The RDA of calcium in Britain varies from 500 mg for adult men (and most adult women) to 1200 mg for pregnant women and nursing mothers, who require additional amounts to prevent loss of calcium from the mother's bones (see Appendix I). The RDA for growing boys or girls is 600 or 700 mg. Recommendations vary widely from one country to another and British recommendations tend to be low compared with those of other countries. They are, however, roughly in line with the recommendations of the Food and Agriculture Organisation (FAO) of the United Nations.

Normally only about 20–30 per cent of the calcium in the diet is absorbed. The rate of absorption may drop to even lower levels if sufficient vitamin D is not available (see p. 249). Phytic acid, which is present in wholemeal bread and wholegrain cereals, may also interfere with the absorption of calcium. If wholemeal bread or wholegrain cereals are eaten regularly, however, the body is able to adapt to their presence and the effect of phytic acid on calcium absorption is less severe than was at one time supposed.

Oxalates, which are present in small amounts in rhubarb and spinach, may react with calcium present in other foods to form insoluble calcium oxalate, thus rendering the calcium unavailable to the body. Luckily, rhubarb and spinach do not constitute a major part of most people's diet, and so we need not be unduly concerned about their detrimental effect on the absorption of calcium.

Some of the calcium absorbed from the diet is subsequently lost in the urine and a smaller amount may be lost in sweat. Such losses must be made up to maintain the concentration of calcium ions in the blood at a level of about 10 mg per 100 ml. If dietary sources of calcium are insufficient the calcium required is taken from the bones and if this continues over a long period serious decalcification may occur.

We tend to regard our bones as a fixed and immutable part of our bodies but, in fact, they are in a state of continuous change. New bone is constantly being formed, and old bone lost, by exchange of calcium ions between bone and blood. While growth is taking place calcium is absorbed at a slightly greater rate than it is lost.

The bones of a newborn baby are soft and consist mainly of collagen. They become hard, or calcified, when minute crystals of calcium phosphate (or hydroxyapatite) are deposited in the soft collagen framework. Bone development continues until *peak bone mass* is achieved, usually sometime between the ages of 20 and 30 years. From this point onwards calcium is gradually lost from the bones which, over a period of time, become less dense. If too much bone loss occurs the condition known as *osteoporosis* – literally, porous bones – may result. This affects many women (and some men) from the age of about 45 onwards. Bone shrinkage and brittle bones which fracture easily are characteristic of osteoporosis.

Osteoporosis was formerly thought to be a consequence of poor absorption of calcium from the diet. It is now known, however, that shortage of the female sex hormone *oestrogen* is the cause in most cases. This is why post-menopausal women, whose bodies secrete less oestrogen, are more likely to suffer from osteoporosis than younger women. The onset of the disease cannot be delayed or prevented by increasing the calcium content of the diet in later life. Achievement of high bone mass through good nutrition and exercise during the period of bone growth, however, is beneficial. Studies have also shown that in areas where water has been fluoridated the incidence of osteoporosis is considerably lower.

Poor calcium absorption by young children may cause the disease *rickets* which is characterized by stunted growth and deformed leg bones. A similar condition in adults is known as *osteomalacia*. This disease was at one time common in underdeveloped countries amongst women whose bones had suffered calcium loss through repeated pregnancies.

The average Briton consumes about 800 mg of calcium per day and it is unlikely that normal diets in Britain or other Western countries will be deficient in the mineral. Calcium deficiency may still occur, however, even when dietary sources are adequate if sufficient vitamin D, which is needed for its absorption, is not available. Calcium is absorbed into the blood through the lining of the small intestine by becoming bound to protein. In the absence of vitamin D the calcium/protein complex cannot be formed and absorption of calcium does not occur.

Phosphorus is found in almost all foods mainly as phosphate. The average daily intake is about 1.5 g and it is unlikely to be lacking in a normal diet. For this reason there are no recommended daily allowance figures.

DIETARY SOURCES OF CALCIUM AND PHOSPHORUS   Only a few foods are rich in calcium and the most important sources in the British diet are milk, cheese, bread and flour. About half the average daily intake comes from dairy foods, a quarter from bread and other cereal foods and the remainder from meat. Bread and flour in Britain (except wholemeal) are fortified with added chalk (see p. 134) to guard against inadequate calcium intake.

The calcium in dairy products is readily absorbed and a pint of milk contains more than the recommended daily intake. The calcium content of some foods is given in Table 11.2.

Table 11.2   Calcium and phosphorus content of foods

Food	Calcium (mg/100 g)	Phosphorus (mg/100 g)
Dairy products and eggs		
Milk	103	88
Butter	15	24
Cheese (Cheddar)	800	520
Eggs (boiled)	52	220
Meat and fish		
Beef (raw)	7	140
Liver (raw)	8	313
Fish (white)	22	171
Sardines (canned)	550	683
Cereals		
Bread (white)	100	79
Bread (wholemeal)	23	210
Flour (70% extraction)	140	84
Rice	4	99
Vegetables and fruit		
Potatoes (old)	8	40
Cabbage (Savoy)	57	54
Spinach	70	93
Apples	4	9
Oranges	41	24

Strange as it may seem, water may be a good source of calcium in those areas where it is 'hard' (see p. 211). For example, a litre of London tap water contains about 200 mg of calcium – 40 per cent of the RDA!

Vegetables cooked in hard water may pick up some calcium from it. When hard water is used for making tea or coffee, however, most of the calcium is deposited as calcium carbonate in the kettle or percolator.

As already explained, there is no danger of the body going short of phosphorus except, perhaps, on a starvation diet. The brain and nervous system are rich in phosphorus and it was at one time thought that foods rich in the mineral would be beneficial to the brain and nerves. This is the origin of the myth that fish is good for the brain.

Iron

Iron accounts for about 0.1 per cent of the mineral elements in the body and the total amount of iron in the body of a healthy adult is only about 4 g. Over half of this is found in the red blood cells in the pigment *haemoglobin* which transports oxygen from the lungs to the tissues. Red blood cells have a life of about four months, and it has been estimated that some ten million of these cells are withdrawn from circulation every second. If the iron contained in these cells passed out of the body it would be difficult to replace it from food. Fortunately, most of the iron released is conserved and it is used to form the new red corpuscles which are produced in the marrow of the bones. In this way the iron present in haemoglobin is used repeatedly.

A small proportion of the iron in the body is present in the muscle protein *myoglobin*; some cell enzymes, such as the *cytochromes*, also contain iron. The remainder of the iron in the body is stored in the liver, spleen and bone marrow in the form of specialized iron-binding proteins called *ferritin* and *haemosiderin*. These iron stores may contain up to 1 g iron in men and about half that amount in women. The iron present can only be released slowly and so it is of no use in combating sudden shortages. The stored iron is of great importance during the first six months of a baby's life because only a small amount of iron is present in milk.

DIETARY SOURCES OF IRON   The main dietary sources of iron are meat, bread and other cereal products, potatoes and vegetables. In some areas water may make a small contribution. However, even the richest dietary sources of iron contain only very small amounts of the element. Lamb's liver, one of the richest sources, contains only about one part of iron in 10 000. The main dietary sources of iron are listed in Table 11.3.

Bread and other cereal-based foods provide about 40 per cent of the total iron intake in an average British diet. Meat and vegetables each provide about 20 per cent and eggs about five per cent. Iron is one of the mineral elements which may be lacking in an average diet and for this reason it is added to all flour in Britain, except wholemeal flour,

Table 11.3   Iron content of foods

Food	Iron content (mg/100 g)	Food	Iron content (mg/100 g)
Liver, lambs, fried	10.9	Eggs, boiled	2.0
Kidney, pigs, fried	9.1	Bread, white	1.7
Flour, wholemeal	4.0	Baked beans	1.4
Spinach, boiled	4.0	Cod, fried	0.5
Oats, porridge	3.8	Cabbage, boiled	0.4
Beef, cooked	3.0	Potatoes, boiled	0.4
Sardines, canned	2.9	Cheese, Cheddar	0.4
Bread, brown or wholemeal	2.5	Apples	0.3
Flour, 72% extraction	2.2	Milk	0.1

so that its iron content is at least 1.65 mg/100 g. Unfortunately, however, studies have shown that most of the iron added to flour passes through the body unabsorbed.

IRON REQUIREMENTS The body utilizes absorbed iron very efficiently and only a minimal amount is needed each day to replace losses. Only a small portion of the iron present in the diet is absorbed by the body, however, and so dietary intake should be much greater than the amount needed to replenish losses.

Adults usually absorb only about ten per cent of the iron in their diet but individuals who have a special need for iron, such as growing children or pregnant women, are able to absorb more. Many factors affect the rate of absorption and the source of iron is amongst them. The iron in meat or offal – haem iron – is organically bound and is absorbed much more readily than the non-haem iron present in plants. It has been shown, for example, that only a small percentage of the iron added to non-wholemeal flour to replace that lost during milling is absorbed. Similarly only 2–3 per cent of the non-haem iron present in vegetables may be absorbed.

The presence of ascorbic acid (vitamin C) in the diet can promote the absorption of non-haem iron because it reduces ferric iron to the absorbable ferrous state and helps to keep it as ferrous iron. Absorption of non-haem iron is also promoted by alcohol, but tea, which forms insoluble tannic acid salts with iron, has the opposite effect.

In Britain the recommended daily intake of iron for an adult male is 10 mg. Women who lose blood through menstruation require a larger intake, probably about 12 mg per day. For pregnant women 13 mg per day is recommended and for nursing mothers 15 mg per day (see Appendix I).

If the amount of iron provided by the diet is insufficient the deficiency is made up from the body's iron stores. With a prolonged shortage the stores of iron will eventually become so depleted that the amount of haemoglobin in the blood will fall below normal levels – a condition known as *anaemia*.

Sodium and potassium

Sodium and potassium are highly reactive metals – so reactive in fact that they combine vigorously with water. In food or in the body, however, they are present as salts such as sodium chloride, NaCl, or potassium chloride, KCl. In these compounds the sodium and potassium are present as cations, Na^+ and K^+, respectively and not in the highly reactive metallic form. Sodium chloride is familiar to us all as common salt and it is mainly in this form that sodium is present in foods. In the body positively charged sodium and potassium ions are accompanied by an equal number of anions, i.e. negatively charged ions, to ensure electrical neutrality. The anions are mainly chloride and phosphate ions but other anions (e.g., carbonate and bicarbonate ions) may occur in food and they also play an important part in body processes.

FUNCTIONS OF SODIUM AND POTASSIUM   Almost all the sodium and potassium in the body is found in the soft body tissues and body fluids. Sodium ions are mainly present in the extracellular tissue fluids and blood plasma, whereas potassium ions are found mainly inside the cells. About 100 g sodium (i.e. about 250 g sodium chloride) and an equivalent amount of potassium salts are present in the body. The volume and osmotic pressure of the blood and tissue fluids are closely related to the concentrations of sodium and potassium ions which are precisely controlled by the body's regulating mechanisms.

The average diet is rich in sodium and potassium and both are easily absorbed by the body. Any excess is removed by the kidneys and excreted in the urine. Some loss also occurs in the form of sweat. When sodium and potassium are removed from the body, water is lost as well. This is why we become thirsty after eating salty food.

Sodium and potassium are crucially involved in the transmission of nerve impulses and in muscle contraction, including the beating of the heart.

DIETARY SOURCES OF SODIUM AND POTASSIUM   The sodium content of most foods in their natural state is usually quite low, but salt is added to many processed foods. Thus, although the sodium content of fresh meat is low, bacon, sausages, pies and most other meat products contain substantial amounts. The same is true for fish: fresh fish contains little sodium but kippers or smoked fish may be very salty. Salt is also added to most butter and margarine, canned vegetables, cheese, bread, and some breakfast cereals. Vegetables cooked in salted water contain much more sodium than fresh vegetables. Marmite is particularly rich in sodium but as it is eaten in small quantities it does not greatly affect the total sodium intake.

Despite its presence in so many foods, most people like even more salt and about one third of the salt in the British diet is added during cooking or at the table. Another third comes from cereals and bread, about one sixth from meat and meat products and the remaining sixth from other foods.

Potassium is present in nearly all foods, especially those of vegetable origin. The main sources in a British diet are vegetables, meat and milk. Fruit and fruit juices contain much more potassium than sodium.

SODIUM AND POTASSIUM REQUIREMENTS   Sodium chloride (or

other sodium salts) must be present in our diet to replace sodium lost through sweating. Sodium excreted in the urine is in excess of the body's requirements and replacement should not be necessary. The daily requirement, therefore, varies according to how much a person perspires. Hot weather, strenuous activity and a tendency to perspire easily all increase the need for salt in the diet. A daily salt intake of 2–3 g should provide sufficient sodium for people in a temperate climate who do not sweat profusely. In extreme conditions. however, as much as 50 g per day can be lost in perspiration. These variable requirements make it difficult to recommend realistic individual daily intake levels.

The average daily salt intake in Britain has been estimated to be about 12 g although, of course, many individuals consume much more or less than this.

Babies less than about one year old are not able to deal effectively with excess sodium because their kidney functions are not fully developed. An infant fed only on breast milk will receive only about 0.5 g salt per day but two or three times as much may be consumed if cow's milk is used. The salt content of powdered milk for babies and other baby foods is kept low to avoid an excessive intake. Salt should never be added to home-prepared baby food and salty spreads such as Marmite should not be given to very young babies.

There is some evidence that there may be an association between high salt intake and the development of high blood pressure, or hypertension, in later life. Other studies indicate that if this is so only about 10–20% of the population, who show special sensitivity to salt, are likely to be affected. High blood pressure is one of the risk factors associated with coronary heart disease and cerebrovascular disease or 'stroke'. The World Health Organization (WHO) does not consider that the evidence is strong enough to justify a recommendation that the population in general should reduce its salt intake. The evidence may be sufficient, however, to justify a reduction in salt intake by those 'at risk' for other reasons. In Britain the 1984 COMA report (see Chapter 15) recommended that average salt intake should not increase further and consideration should be given to ways of decreasing it. The 1983 NACNE report (see Chapter 15) on the other hand, recommended that the average salt intake should be reduced by 3 g per day.

Potassium is present in such a wide variety of foods and is so readily absorbed and excreted that it is not necessary to specify a recommended daily amount. The average daily diet contains about 3 g of potassium.

Magnesium

A human body contains about 20–25 g magnesium and most of it is found in the bones as magnesium phosphate. Magnesium is also present in ionic form in all tissues where it plays a part in many reactions involved in energy utilization.

Magnesium occurs widely in foods. It is present in green vegetables as a part of the chlorophyll molecule and vegetables provide two-thirds of the magnesium in an average British diet. Meat is also a good source as a consequence of animals eating grass and other vegetation.

Magnesium deficiency is rare and is normally caused by illness and not because of low intake.

Is has been estimated that adults require 200–300 mg magnesium per day and the average British diet provides this adequately.

Zinc

An adequate intake of zinc is essential for the maintenance of good health. It forms part of the enzyme *carbonic anhydrase* found in red blood cells, which assists in releasing carbon dioxide from venous blood passing through the lungs. Zinc is also important as a constituent of several other enzymes and it plays a part in protein and carbohydrate metabolism. Prolonged shortage of zinc can lead to retarded physical and mental development in adolescents.

Zinc is present in a wide variety of foods including meat and meat products, milk, bread and other cereal products. The estimated daily intake in Britain is 15 mg. There are no recommended daily intake figures in Britain but the Food and Nutrition Board in the USA has recommended a daily intake of 15 mg. The Food and Agriculture Organization of the United Nations suggest a daily adult intake of 22 mg with the proviso that growing children and pregnant or lactating women would need more.

It is likely that less than half the zinc in the diet is absorbed.

TRACE ELEMENTS

In addition to the major mineral elements already considered, the body also requires minute quantities of certain other elements known as *essential trace elements* which are necessary for human life. These are as follows: cobalt, copper, manganese, molybdenum, selenium, fluorine, iodine and chromium. The exact role that these elements play in the body is often not fully known, though most of them form part of vitamin, enzyme or hormone molecules. Only minute amounts are required in the diet and any normal diet will contain sufficient for the body's needs. Table 11.4 summarizes the main facts about the essential trace elements.

The metallic trace elements are particularly associated with enzyme activity and are usually an integral part of the enzyme. Copper forms part of several enzyme systems including *cytochrome oxidase*, and *tyrosinase*. Copper and all the other metallic essential trace elements have a variable valency and this enables them to take part in oxidation–reduction reactions in the body. In cytochrome oxidase copper is associated with iron and catalyses oxidation–reduction mechanisms concerned with tissue respiration. Tyrosinase is concerned with the oxidation of tyrosine and seems to be unique in that

Table 11.4 Some essential trace elements

Element	Approximate average daily intake	Approximate adult body content	Main food sources	Functions in body
Cobalt (Co)	0.3 mg	1.5 mg	Liver and other meat	Required for formation of red blood cells
Copper (Cu)	3.5 mg	75 mg	Green vegetables, fish and liver	Component of many enzymes. Necessary for haemoglobin formation
Chromium (Cr)	0.15 mg	1 mg	Liver, cereals, beer, yeast	Contained in all tissues. May be involved in glucose metabolism
Fluorine (F)	1.8 mg	2.5 g	Tea, sea food, water	Required for bone and tooth formation
Iodine (I)	0.2 mg	25 mg	Milk, sea food, iodized salt	Component of thyroid hormones
Manganese (Mn)	3.5 mg	15 mg	Tea, cereals, pulses, nuts	Forms part of some enzyme systems
Molybdenum (Mo)	0.15 mg	?	Kidney, cereals, vegetables	Enzyme activation
Selenium (Se)	0.2 mg	25 mg	Cereals, meat, fish	Present in some enzymes. Associated with vitamin E activity

Nickel, silicon, tin and vanadium may also have some part to play in human metabolism it can catalyse the reaction even when tyrosine is part of an intact protein. Manganese activates the enzymes *alkaline phosphatase* and *arginase* which are concerned with bone and urea formation respectively.

Cobalt forms part of the vitamin B_{12} molecule and the element can only be utilized by man in this form.

The amount of essential trace elements in the body is very small indeed, and the amount needed in the diet is minute. For example, the total amount of cobalt in the body is about 1.5 mg and the amount required daily is only a few millionths of a gram. Normal diets supply a sufficiency of all the essential trace elements, except possibly iodine and fluorine. For this reason these two elements warrant separate consideration.

IODINE   Iodine is the heaviest member of the halogen group which comprises the chemically related elements fluorine, chlorine, bromine and iodine. They all occur in nature in the form of salts and are all found in sea-water. All the halogens are found in the body and they are all essential except bromine. Iodine is carried round the body in blood as iodide and is absorbed in the thyroid gland in the neck where it is converted to the hormones *thyroxine* and *tri-iodothyroxine*. These two important hormones are concerned with the general metabolic activity of the body and control the rate of energy production in all cells.

The body normally contains only about 20–50 mg iodine and the amount required daily in the diet is very small indeed, about 0.15 mg being sufficient for normal needs. When the diet provides insufficient iodine the thyroid gland may increase in size in an attempt to compensate for the deficiency. The characteristic swollen neck, or goitre, was at one time known in Britain as 'Derbyshire neck' because of the prevalence of the condition there. Goitre still occurs in some parts of the world, especially mountainous and inland areas, where iodine levels in the soil and hence in vegetation is low. Its incidence in developed countries, where preventive measures can be taken, is low.

Small amounts of iodine may be present in drinking water, and it is also obtained from food, sea foods being the richest source. Thus cod, salmon and herring are all useful sources of iodine, although the best marine source is cod-liver oil. Milk and other dairy products are important dietary sources of iodine in Britain as a result of iodine-enrichment of cattle feed. Vegetables grown on iodine-rich soil also contain available iodine but most cereals, legumes and roots have a low iodine content.

Some vegetables are said to be *goitrogenic*, i.e. capable of causing goitre. Cabbage, cauliflower and Brussels sprouts can all interfere with the uptake of iodine by the thyroid gland and thus cause goitre.

This is only likely to occur, however, if substantial amounts are eaten and the iodine content of the diet is very low.

Some seaweeds concentrate iodides from sea-water and are, therefore, a useful reservoir of combined iodine. In some parts of the world certain seaweeds are regarded as valuable foods for man and cooked seaweed, known as laverbread, is eaten in southwest Wales.

In areas where the iodine content of the diet is low a satisfactory way of increasing the intake of iodine is the use of 'iodized salt'. This is prepared by adding about one part of potassium iodide to about 40 000 parts of salt. Potassium iodide is soluble in water and is rapidly absorbed into the blood, any surplus being quite harmless. The use of iodized salt is a simple and harmless way of supplementing the iodide obtained from food.

FLUORINE Fluorine is the lightest and most reactive member of the halogen group and in the body it is found in bones and teeth, although there is as yet no direct evidence to prove that it is essential to the body. It appears to harden tooth enamel by combining with calcium phosphate in teeth. It has been shown that traces of combined fluorine are beneficial in protecting teeth against decay, the protective effect being most noticeable in children under eight years of age.

Minerals containing combined fluorine have a wide distribution in nature, though they only occur in small quantities. Small amounts of fluorine compounds are, therefore, usually present in natural water. Water in Britain nearly always contains traces of fluorides, though their concentration rarely exceeds one part per million (ppm) parts of water. Drinking water is the main source of fluorine in the diet as very few foods contain more than 1 ppm, the main exceptions being sea fish, which contain 5–10 ppm, and tea which may contain up to 100 ppm.

The average daily intake of fluoride in Britain from food and beverages is about 0.6–1.8 mg. In areas where drinking water contains very little or no fluoride the intake of fluoride may be too low to provide protection against dental decay. This situation can be remedied if controlled amounts of fluoride are added to drinking water. This may be done by adding sodium fluoride or sodium silicofluoride, Na_2SiF_6, or hydrofluosilicic acid in such quantities that the total fluoride content of the water is 1 ppm. The amount of fluoride added must be carefully controlled because if it exceeds 1.5 ppm teeth may become mottled in appearance.

Experiments carried out over many years in the USA, Britain and other countries prove beyond reasonable doubt that fluoridation of water to increase the fluoride content to 1 ppm reduces the incidence of dental caries, particularly in young children. An authoritative report by the Royal College of Physicians (1976) concluded that:

1. The presence of fluoride in drinking water substantially reduces dental caries throughout life.
2. A level of 1 ppm of fluoride in drinking water is completely safe irrespective of the hardness of the water.
3. Alternative methods of taking fluoride, such as toothpaste or tablets, are less effective than fluoridation of water supplies.
4. Fluoridation does not harm the environment.
5. Where the natural fluoride content of water supplies is less than 1 ppm, fluoride should be added to bring it to this level.

TRACE ELEMENTS AND FOOD PROCESSING　Trace elements may become incorporated into food during processing and although minute amounts of such elements may be beneficial or even essential to the body, slightly larger amounts are frequently toxic. Modern analytical techniques such as flame photometry and atomic absorption spectrophotometry, which enable concentrations of less than 1 ppm of many elements to be determined, have paved the way for legislation which lays down limits for a number of elements. The ability to measure concentrations of less than 1 ppm and the improvement in processing techniques is making it possible to reduce such limits. Table 11.5 summarizes the position and gives examples of specific and general limits that are applied.

The inclusion of traces of some elements in foods during processing can have an adverse effect on quality and this is an additional reason

Table 11.5　Limits for trace elements in food

| Element | General limit | | Examples of specific limits (ppm) |
	Statutory or recommended	Limit in ppm	
Lead	Statutory	1	Dried herbs and spices, 10; corned beef, tea, 5; canned fish, 3; canned foods, 2; baby foods, 0.2
Arsenic	Statutory	1	Spices, 5; chicory, 4; ice-cream, soft drink concentrates, fruit juices and beer, 0.5
Fluorine	–	–	Baking powder, 15; self-raising flour, 3
Copper	Recommended	20	Tea, 150; chicory and gelatine, 30; beer and wines, 7
Zinc	Recommended	50	Gelatine, 100; ready to drink beverages, 5
Tin	Recommended	250	Canned foods only

for measuring and controlling the amounts of trace elements entering food in this way. Copper, for example, even in concentrations of less than 0.1 ppm, can lead to the development of rancidity in milk and butter.

SUGGESTIONS FOR FURTHER READING

BRITISH NUTRITIONAL FOUNDATION (1985). *Salt in the Diet*. Briefing Paper No. 7. BNF, London.

DAVIES, L.S. AND HOLDSWORTH, M.D. (1985). *Prevention of Osteoporosis*. Symposium report. WHO, Geneva.

NATIONAL DAIRY COUNCIL (1986), *Calcium and Health*, Fact File 1. NDC, London.

ROYAL COLLEGE OF PHYSICIANS (1976). *Fluoride, Teeth and Health*, Pitman Medical, London.

WHO (1973). Report on *Trace Elements in Human Nutrition*. HMSO, London.

CHAPTER 12
Vitamins

Vitamins are organic compounds found in small amounts in many foods; their presence in the diet is essential because, in most cases, the body is unable to synthesize them from other nutrients. Various diseases, called *deficiency diseases*, are associated with shortages of specific vitamins. Deficiency diseases have caused much suffering and death in the past but today they can be prevented and cured by ensuring that the diet contains a sufficient quantity and variety of vitamins.

Most vitamins have complex chemical structures; they do not belong to one chemical family but are all quite different from each other. However, the structures of all of them are known and, with one exception, they can be prepared synthetically. Before their structures were determined the vitamins were designated by letters as vitamin A, vitamin B and so on. They are now known by names which give some indication of their chemical structure, and in general these names are used in preference to the letters. In some cases, however, the letters by which they were originally known are still used.

Foods contain only very small quantities of vitamins, but these small amounts carry out some most important tasks in the body. Members of the B group of vitamins, for example, form part of several coenzyme molecules which are necessary for the maintenance of good health. The other vitamins are equally essential, although in some cases their exact job in the chemistry of the body is not known.

Only small amounts of vitamins are needed by the body and the quantities present in foods are usually sufficient for people's needs. They are, however, distributed among many types of food, and to ensure that all the vitamins are present in the diet it is important that a variety of different foods is eaten. The vitamin content of a food can vary quite considerably. This is especially so with fruit and vegetables, where the vitamin content depends, among other things, on the freshness and variety of the fruit or vegetable and climatic conditions during its growth. The figures for the vitamin content of foods given in this chapter are average values and this must be borne in mind when consulting them.

It is so important that sufficient quantities of vitamins are consumed that in some cases extra vitamins are added to food. Examples have already been encountered in connection with flour, to which the vitamins thiamine and niacin are added to replace losses which occur during milling, and margarine, to which vitamins A and D are added. Similarly, most brands of cornflakes are 'fortified' with added B group vitamins to compensate for the low vitamin content of the maize from which they are made.

When vitamins are added to foods synthetic vitamins (i.e. man-made vitamins) are often used rather than the naturally occurring compounds. Synthetic vitamins are identical in structure to the naturally occurring vitamins and behave in the body in the same way. They should not be regarded as substitutes.

The main sources and functions of the vitamins are summarized in Table 12.1. It is convenient to divide the vitamins into a fat-soluble or water-soluble group. The fat-soluble vitamins, namely vitamins A, D, E and K, are mostly found in fatty foods and fish-liver oils are particularly rich in vitamins A and D. These two vitamins are also found in human liver and if the diet contains more vitamin A or D

Table 12.1 Vitamins (the RDA of vitamins are shown in Appendix I)

Name	Main sources	Functions in the body and effect of shortage
Fat-soluble vitamins		
Vitamin A or retinol	Milk, dairy products, margarine, fish-liver oil. Also made in the body from carotenes found in green vegetables and carrots.	Necessary for healthy skin and also for normal growth and development. Deficiency will slow down growth and may lead to disorders of the skin, lowered resistance to infection and disturbances of vision such as night blindness.
Vitamin D or cholecalciferol	Margarine, butter-milk, fish-liver oils, fat fish.	Necessary for the formation of strong bones and teeth. A shortage may cause bone diseases or dental decay.
Vitamin E or tocopherols	Plant-seed oils.	Not known.
Vitamin K or naphthoquinones	Green vegetables.	Assists blood clotting.

continued

Table 12.1 *Continued*

Name	Main sources	Functions in the body and effect of shortage
Water-soluble vitamins		
The B vitamins		
Thiamine, B_1 Riboflavin, B_2 Niacin Pyridoxine, B_6 Pantothenic acid Biotin	Bread and flour, meat, milk, potatoes, yeast extract, fortified cornflakes (See text for details).	Function as co-enzymes in many of the reactions involved in making use of food. Shortage causes loss of appetite, slows growth and development and impairs general health. Severe deficiency disease such as pellagra or beriberi.
Cobalamin, B_{12} Folic acid	Offal, meat, milk, fortified cornflakes. Potatoes, offal, green vegetables, bread, Marmite, fortified cornflakes.	Necessary for formation of nucleic acids and red blood cells. Shortage may lead to megaloblastic anaemia and (for cobalamin) to pernicious anaemia (see text).
Vitamin C or ascorbic acid	Green vegetables, fruits, potatoes, blackcurrant syrup, rosehip syrup.	Necessary for the proper formation of teeth, bones and blood vessels. Shortage causes a check in the growth of children and if prolonged may lead to the disease scurvy.

than is immediately required the surplus is stored in the liver. Enough of these vitamins are stored in the liver of a well-nourished person to satisfy the body's needs for several months if they are absent from the diet. If the diet contains too much vitamin A or D, however, the surplus will accumulate in the liver and may be harmful. Such excessive intakes are most unlikely to result from over-eating but may occur through over-enthusiastic use of vitamin pills or dietary supplements.

The water-soluble group of vitamins is made up of several B vitamins and ascorbic acid (vitamin C). The body is unable to build up a store of these vitamins and if the diet contains more than is immediately required the surplus is excreted in the urine. Nevertheless, a well-nourished person can remain healthy for several months on a diet which contains little vitamin C and for several years on a diet which contains little or no vitamin B_{12}. In general, however, adequate and regular dietary supplies of the water-soluble vitamins are necessary for the maintenance of good health.

The optimum daily intake for each vitamin cannot be stated with certainty. Precise requirements vary from person to person and with the nature of the rest of the diet. Despite these uncertainties various national and international bodies have recommended desirable vitamin intakes for certain groups of the population. These are published as *recommended daily amounts* (RDAs) which are at such a level as to ensure that the requirements of almost all healthy persons in the group concerned will be satisfied. In Britain the Department of Health and Social Security (DHSS) published tables of RDAs (1979, see Appendix I) and recommendations are also published by the World Health Organization (WHO). It should, perhaps, be emphasized that despite the word 'daily' it is not necessary to regard RDAs as strict *daily* allocations. To begin with, as has already been explained, the amounts specified can only be approximate; the precise amount required will vary from person to person and from time to time. Secondly, the body has a sufficient reserve of vitamins to be able to cope easily with day-to-day variations. As far as vitamins are concerned RDAs are best regarded as desirable *average* intakes and daily variations (up or down) from the RDA will generally be of no significance.

FAT-SOLUBLE VITAMINS

Retinol or vitamin A

Retinol is a pale yellow solid which dissolves freely in oils and fats but is only very slightly soluble in water. It is a fat-soluble vitamin and so it is found in the fatty parts of foods, for example, in the fat of milk and butter, fish-liver oils and the small amount of fat present in green vegetables and carrots.

Retinol is a fairly complex unsaturated alcohol of molecular formula $C_{20}H_{29}OH$. It behaves chemically in the same way as other alcohols. In animal tissues it is stored and transported as an ester formed with a long chain fatty acid such as stearic acid or palmitic acid which is bound to a protein.

Retinol has been synthesized and is now produced industrially on a fairly large scale for the enrichment of margarine.

Vegetables contain no retinol as such, but pigments called *carotenes* are present. These are converted to retinol in the wall of the small intestine during absorption and hence vegetables have considerable vitamin A activity. Several carotenes are known; the most important is β-carotene which is often referred to as provitamin A. β-carotene is a red solid which was first isolated from carrots; indeed, it owes its name to this relationship. Solutions of β-carotene are yellow in colour and it is used for colouring margarine. The molecule of β-carotene is almost exactly twice as big as that of vitamin A, but it is an unsaturated hydrocarbon, not an alcohol.

The vitamin A activity of carotenes is not as great as that of retinol itself. β-carotene, for example, is only about one-sixth as effective as an equal weight of retinol. Other carotenes present in vegetable foods are converted to retinol even less efficiently and have half the activity of β-carotene. To allow for this variation in availability the vitamin A activity of foods is usually expressed in *retinol equivalents.*

SOURCES OF RETINOL AND CAROTENES   Retinol is found in animal tissues (especially liver) and dairy products. Fish-liver oils are the richest source and consumption of cod-liver or halibut-liver oil is a simple way of ensuring that a sufficient supply of the vitamin is obtained.

Carotenes are found in plant tissues and about one-third of the vitamin A activity of the average British diet is contributed by carotenes. Carrots, dark green vegetables and yellow fruits are all good sources of carotenes. Some green vegetables are rather poor sources however. Lettuce, with its pale green leaves, cabbage (especially the paler inner parts) and peas are not good sources, but spinach with its dark green leaves is well endowed with carotene. When vegetables are eaten not all the carotene they contain is absorbed and only a fraction of that absorbed may be converted into retinol. The source of carotene may affect its availability; for example, carotene is obtained from green vegetables more easily than from carrots, which have a comparatively fibrous structure.

Milk and milk products are also good sources of vitamin A, but the amount present depends upon the amount of carotene or retinol in the food eaten by the cow and so dairy produce is usually a richer source of the vitamin in summer, when fresh grass is available, than in winter. Cod-liver oil or synthetic retinol is incorporated into many animal foodstuffs, however, and so the difference is not as great as one might expect.

Table 12.2 shows the main sources of vitamin A activity in the diet. Retinol and carotenes are highly unsaturated and so they are easily destroyed by oxidation, especially at high temperatures. They are much more susceptible to oxidation after extraction from foods than when in animal or plant tissues. Losses due to oxidation during normal cooking processes are small, but considerable loss may occur during storage of dehydrated food if precautions are not taken to exclude oxygen. Apart from this sensitivity to oxidation, retinol and carotenes are reasonably stable and are only slowly destroyed at the temperatures used in cooking food. They are also almost insoluble in

Table 12.2 Average values for vitamin A content of foods in μg/100 g

Food	Retinol equivalents
Foods supplying retinol	
Halibut-liver oil	900 000
Cod-liver oil	18 000
Liver, lambs, fried	30 500
Herring	45
Sardines, canned	7
Butter	1 000
Margarine	985
Cheese, Cheddar	363
Eggs, boiled	190
Milk	44
Foods supplying carotene	
Red palm oil	20 000
Carrots	2 000
Spinach, boiled	1 000
Lettuce	167
Tomatoes	100
Bananas	33

Foods with negligible vitamin A activity

Potatoes
Cooking fats, lard and suet
Bacon, pork, beef (trace) and mutton (trace)
Bread, flour and other cereals
Sugar, jams and syrups
White fish

water and so there is little or no loss by extraction during the boiling of vegetables.

RECOMMENDED INTAKES OF VITAMIN A   The amount of vitamin A required for the maintenance of normal health depends largely upon age and the recommended daily intakes are shown in Appendix I. The recommendations are expressed in terms of retinol equivalents to allow for the fact that much of the vitamin A activity of a mixed diet is provided by carotenes. The RDA for adults is 750 μg of retinol equivalents daily. During lactation this should be increased to 1200 μg daily. A smaller RDA is recommended for younger people.

Retinol is not soluble in water and an excess above the body's need is not excreted in the urine but accumulates in the liver. This is why animal liver is such a valuable source of the vitamin. The liver of a well-nourished person may contain sufficient retinol to permit subsistence for several months without further intake of retinol or carotene.

Because retinol accumulates in this way an excessive intake should be avoided. Mothers who give their babies vitamin A supplements in the form of fish-liver oil should take particular care not to exceed the recommended dose. Adults are not immune from the ill-effects of grossly excessive vitamin A intake: vitamin pill and health food addicts should beware!

Recommended intakes of vitamin A are expressed in terms of a daily amount, as with other nutrients. Day-to-day deficiencies of the vitamin are normally of no consequence, however, because of the liver's capacity to store the vitamin. A normal mixed diet which provides adequate amounts of other nutrients will almost certainly provide sufficient vitamin A to meet the recommended intake levels over a period, even if it does not do so every day.

EFFECTS OF RETINOL DEFICIENCY   A long-term deficiency of vitamin A may lead to a condition known as 'night blindness' which makes it difficult to see in a dim light. Normally one's eyesight is able to adapt to changes in illumination. This is why one is able to see one's surroundings after a short while in a cinema which at first seems very dark. Night blindness is caused by a shortage of a retinol derivative called *rhodopsin* (or visual purple) which is essential for the proper functioning of the retina at the back of the eye. Night blindness is common in some parts of Asia and Africa where the diet is deficient in vitamin A.

An adequate intake of vitamin A is essential for the maintenance of healthy skin and other surface tissues such as the mucous membranes. Long-term deficiency may cause an eye disease known as *xerophthalmia* in which dead cells accumulate on the surface of the eyes causing them to become dry and opaque. The cornea may become ulcerated and infected– a condition known as *keratomalacia* – and blindness is a common sequel. Although the cause of the condition is known and preventive measures can easily be taken it is estimated that up to 20 000 children become blind in this way every year in underdeveloped countries.

Shortage of vitamin A in infancy during the formation of teeth may produce poor teeth and even after they have been formed a lack of vitamin A may affect the enamel.

Cholecalciferol or vitamin D

Naturally occurring vitamin D is more precisely referred to as vitamin D_3 or *cholecalciferol*. Another form of vitamin D, known as vitamin D_2 or *ergocalciferol*, can be made by exposing the compound ergosterol, which is found in fungi and yeasts, to ultraviolet light. The name vitamin D_1, which was at one time used for a mixture of substances displaying vitamin D activity, is not now used. Cholecalciferol, vitamin D₃, is the only one of these compounds of dietary importance and in what follows it will be referred to simply as vitamin D.

Vitamin D is a white crystalline solid which, like vitamin A, is freely soluble in oils and fats but insoluble in water. Vitamin D is stored in the liver and since it is insoluble in water an excess cannot easily be removed in the urine and it accumulates in the liver. Too much vitamin D can be harmful and in this respect also it resembles vitamin A.

SOURCES OF VITAMIN D Vitamin D is not widely distributed and it is found almost exclusively in animal foods. Fish-liver oils are the richest natural source. The main sources in an average British diet are margarine, vitamin-fortified breakfast cereals, butter, eggs and fatty fish. Synthetic vitamin D is added to all margarine sold in Britain, except that used by the baking industry, and also to many baby foods. Table 12.3 summarizes the important sources of vitamin D.

Vitamin D is not destroyed to any significant extent during normal cooking processes.

As well as being obtained from the diet vitamin D is produced in the body by the action of sunlight on a chemically related compound present in the skin.

FUNCTIONS AND RECOMMENDED INTAKES OF VITAMIN D Vitamin D is concerned with the absorption of calcium and phosphorus by the body. In the absence of the vitamin the body is unable to make use of these elements and they are lost in the faeces. Phosphorus and calcium are both needed to form bones as explained earlier (see p. 228). Deficiency of vitamin D causes rickets in the young and the related bone disease osteomalacia in those in whom bone growth has ceased. Rickets is characterized by curvature of the bones in the limbs and other distressing symptoms of improper bone formation. It was formerly very common in Britain but is now rare except in children of Asian immigrants amongst whom it is a continuing problem.

It has been recognized for about 90 years that rickets is more prevalent in industrial areas in the temperate regions where sunlight is

Table 12.3 Average values for vitamin D content of foods in μg/100 g

Halibut-liver oil	up to 10 000	Butter	up to 2
Cod-liver oil	200–750	Eggs, whole	1–1.5
Herrings, sardines	5–45	Cheese	0.3
Salmon	4–30	Milk (Summer)	0.1
Cornflakes (fortified)	2.8	Egg-yolk	4–10
Margarine	2–2.5		

deficient. The disease has been successfully treated by exposure to the maximum amount of sunlight and later it was shown that any other source of ultraviolet light (e.g., a 'sun-lamp') is also efficient. The reason for this is now clear; it is because the ultraviolet light converts a provitamin present in the tissues of the skin into the vitamin, which is then able to carry out its function.

No amount of vitamin D or exposure to sunlight will prevent the development of rickets if the diet contains insufficient calcium. The disease may be caused by calcium shortage, vitamin D shortage or lack of sunlight.

Rickets is often found in conjunction with dental caries because vitamin D is also necessary for proper calcification of teeth. Not only does vitamin D assist in the formation of healthy teeth but it can also help to prevent the development of dental caries in existing teeth (although other factors are also involved). Adults who do not receive sufficient vitamin D may develop osteomalacia. In this disease calcium is lost from the bones and softer tissue takes its place.

Most people obtain sufficient vitamin D through the action of sunlight on their skin. There are no recommended daily amounts for adults. There are, however, certain vulnerable groups for whom RDA are given as a safeguard against insufficient exposure to sunlight. For babies under one year old the RDA is 7.5 μg daily, and for children 1–4 years old and for pregnant and nursing women the RDA is 10 μg daily; the same amount is suggested for those who are housebound.

A high intake of vitamin D can be harmful. Too much calcium may be absorbed from the diet and the excess is deposited in the kidneys where it causes damage and eventually death may result. There is a particular danger that babies who are given extra vitamin D in the form of fish-liver oil could receive an excessive intake unless the recommended dosage is carefully observed.

Tocopherols or vitamin E

Vitamin E is the name given to α-*tocopherol* (a light yellow oil of formula $C_{29}H_{50}O_2$) and a group of fat-soluble saturated and unsaturated alcohols closely related to it.

Most plant tissues contain some vitamin E and vegetable oils such as maize oil, soya bean oil and, especially, wheatgerm oil are good sources. Meat and other foods of animal origin are poor sources.

Vitamin E is a natural antioxidant and tocopherols are permitted food additives (E307–309; see Chapter 18). Its presence in unsaturated vegetable oils probably indicates that its function in plant tissues is to protect these easily oxidized oils. It may also perform a similar function in the human body by protecting easily oxidized nutrients such as unsaturated fatty acids, retinol and vitamin C from oxidation.

There is evidence that vitamin E can help to prevent the occurrence of a serious eye disease called retrolental fibroplasia which affects premature babies. This disease is caused by the action of oxygen on the developing blood vessels in the baby's eyes.

Lack of vitamin E renders male rats sterile: female rats deficient in the vitamin can conceive but the pregnancy is interrupted and no live young are born. There is no conclusive evidence that vitamin E influences human fertility, although various claims have been made. Vitamin E supplements are sometimes taken in the optimistic belief that the vitamin will delay ageing, improve the condition of the skin or enhance sexual capabilities. All of these hopes are unfounded: a condition of vitamin E deficiency has not yet been recognized in man and there is no recommended daily intake.

Naphthoquinones or vitamin K

Vitamin K comprises several closely related fat-soluble compounds derived from menadione (2-methyl-1,4-naphthoquinone) all of which display vitamin K activity. It is essential for normal clotting of the blood; without vitamin K the liver is unable to synthesize *prothrombin* which is the precursor of the blood-clotting enzyme *thrombin*.

Vitamin K is present in most foods but green leafy vegetables are the richest source. Bacterial synthesis in the bowel provides humans with vitamin K in addition to that obtained from foodstuffs.

In most cases the amount made available in this way is sufficient to supply the body's requirements. There is little danger of vitamin K deficiency in a person who eats a normal diet and there are no recommended daily intakes.

WATER-SOLUBLE VITAMINS

The B group of vitamins

The B group of vitamins comprises several vitamins which have similar functions and which are often found together in foods. In the body they are largely concerned with the release of energy from food. They are all soluble, to a greater or lesser extent, in water and since the body lacks the capacity for storing them any excess over immediate requirements is excreted in the urine. The members of the B group of vitamins are:

Thiamine, or vitamin B_1
Riboflavin, or vitamin B_2
Niacin (nicotinic acid and nicotinamide)
Pyridoxine, or vitamin B_6
Pantothenic acid

Biotin
Cobalamin, or vitamin B_{12} (formerly called *cyanocobalamin*)
Folic acid

Thiamine or vitamin B_1

This is a white, water-soluble crystalline solid. The thiamine molecule has a complex structure which includes an amino ($-NH_2$) group and a hydroxyl group. Like all amines, thiamine forms salts with acids. Thiamine hydrochloride is made on quite a large scale for use in fortifying flour. The hydroxyl group can be esterified and thiamine is found in foods and in the body as its pyrophosphate ester.

SOURCES OF THIAMINE   Thiamine plays an essential part in the utilization of carbohydrates by living cells. As a result it is present in all natural foods to some extent. Unfortunately, it is often absent from processed foods because it has been removed or destroyed in the preparation of the food for the market. Polished rice and low extraction rate flour from which thiamine has been largely removed, sugar, refined oils and fats and alcoholic beverages are examples of foods which contain little or no thiamine. Nevertheless, thiamine is still present in a fairly wide range of foodstuffs as Table 12.4 shows. Most of the thiamine in an average British diet comes from fortified bread and cereal products, potatoes, milk and meat.

Because thiamine is so soluble in water, as much as 50 per cent may be lost when vegetables are boiled. Potatoes boiled in their skins retain up to 90 per cent of their thiamine compared with a retention of about 75 per cent in the case of boiled peeled potatoes. Thiamine decomposes on heating though it is fairly stable at the boiling point of water and little loss occurs at this temperature in acid conditions. In neutral or alkaline conditions breakdown is more rapid. Foods which have been subjected to higher temperatures, as in roasting, or in 'processing' during canning, may have a large proportion of their thiamine destroyed. Meat loses about 15–40 per cent of its thiamine when boiled, 40–50 per cent when roasted and up to 75 per cent when canned. When bread is baked some 20–30 per cent of its thiamine

Table 12.4   Average values for thiamine content of foods in mg/100 g

Marmite	5.00	Bread (wholemeal)	0.26
Cornflakes (fortified)	1.00	Bread (brown)	0.24
Oats (porridge)	0.90	Bread (white)	0.18
Pork (cooked)	0.60	Mutton or lamb (cooked)	0.10
Bacon (fried)	0.57	Eggs (raw or boiled)	0.09
Kidney (pigs, fried)	0.41	Potatoes (boiled)	0.20
Peas (frozen, boiled)	0.30	Milk (pasteurized)	0.05
		Cheese (Cheddar)	0.04

present in the flour may be destroyed by the moist heat. In cakes made with baking powder all the thiamine may be destroyed by reaction with the baking powder. Some preservatives also destroy thiamine and sulphites, those which are used in sausages are particularly prone to cause thiamine breakdown.

The deficiency disease beriberi, which is caused by a deficiency of thiamine, is almost unknown in Britain, but it is still common in Far Eastern countries, where the standard of living is very low. In these countries the main article of diet is polished rice, which contains little thiamine. The husk and silverskin of the rice grains are removed to improve its palatability and keeping properties. Unfortunately, however, this process also removes most of the thiamine. If the rice grains are parboiled before the husks are removed much of the thiamine is absorbed and retained by the endosperm. Loss of thiamine occurs in the same way during the milling of wheat to produce flour of low extraction rate and it is for this reason that synthetic thiamine is now added to all such flour produced in Britain. Enriched flour provides about 25 per cent of the thiamine in the average diet in Britain.

FUNCTIONS AND RECOMMENDED INTAKES OF THIAMINE Thiamine is esterified with pyrophosphoric acid in the body to give *thiamine pyrophosphate*, or cocarboxylase, which is an essential coenzyme involved in the utilization of carbohydrates. In the absence of sufficient thiamine, carbohydrate metabolism is upset and this produces a check in the growth of children together with a loss of appetite and other symptoms such as irritability, fatigue and dizziness. Prolonged and severe deficiency can cause the disease beriberi. Several types of this disease are known but all of them are associated with loss of appetite, leading to reduced food intake and, in time, emaciation and an enlargement of the heart occur. The nervous system is badly affected and this may produce partial paralysis and muscular weakness.

The amount of thiamine needed in the diet depends upon the quantity of carbohydrate consumed. A diet which consists largely of carbohydrate requires more thiamine than more varied diets, and because of this dependence on carbohydrate intake it is difficult to estimate the daily requirements of thiamine. The situation is made more complicated because some thiamine may be synthesized by bacteria in the large intestine and the amount available from this source varies from person to person and from time to time. The WHO advocates that thiamine intake should be related to the total energy content of the diet and that the appropriate relationship is 96 μg/MJ. The British RDAs (see Appendix I) conform closely to this numerical relationship. They vary from 0.3 mg for children under one year old to 1.3 mg for very active men.

In common with other water-soluble vitamins thiamine is not stored by the body and any excess over immediate requirements is rapidly excreted in the urine. A *regular* and adequate supply of the vitamin is thus essential.

Riboflavin or vitamin B₂

This is a yellowish-green fluorescent solid which, like thiamine, has a complex chemical structure. The molecule contains a complex heterocyclic ring system of carbon and nitrogen atoms combined with the sugar ribose. Riboflavin is synthesized commercially and is used in some countries for enrichment of food.

SOURCES OF RIBOFLAVIN   Riboflavin is widely distributed in plant and animal tissues and the more important sources are shown in Table 12.5. The main sources of riboflavin in the British diet are milk, meat, fortified breakfast cereals and eggs. It is found in beer but it is present in such small amounts that about nine pints would be required to provide the RDA for a man.

Riboflavin is only very slightly soluble in water and losses by solution during cooking are small. Heating causes little breakdown of riboflavin and little or no loss occurs during canning. Meat loses about one-quarter of its riboflavin during roasting. Greater losses occur if riboflavin is heated under alkaline conditions such as occur when sodium bicarbonate is added to the water used for boiling vegetables.

Although riboflavin is very stable to heat it is sensitive to light. This is not important with solid foods such as meat but serious losses can occur in milk. Up to three-quarters of the riboflavin present in milk may be destroyed by exposure to direct sunlight for three and a half hours. The substances produced when riboflavin breaks down in this way are oxidizing agents capable of reacting with, and totally destroying, the ascorbic acid (vitamin C) present in the milk. In addition the fats present in milk may be partially oxidized with the production of unpleasant 'off' flavours. Obviously it is not good practice to allow bottles of milk to remain for too long on the doorstep or, for that matter, in a brightly illuminated supermarket display cabinet.

Table 12.5 Average values for riboflavin content of foods in mg/100 g

Marmite	6.75	Milk	0.17
Liver (lambs, fried)	5.65	Bread (wholemeal)	0.06
Kidney (pigs, fried)	3.70	Bread (brown)	0.06
Cornflakes (fortified)	1.5	Bread (white)	0.03
Cheese (Cheddar)	0.50	Beer (keg biter)	0.03
Eggs (boiled)	0.47	Cabbage (boiled)	0.03
Beef (cooked)	0.33	Potatoes (boiled)	0.02

FUNCTIONS AND RECOMMENDED INTAKES OF RIBOFLAVIN   In the body, riboflavin is esterified with phosphoric acid or pyrophosphoric acid and forms part of two coenzymes concerned in a variety of oxidation–reduction processes in living cells. A deficiency of riboflavin produces a check in the growth of children and lesions on the lips and cracking and scaliness at the corners of the mouth may occur. The tongue and eyes may also become irritated.

When riboflavin is eaten it is stored temporarily in the liver until it is needed by the body. It is not possible to store large amounts in this way, however, and it is necessary for regular and adequate amounts to be eaten. The quantity of riboflavin needed for the maintenance of health is not known with certainty but it is believed to be related to the basal metabolic rate rather than (as with thiamine) to total energy content of the diet. The RDAs given in Appendix I have been calculated on this basis and should prove ample in all circumstances. The RDA for an adult man is 1.6 mg and for an adult woman 1.3 mg. During pregnancy and lactation the RDA is increased by 0.3 mg and 0.5 mg respectively. The riboflavin in the diet may be supplemented to a small extent by that produced by bacterial synthesis in the large intestine but not all of this is absorbed.

Niacin (nicotinic acid and nicotinamide)

The B group vitamin known as niacin exists in two forms; a pyridine carboxylic acid, nicotinic acid, and its amide, nicotinamide. Unlike most other members of the B group of vitamins these two substances have simple chemical structures. The acid and the amide are equally effective as vitamins.

Nicotinic acid   Nicotinamide

Nicotinic acid was first prepared, long before its importance as a vitamin was realized, from nicotine. In food, however, it is not derived from nicotine nor is it produced during tobacco smoking. It was thought that the name nicotinic acid might give the general public an undesirable impression of close relationship to nicotine and this is why the name niacin has been adopted. It is used to refer to both nicotinic acid and nicotinamide, although the latter is also sometimes called niacinamide. Niacin is manufactured on a fairly large scale for use in enriching flour.

SOURCES OF NIACIN   Niacin is found in both animal and vegetable tissues. The main sources of the vitamin in the average British diet are meat and meat products, potatoes, bread and fortified breakfast cereals. In Britain all flour (except wholemeal) is 'fortified' with added niacin (see p. 134) and about one-seventh of the RDA is obtained from bread and other flour products.

Some cereal products are fairly rich in niacin but unfortunately it is bound up in a complex with hemicelluloses called *niacytin*. This is not broken down during digestion and so the niacin is not available to the body. For this reason, therefore, unfortified cereal products must be regarded as poor sources of niacin. The amino acid tryptophan, which is present in cereals, can be converted by the body to niacin but the amount made available from cereals in this way is normally small.

Milk and eggs contain little niacin but their proteins are especially rich in tryptophan and so they serve as good sources of the vitamin. To allow for the presence of tryptophan in the diet the niacin content of foods is conveniently expressed in terms of *niacin equivalents* and for this purpose 60 mg of tryptophan are taken to be equivalent to 1 mg of niacin. The values given in Table 12.6 are expressed in this way.

Niacin is not easily decomposed by heating and it is only moderately soluble in water so that losses in cooking are small.

FUNCTIONS AND RECOMMENDED INTAKES OF NIACIN   Nicotinamide occurs in the body as part of two essential coenzymes concerned in a large number of oxidation processes involved in the utilization of carbohydrates, fats and proteins. The amount required is difficult to estimate because some niacin is produced and made available to the

Table 12.6   Average values for niacin equivalent content of foods in mg/100 g

Yeast extract (Marmite)	75.0	Cheese (Cheddar)	6.2
Beef extract (Bovril)	67.7	Cod (fried)	4.9
Liver (lambs, fried)	24.7	Eggs (boiled)	3.7
Cornflakes	21.9	Peas (frozen)	2.6
Kidney (pigs, fried)	20.1	Bread (brown)	1.9
Tuna (canned)	17.2	Bread (white)	1.8
Chicken (roast)	12.8	Bread (wholemeal)	1.7
Sardines (canned)	12.6	Potatoes	1.5
Bacon (grilled)	12.5		
Lamb (roast)	11.0		
Pork chop (grilled)	11.0		
Beef (roast)	10.2		
Corned beef	9.1		

body by microorganisms present in the large intestine. The amount of niacin required for the maintenance of good health is believed to be related to the basal metabolic rate and the RDA is about 11 times that for riboflavin after allowing for the likely contribution by tryptophan (see Appendix I). The RDA for an adult man is 18 mg and for an adult woman 15 mg. During pregnancy and lactation increased intakes of 3 mg and 6 mg respectively are recommended.

A severe deficiency of niacin can cause the disease pellagra which is characterized by dermatitis, diarrhoea and symptoms of mental disorder. Less severe deficiencies can produce one or more of these symptoms. Pellagra has long been associated, like many other deficiency diseases, with a low standard of living. In particular, pellagra results from subsistence on a diet consisting mainly of maize, and for this reason it was in the past particularly prevalent in the southern States of the USA, where maize was a staple food. The niacin present in maize is not available for the reasons already given and its proteins are deficient in tryptophan. Pellagra is not associated solely with consumption of maize, however, and it may occur whenever the intake of nicotinamide, or its precursor tryptophan, is insufficient. Some nicotinamide may be synthesized by microbial action in the large intestine but the amount absorbed is small.

Pyridoxine or vitamin B$_6$

Pyridoxine or vitamin B$_6$ is the name given to a group of three simple pyridine derivatives which have the structures shown:

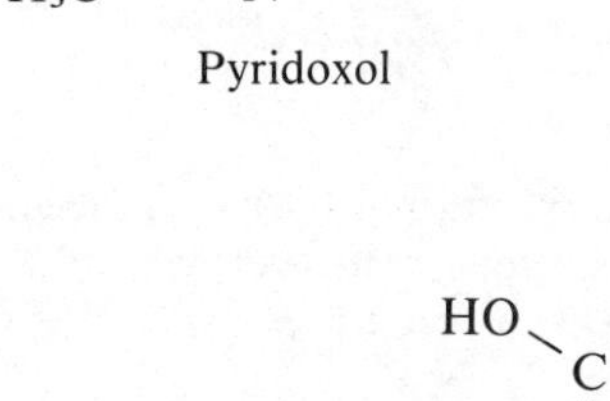

Pyridoxol

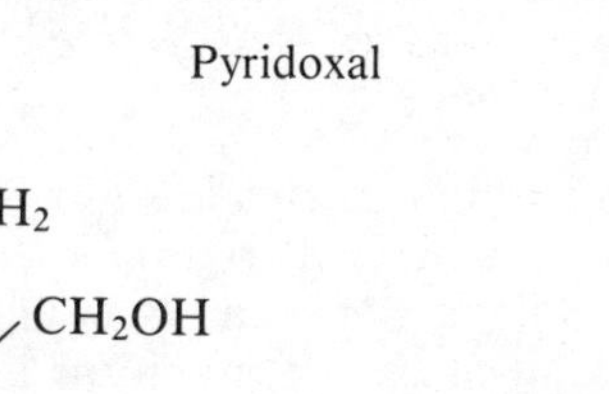

Pyridoxal

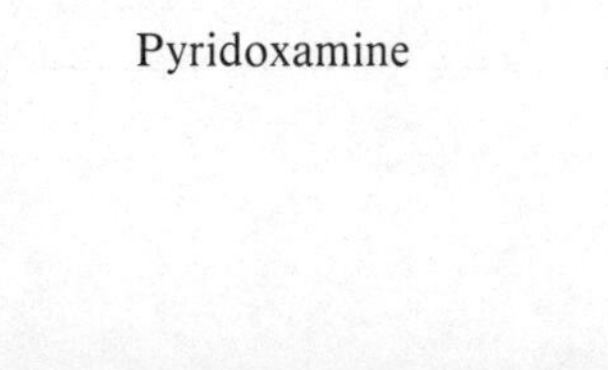

Pyridoxamine

All three compounds have been synthesized and are equally potent as vitamins. Vitamin B_6 is found in foods which contain the other B vitamins. The main sources in the diet are potatoes and other vegetables, milk and meat. The vitamin B_6 content of some foods is shown in Table 12.7.

Table 12.7   Average values for vitamin B_6 content of foods in mg/100 g

Wheat germ	0.95	Potatoes	0.25
Bananas	0.51	Baked beans	0.12
Turkey	0.44	Bread (wholemeal)	0.12
Chicken	0.29	Peas (frozen)	0.10
Fish (white)	0.29	Bread (white)	0.07
Brussels sprouts	0.28	Milk	0.06
Beef	0.27	Oranges	0.06

Symptoms of vitamin B_6 deficiency in animals can be produced by feeding them with a diet devoid of the vitamin. It is not easy, however, to do the same thing with humans although various skin lesions are reputed to be caused by vitamin B_6 deficiency. Infants fed on milk powders deficient in vitamin B_6 were found to suffer from convulsions but responded rapidly to treatment with the vitamin. It is believed that pyridoxine and pyridoxamine are converted in the body to pyridoxal, the phosphate of which functions as a coenzyme in protein metabolism. However, the precise function of vitamin B_6 in the maintenance of good health is not well established.

It is difficult to stipulate a recommended daily intake of vitamin B_6 because so little is known of its functions and the effects of a deficiency. However, it is distributed widely and a varied diet should provide 1 to 2 mg per day which appears to be sufficient for most people.

Pantothenic acid

This vitamin is a pale yellow oil with the structure shown:

$$HOCH_2C(CH_3)_2CHOHCONHCH_2CH_2COOH$$

Pantothenic acid

It is found in a wide variety of plant and animal tissues, indeed the name is derived from greek words meaning 'from everywhere'. It is soluble in water and is rapidly destroyed by treatment with acids or alkalis or by heating in the dry state.

Pantothenic acid is an essential constituent of coenzyme A which is concerned in all metabolic processes involving removal or addition of an acetyl group ($-COCH_3$). Such processes are of great importance in the many complex transformations occurring within the human body, especially those concerned with the release of energy from carbohydrate and fat.

Pantothenic acid is undoubtedly of fundamental importance as a coenzyme and symptoms of deficiency of the vitamin have been produced in numerous species of animals and also in man by diets devoid of pantothenic acid. The daily requirements of pantothenic acid are thought to be about 5–10 mg per 10 MJ of energy intake but no hard and fast recommendations are given. It is so widely distributed that a normal daily diet, which contains 10–20 mg, should be adequate and there is no danger of a deficiency.

Biotin

Biotin is another widely distributed vitamin which is required in minute amounts as a coenzyme involved in the metabolism of fats and carbohydrates. Many foods contain biotin. Liver and kidney are good dietary sources (but see below) and smaller amounts are found in egg-yolk, milk and bananas.

Such small amounts of biotin are required by the body that sufficient may be produced by the microorganisms present in the large intestine. Consequently dietary sources are not of great importance and there is no RDA.

Raw egg-white contains a protein or protein-like substance called *avidin* which combines with the biotin of the yolk to form a stable compound. This is not absorbed from the intestinal tract and so the biotin is not available to the body. Avidin can also render unavailable the biotin in other foods. This avidity for biotin is not shown by cooked egg-white.

Cobalamin or vitamin B₁₂

Cobalamin is a deep red crystalline substance of molecular formula $C_{63}H_{90}O_{14}N_{14}PCo$, and it has by far the most complex chemical structure of any vitamin. The presence of a cobalt atom in the molecule is a noteworthy feature.

Cobalamin is found in small amounts in all animal tissues but it is absent from foods of vegetable origin. It is required by the body in extremely minute amounts and vegetarians usually obtain sufficient from eggs and milk. Vegans, who abstain completely from foods of animal origin, including dairy foods, may suffer from a deficiency. Fortunately, cobalamin can be made from a mould used to produce the antibiotic streptomycin and supplies are available for vegans from this source.

Cobalamin plays a part in the production of nucleic acids and in the complex process of cell division in the body. It is especially important, in conjunction with folic acid and iron, for the formation of red blood cells. It is also involved in the formation of the *myelin* tube or sheath which surrounds each nerve fibre.

The amount of cobalamin required for the maintenance of good health is not known with complete certainty and there is no RDA specified in Britain. Only minute amounts are needed, however, and as little as 1 μg per day is thought by the WHO to be sufficient for adults. This very small amount is almost certain to be present in all diets except the rigorous vegetarian dietary regime of vegans. Milk, for example, which is not a particularly rich source, contains over 2 μg per pint. The body holds good reserves of cobalamin and it has been estimated that enough is available to last for up to five years in the absence of a dietary intake.

Table 12.8 shows the average cobalamin content of some foods: the main sources in the average British diet are meat, offal and milk. Cobalamin is fairly stable to heat and is only slightly soluble in water so losses during cooking are small.

Some people who are unable to absorb cobalamin from their diet suffer from a serious disease known as *pernicious anaemia*, in which extreme anaemia is accompanied by degeneration of the nerve tracts in the spinal cord. At one time this disease was invariably fatal but it is now treated very successfully by injection of hydroxocobalamin at three-monthly intervals. Pernicious anaemia is caused by the absence from the gut of an *intrinsic factor* which is essential for the absorption of cobalamin. It is *not* a deficiency disease because, if the intrinsic factor is absent, it will occur even when the diet contains sufficient cobalamin.

Table 12.8 Average values for cobalamin content of foods in μg/100 g

Liver (lambs)	54.0	Eggs	1.7
Liver (pigs)	23.0	Cheese (Cheddar)	1.5
Marmite	8.0	Milk	0.4
Beef, lamb, pork	2.0		
Fish (white)	2.0		
Cornflakes (fortified)	1.7		

Folic acid

Folic acid is the name given to a group of closely related compounds derived from pteroylglutamic acid. It is involved in the body, in conjunction with cobalamin, in the production of nucleic acids and, in particular, in the formation of red blood cells.

A deficiency of folic acid causes a particular type of anaemia called *megaloblastic anaemia*. This is similar to the anaemia caused by non-absorption of cobalamin but it is not accompanied by the degeneration of the nerve cells which is a feature of pernicious anaemia. Pregnant women are prone to develop this type of anaemia and it has been reported that it occurs in up to 25 per cent of pregnant women in Britain. Folic acid deficiency during pregnancy is likely to lead to premature birth and low birthweight. If a mother's diet is deficient in folic acid before conception or during the early stages of pregnancy there is evidence of an increased risk that the baby will be born with neural tube defects such as spina bifida.

Surprisingly, there are no British RDAs for folic acid. The WHO recommends a daily intake of 170 μg for adult women and 200 μg for adult men.

Folic acid is found in small amounts in a wide variety of foods; liver, green vegetables, potatoes, Marmite and fortified cornflakes are good sources. Fruit and vegetables contribute about 40 per cent of the folic acid in an average British diet and bread about 20 per cent. Folic acid is easily destroyed during cooking and a good deal can be lost in the water used for cooking vegetables. Even greater losses occur if sodium bicarbonate is added to the water to preserve the colour of green vegetables.

Ascorbic acid or vitamin C

Ascorbic acid, or vitamin C, is a white water-soluble solid of formula $C_6H_8O_6$. In spite of the name ascorbic acid the molecule does not contain a free carboxyl group. It is really a *lactone* formed from the free acid by loss of water between a carboxyl group on one carbon atom and a hydroxyl group on another. It has the structure shown:

$$
\begin{array}{cc}
\begin{array}{c}
\text{HO} \quad \text{OH} \\
| \quad\quad | \\
\text{C} = \text{C} \\
| \quad\quad | \\
\text{HOCH}_2\text{CHOH} - \text{HC} \quad \text{CO} \\
\diagdown \quad \diagup \\
\text{O}
\end{array}
&
\begin{array}{c}
\text{HO} \quad \text{OH} \\
| \quad\quad | \\
\text{C} = \text{C} \\
| \quad\quad | \\
\text{HOCH}_2\text{CHOH} - \text{HC} \quad \text{CO} \\
| \quad\quad | \\
\text{HO} \quad \text{OH}
\end{array}
\\
\text{Ascorbic acid}
&
\begin{array}{c}
\text{Free acid corresponding to} \\
\text{ascorbic acid}
\end{array}
\end{array}
$$

Lactones behave much like acids and for many purposes can be regarded as acids. Ascorbic acid has the sharp taste usually associated with acids and will form salts. It is optically active and is dextrorotatory; laevorotatory ascorbic acid is also known but it has little or no vitamin activity. Ascorbic acid is a good reducing agent and consequently it is easily oxidized. The oxidation product *dehydroascorbic acid* is easily reconverted into ascorbic acid by mild reducing agents, and because this reduction can be accomplished by the body it is as active as ascorbic acid itself. It is, however, less stable than ascorbic acid and only small amounts are present in foods.

Of all the vitamins, ascorbic acid is the most easily destroyed by oxidation, and in extracts, juices and foods with cut surfaces, it may be oxidized by exposure to air. The oxidation is enzymically catalysed by oxidases which are contained within the cells of foodstuffs and are set free on cutting, chopping or crushing. The rate of oxidation is greatly accelerated by heat (if the temperature is not high enough to destroy the oxidases), by alkalis and especially by traces of copper which catalyses the oxidation. The rate of oxidation is diminished in a weak acid solution and by storage in the cold.

SOURCES OF ASCORBIC ACID   Ascorbic acid occurs mainly in foods of plant origin. Fruits are usually good sources but many popular eating apples, pears and plums supply negligible amounts. Green vegetables and potatoes are the most important sources of ascorbic acid in the British diet. Table 12.9 lists the average ascorbic acid content of the more important sources of this vitamin.

The amount of ascorbic acid present in vegetables is greatest in the periods of active growth during spring and early summer. Storage decreases the ascorbic acid content and this can be clearly seen in the figures given for potatoes in Table 12.9.

Table 12.9   Average values for ascorbic acid content of foods in mg/100 g

Blackcurrants	200	Potatoes (raw, new)	30
Rosehip syrup	175	,, raw, Oct–Nov	20
Sprouts (raw)	87	,, ,, Dec	15
(boiled)	41	,, ,, Jan–Feb	10
Cauliflower (raw)	64	,, ,, March on	8
(cooked)	20	,, (boiled)	6
Cabbage (raw)	55	Apples, eating, (raw)	5
(boiled)	20	Lettuce (raw)	15
Spinach (raw)	60	Bananas (raw)	10
(boiled)	25	Beetroot (boiled)	5
Watercress (raw)	60	Onions (raw)	10
Strawberries	60	(boiled)	6
Oranges (raw)	50	Carrots (raw)	6
Lemons (juice)	50	(boiled)	4
Grapefruit (raw)	40	Plums (raw)	3
Peas (raw)	25	(stewed)	2
(boiled)	15	Pears (raw)	3
(dried boiled)	trace	(stewed)	2
Tomato (raw or juice)	20	(canned)	1
Liver (lambs, fried)	19	Cow's milk (fresh)	2
		Human milk	5

Potatoes contain less ascorbic acid than green vegetables but such large quantities of them are eaten that they constitute an important source of this vitamin. A normal serving of boiled new potatoes provides about 90 per cent of the recommended daily intake of ascorbic acid and about one-quarter of the ascorbic acid in the British diet is derived from potatoes – roughly as much as that from all other vegetables. For many years potatoes were the main source of ascorbic acid in Britain but fruit and fruit juices are now the main source.

As much as 75 per cent of the ascorbic acid present in green vegetables may be lost during cooking. This loss can be avoided by eating raw green vegetables in salads but the amounts which can conveniently be eaten in this way are comparatively small and more ascorbic acid may be obtained by eating a larger quantity of cooked vegetables. For example, 25 g of lettuce, which is a convenient serving, provides 4 mg of ascorbic acid compared with 23 mg provided by 100 g of cooked cabbage. Raw cabbage is a much better source of ascorbic acid than lettuce, 25 g provides about 13 mg of the vitamin.

Cow's milk has only about one-quarter or one-third the ascorbic acid content of human milk and some of this is destroyed during pasteurization. Exposure of milk to sunlight also causes diminution of its ascorbic acid content and this change is brought about by the breakdown products of riboflavin (see p. 254). It is important that babies, and particularly those fed on cow's milk which has been boiled, should be provided with other sources of the vitamin. Concentrated orange juice, rose-hip syrup or blackcurrant juice are attractive additional sources of the vitamin. When babies progress to a mixed diet there is less need for such supplements and at two years of age a normal diet should provide sufficient ascorbic acid.

Canned fruits and vegetables vary considerably in their ascorbic acid content but some, for example, tomatoes, are good sources of the vitamin. Some loss of ascorbic acid is inevitable in canning but good quality canned fruits and vegetables often contain more of this vitamin than 'fresh' fruit and vegetables cooked at home. This is because they are canned while fresh and cooked under carefully controlled conditions.

Foods such as yeast, egg-yolk, meats and cereals, which are rich in B vitamins are usually devoid of ascorbic acid, but liver and kidney are exceptions.

LOSS OF ASCORBIC ACID IN COOKING   It has already been mentioned that losses of ascorbic acid occur during storage of fruits and vegetables. Some loss also occurs during the preparation and cooking of foods. This is due partly to oxidation and partly to solution in the water used for cooking. To avoid undue losses of this vitamin, vegetables and fruit should not be crushed or finely chopped before cooking, as this sets free oxidases which catalyse the oxidation of ascorbic acid by the air. There is evidence to show that loss of ascorbic acid is greater when vegetables are cut with a blunt knife than with a sharp one, presumably because the former ruptures more cells by crushing and hence sets free more oxidases than the latter.

If vegetables are put in cold water which is brought to the boil the dissolved oxygen in the water will, in the presence of the oxidases, destroy a substantial amount of the ascorbic acid present. The oxidases are most active at about 60–85°C and above this temperature are quickly inactivated. It is best to place vegetables in boiling water because this contains no dissolved oxygen and the oxidases are rapidly destroyed at this temperature. With potatoes, for example, it has been found that cooking in this way causes only half the reduction in ascorbic acid content experienced with the normal method of cooking.

A minimum quantity of water should be used when cooking vegetables so that large amounts of the vitamin are not dissolved. This is particularly important with leafy vegetables, such as cabbage, because of the large surface area from which vitamin losses may occur. Vegetables which are completely immersed in water when being cooked may lose up to 80 per cent of their ascorbic acid. If they are only one-quarter covered by water only about half as much ascorbic acid is lost. With potatoes, which have a smaller surface area per unit weight, and in which gelatinization of the starch prevents diffusion of the ascorbic acid, the quantity of water used in cooking does not greatly affect the amount of ascorbic acid lost. Alkaline conditions should be avoided during cooking and the addition of bicarbonate of soda to green vegetables to preserve or intensify their colour should be avoided, although small quantities may be used to reduce the sourness of acid fruits without serious diminution in the ascorbic acid content.

Ascorbic acid is very rapidly oxidized in the presence of trace amounts of copper. For this reason copper or copper-alloy cooking vessels should not be used for cooking foods rich in vitamin C.

Finally, cooked food should not be kept hot longer than is absolutely necessary before serving because this can destroy almost all the ascorbic acid. It has been shown that a loss of 25 per cent of the ascorbic acid of hot cooked food occurs in 15 minutes and 75 per cent in 90 minutes. This is likely to be the greatest cause of loss of ascorbic acid in restaurants and hotels and in canteens where cooked food is received from a central kitchen.

FUNCTIONS AND RECOMMENDED INTAKES OF ASCORBIC ACID   The functions of ascorbic acid in the body are not known with certainty but it has been shown to be necessary for the formation of the intercellular connecting protein collagen. The cells of the body concerned in the formation of bone and the enamel and dentine of teeth lose their normal functional activity in the absence of ascorbic acid.

A lack of ascorbic acid in the diet causes a condition known as scurvy, which is characterized by haemorrhages under the skin and in other tissues and swollen and spongy gums from which the teeth are easily dislodged or may fall out. Scurvy in infants is associated with great tenderness and pain in the lower limbs together with changes in the bone structure which are not found in adult scurvy. The disease has been known for hundreds of years and was formerly rife among sailors and others whose diet was deficient in ascorbic acid. The cause of the disease was not known but in the course of time it was noted that consumption of fresh foods, particularly vegetables and fruits, could prevent and cure it. Scurvy is almost unknown in Britain at present although it still occurs occasionally.

A deficiency of ascorbic acid not severe enough to cause scurvy is thought to increase the susceptibility of the mouth and gums to infection and to slow down the rate at which wounds and fractures heal. Increased susceptibility to many kinds of infection, including the common cold, has been attributed to a shortage of ascorbic acid, but though this is known to be true for guinea-pigs it is not at all certain that the same applies to human beings. It is known, however, that increased amounts of the vitamin are needed by the body when suffering from infectious diseases. Ascorbic acid also assists in the absorption of iron by promoting its reduction to the ferrous state.

The amount of ascorbic acid needed for the maintenance of good health has been a subject of much controversy. An intake of 10 mg per day has been found to be sufficient to protect adults against symptoms of scurvy and even to cure the disease. If more than 10 mg is taken the body goes on absorbing it up to about 100 mg per day. Beyond this point, the so-called saturation level, the excess is excreted in the urine. The British RDA for adults is 30 mg and this is a compromise between these two extremes. During pregnancy and lactation the RDA is increased to 60 mg. The RDA for other age groups is shown in Appendix I.

SUGGESTIONS FOR FURTHER READING

BIRCH, G.G. AND PARKER, T. (Eds) (1974). *Vitamin C* (Symposium Report). Applied Science, Barking.

MARKS, J.A. (1945). *Guide to the Vitamins – their role in health and disease.* Medical and Technical Publishing Co., Lancaster.

STEIN, M. (Ed.) (1971). *Vitamins.* Churchill Livingstone, Edinburgh.

CHAPTER 13

Fruit and vegetables

Fruit and vegetables form an important part of the diet and are usually regarded as 'good' foods. They are major sources of vitamin C, folic acid and dietary fibre but, as Tables 13.1 and 13.2 show, they are not, in general, rich in other nutrients.

FRUIT

Botanically, a fruit is a matured ovary of a flower, including the seed (or seeds) and any part of the flower remaining attached to it. Fruits which conform strictly to this definition are called *true fruits* and they include nuts, legumes, berries and drupes (see below). Fruits such as apples, pears and strawberries which include some other part of the flower which has enlarged as well as the matured ovary are called *false fruits*.

Botanists regard peas, tomatoes, peppers, aubergines and cucumbers as fruits. Most people, however, have a more restricted idea of what constitutes a fruit and look upon them as vegetables. We have come to regard as fruits the succulent parts of plants which are characterized by a sweet or acid taste and a distinct flavour. This somewhat vague description — it hardly merits being called a definition — is more suitable for our hybrid scientific/culinary purposes than the more restrictive botanical definition.

Types of fruit

Hundreds of different fruits are known and each one may exist in a number of varieties. There are over 6000 different varieties of apples, for example, all originating from the wild crab-apples *Malus pumila*, and *Malus silvestris*.

Fruits can be classified in a number of ways but a simple method, which is satisfactory for our purposes, is based on their physical characteristics.

Pomes have a seed-bearing compartmented core surrounded by a firm fleshy body: apples and pears are pomes.

Drupes have a stone or nut embedded in edible flesh: cherries, plums, peaches and olives are obvious examples.
Berries have seeds enclosed in a pulp: grapes and oranges are berries.

Pomes

The most important members of this class of fruit are apples and pears, both of which are members of the same botanical family – the rose family or *Rosaceae*. The core of an apple or pear is really the matured ovary consisting of five carpels each containing two seeds or 'pips'.

Apples are grown almost everywhere except in the very hottest and coldest areas of the world. Most apples are eaten raw and dessert apples, or eating apples, are usually small and sweet with a fragrant aroma and flavour. Cooking apples are usually large and green; they have a sharper taste than eating apples owing to the presence of more acid and they pulp easily on cooking. Apples are not particularly good sources of nutrients but they contain some vitamin C and about two per cent dietary fibre. About one-third of the dietary fibre is in the form of pectin which is commercially extracted from apple pulp for use as a gelling agent (see p. 108). Large quantities of apples are converted to cider.

Pears are closely related to apples but they differ in flavour and often have gritty particles or 'stone cells' embedded in their flesh. Pears are roughly equivalent to apples nutritionally. Just as apple juice is used for making cider, pear juice is made into perry.

Drupes

Drupes are juicy fruits containing one seed surrounded by a hard woody layer which together form a stone.

Plums grow in most parts of the world and are eaten as a fresh dessert fruit and used for jam manufacture or canning. Dried plums are known as *prunes*.

Cherries vary in colour from yellow to black and in acidity from sour to sweet.

Peaches occur in two main varieties. Freestone peaches have a stone which can be easily separated from the flesh. The more popular cling peaches are firmer, more deeply coloured and less easily damaged than the freestone variety. The *nectarine* is a type of peach but it is smaller and hairless.

Blackberries, *raspberries* and *strawberries* are usually grouped together as soft fruits. Despite their names they are not berries. The first two are drupelets, i.e. aggregates of several 'mini-drupes' which are attached to the swollen tip, or receptacle, of the fruit stalk. Strawberries are false fruits: the true fruits are the tiny seed-like *achenes* or 'pips' on the outside of the fruit.

Blackberries and raspberries grow wild in most parts of Europe but they are also cultivated commercially. They are used mainly in cooking and for jam and jelly manufacture. Strawberries have a high vitamin C content. Fresh strawberries are usually eaten raw but jam manufacturers use large quantities of the preserved fruit.

Berries

CITRUS FRUITS   The more important citrus fruits are oranges, lemons, limes and grapefruit. Their seeds are contained in segmented sections or *carpels* of juicy flesh which are surrounded and protected by a tough skin or peel. Citrus fruits are rich in vitamin C, particularly in the pithy white layer or *albedo* found under the peel. Orange juice contains only about 20–30 per cent of the fruit's vitamin C and grapefruit juice has an even smaller proportion. Citrus fruits have a refreshing taste because of their high water content and the presence of citric acid and sucrose. The relative amounts of these two substances determines whether the fruit is sharp or sweet to the taste.

Oranges are by far the most important citrus fruit and orange juice is a major important source of vitamin C in the British diet. Oranges can be grown in tropical and subtropical countries but they develop their characteristic colour only where night temperatures are below 10°C during the ripening period. Green (but ripe) oranges may be treated with ethylene gas which artificially stimulates them (see p. 272) and brings out their orange colour. Bitter oranges, or Seville oranges, are used for marmalade manufacture. There are many different varieties of orange and also orange-like fruits such as tangerines, mandarins and satsumas which have characteristic flavours and a soft, loosely adherent skin. Some 'soft-citrus' fruits are hybrids of oranges and tangerines or mandarins.

Lemons are not eaten as dessert fruits because of their high citric acid content and consequent sharp taste.

Grapefruit derives its name from the fact that the fruit grows in vinelike clusters containing 3–18 fruits. Over 25 grapefruit varieties are known, some of which have a pink or reddish flesh which is caused by the presence of the carotenoid *lycopene*.

OTHER BERRIES   *Blackcurrants* are mainly notable for their high vitamin C content and one of their main uses is for the manufacture of blackcurrant juice, for use in fruit drinks rich in this vitamin. Large quantities are also made into jam or sun-dried to produce fruit for use in cakes. Currants also occur in red and white varieties.

Grapes are probably the world's most widely cultivated fruit crop. Only a small proportion of the grapes grown is used as food, the bulk being converted to wine by fermentation of the juice as explained in Chapter 10. Most of the grapes grown are varieties of a single European grape *Vitis vinifera*. Sun-dried grapes or *raisins* are usually made from seedless grapes. Grapes are not rich in nutrients and they contain only about one-tenth as much vitamin C as oranges.

Bananas are cultivated in most tropical countries but those imported into Britain come mainly from Central America and the Caribbean islands. The fruit is picked while it is unripe and during transportation to Britain by ship it is kept at about 12°C in ventilated holds to retard ripening. Before sale the fruit is ripened in temperature-controlled warehouses at 15°C. During ripening the banana skin changes from green to yellow in colour and the fleshy part of the fruit becomes softer and sweeter as its starch is almost completely converted to sugar.

Bananas are not particularly important as a food in Britain and they are normally eaten raw as a sweet dessert fruit. In some other parts of the world, however, boiled or steamed green bananas serve as an important source of starch.

Ripe bananas are little more than easily digested sources of carbohydrate. They provide energy and some vitamin C but very little else. They are roughly equal in nutrient density to potatoes but as much smaller quantities are eaten they are not important as providers of nutrients.

Nutritional value of fruit

Fruit is refreshing to eat and adds colour and flavour to the diet. Most fruits consist largely of water, however, and hence their nutrient content is low, as Table 13.1 shows. Their main importance is as a source of vitamin C and dietary fibre. About 40 per cent of the vitamin C content of the average British diet is provided by fruit and fruit juices. Most of this comes from citrus fruit and, especially, fruit juices and little from bananas, apples and pears which are the most popular fruits. Some fruits also contain vitamin A and they also make a small contribution to the mineral content of the diet. Despite their sweetness fruits contribute only about five per cent of the energy content of the average British diet.

Fruit contains dietary fibre in the form of cellulose, hemicellulose, pectin and protopectin. In fact, a good deal of what is left after the juice has been completely expressed from a fruit consists of dietary fibre. If we discount those parts not normally eaten – such as pips, cores and the peel of citrus fruits – the actual dry weight of fibre is quite small and usually amounts to less than two per cent of the weight

Table 13.1 Average values for nutrient content of fruit per 100 g of edible portion

Type of fruit	Energy (kJ)	Protein (g)	Carbohydrate (as monosaccharide) (g)	Water (g)	Calcium (mg)	Iron (mg)	Sodium (mg)	Vitamin A (retinol equiv.) (µg)	Thiamine (mg)	Riboflavin (mg)	Niacin (equiv.) (mg)	Vitamin C (mg)
Apples	196	0.3	11.9	84	4	0.3	2	5	0.04	0.02	0.1	5
Bananas	326	1.1	19.2	71	7	0.4	1	33	0.04	0.07	0.8	10
Blackcurrants	121	0.9	6.6	77	60	1.3	3	33	0.03	0.06	0.4	200
Cherries	201	0.6	11.9	82	16	0.4	3	20	0.05	0.07	0.4	5
Dates (dried)	1056	2.0	63.9	15	68	1.6	5	10	0.07	0.04	2.9	0
Figs (dried)	908	3.6	52.9	17	280	4.2	87	8	0.10	0.08	2.2	0
Gooseberries (cooked)	62	0.9	2.9	90	24	0.3	2	25	0.03	0.03	0.5	31
Grapes	268	0.6	16.1	79	19	0.3	2	0	0.04	0.02	0.3	4
Grapefruit	95	0.6	5.3	91	17	0.3	1	0	0.05	0.02	0.3	40
Melon	97	0.8	5.2	94	16	0.4	17	175	0.05	0.03	0.3	50
Oranges	150	0.8	8.5	86	41	0.3	3	8	0.10	0.03	0.3	50
Orange juice	161	0.6	9.4	88	12	0.3	2	8	0.08	0.02	0.3	25–45
Peaches	156	0.6	9.1	86	5	0.4	3	83	0.02	0.05	1.1	8
Pears	175	0.3	10.6	83	8	0.2	2	2	0.03	0.03	0.3	3
Pineapple (canned in juice)	194	0.5	11.6	77	12	0.4	1	7	0.08	0.02	0.3	20–40
Plums	137	0.6	7.9	85	12	0.3	2	37	0.05	0.03	0.6	3
Prunes	686	2.4	40.3	23	38	2.9	12	160	0.10	0.20	1.9	0
Raspberries	105	0.9	5.6	83	41	1.2	3	13	0.02	0.03	0.5	25
Strawberries	109	0.6	6.2	89	22	0.7	2	5	0.02	0.03	0.5	60

of the fruit. Nevertheless, about ten per cent of the dietary fibre in the average British diet comes from fruit. Consumption of additional fruit is a pleasant (but expensive) way of increasing the intake of dietary fibre.

When fruit is stored there is a progressive loss of vitamin C and up to 20 per cent of that present in citrus fruit may be lost in one month. There is usually also a loss of thiamine but as only small amounts are present initially the loss is not nutritionally significant. Small amounts of carotene may also be lost.

Canned fruit can be a good source of vitamin C, despite losses of 20–30 per cent which may occur during the canning process. This is because canning takes place promptly after picking when vitamin C content is at a peak. The carotene and thiamine content of fruit is not much reduced during canning. If canned fruit is stored for a long period some loss of vitamin C may occur but less than 15 per cent is usually lost during one year's storage.

Frozen fruit retains most of its vitamin C during the freezing process and subsequent storage. Some losses occur during the preliminary trimming, washing and blanching.

When fruit is dried to produce raisins, sultanas, prunes or dried currants its vitamin C content and about 50 per cent of its thiamine are destroyed. The carotene content is little affected.

Ripening and storage of fruit

RIPENING Most of our food can be eaten at any time in its life cycle: it does not pass through a period when its appeal to the palate is outstandingly greater than at other times. This is not so with fruit, however. Unripe fruit is practically inedible but when it ripens it changes dramatically to a condition in which flavour, colour and texture are all at a peak. Unripe fruit is often green but during ripening the green colour may be replaced by a yellow or reddish hue. The flesh softens and becomes sweeter and juicier and a characteristic 'ripe' flavour and odour develops.

The changes which occur during ripening are, in part, consequences of enzymic conversion of complex substances to simpler ones. Hard starch-packed cells are softened by conversion of starch to sugars which dissolve and enhance the sweetness and juiciness of the fruit. At the same time insoluble protopectin, which cements the plant cells tightly together, is converted to soluble pectin which connects them more flexibly.

When a fruit is harvested it is cut off from a supply of nutrients and growth ceases. Ripening may continue, however, and sometimes the fruit ripens more rapidly than if it had been allowed to continue growing.

Citrus fruits and other fruits which do not store starch obtain their sugar from the leaves of the plant on which they grow. Thus they do not become sweeter after picking.

Some fruit produces minute amounts of the unsaturated hydrocarbon ethylene (or ethene), $H_2C=CH_2$, during its growing period. Larger amounts are produced when the fruit is passing through the critical ripening period when cell activity is at a maximum. Not only is ethylene produced during ripening but it has been shown that it actively promotes ripening. If unripe fruit is exposed to small amounts of ethylene in air (less than one part per million) it initiates ripening and, in addition, stimulates the fruit to produce its own ethylene. This has been commercially exploited by using the gas to accelerate the ripening of fruit which has been stored in conditions which retard ripening. Citrus fruits do not produce ethylene when they ripen naturally but even these can be artificially ripened by using an ethylene-enriched atmosphere. In this way green but ripe oranges which, understandably enough, have little market appeal can be converted into the familiar orange-coloured fruit. Green bananas can be ripened in the same way.

STORAGE   Ripeness marks the end of a fruit's growth and the beginning of its death. Fruits which have a soft flesh and a thin skin pass rapidly from ripeness to rottenness and they can hardly be said to have a storage life at normal temperatures. Harder fruits, however, and those protected by a tough skin, can remain in good condition for several months.

Stored fruit deteriorates through normal ageing and shrinkage caused by loss of water. Undamaged fruit may remain edible for some time but eventually it will decay as a result of the continued activity of its own enzymes and attack by microorganisms. Soft fruit usually becomes inedible owing to the growth of moulds and yeasts on its surface. Mould and yeast spores are always present in the air (see Chapter 16) and on the skin of fruit and they are especially numerous near other mouldy fruit. As fruits age and shrink, especially if the skin has been damaged, sugary juice escapes and coats the skin with an ideal growth medium for moulds and yeasts. In general, fruit is not attractive to bacteria because of its acidity and lack of protein but mould growth occurs with ease.

Dark patches may appear on the surface of fruit which has been roughly handled. Ruptured cell membranes allow the contents of cells to mix and soft spots develop. The enzyme *phenolase* (or *polyphenoxidase*) oxidizes phenolic cell components to compounds which polymerize to dark-coloured polymers. This is how the familiar brown 'bruises' on the surface of damaged fruit are produced. Phenolase is not present in citrus fruits, melons or tomatoes and thus they do not develop brown bruises. Nevertheless, when cell rupture occurs and juice leaks they are prone to attack by moulds and can rapidly become inedible.

Fruit should be stored in conditions which slow down the action of its own destructive enzymes and the growth of moulds. It should be kept cool and in conditions where loss of moisture is minimized. Humidity should be high enough to prevent undue loss of water but not so high as to encourage mould growth. Temperatures just above freezing point are best for slowing down enzyme action and mould growth but are too low for many tropical fruits. Bananas, for example, can be irreversibly damaged if kept below 10°C for more than a few hours. The 'crisper' compartment of a refrigerator, where neither temperature nor humidity is too low, provides good conditions for storing most fruits.

Unblemished apples can be kept for long periods in cool surroundings but some varieties are damaged if the temperature falls below 3°C. Commercially their storage life can be considerably extended if they are kept in an atmosphere containing little oxygen (1–3 per cent) and added carbon dioxide (1–5 per cent). Ethylene production is suppressed and post-harvest ripening is delayed by this treatment.

VEGETABLES

The word 'vegetable' is used in several ways. In one sense – as employed in the phrase 'animal, vegetable or mineral' – it includes all substances of plant origin, including fruit. Here, however, we shall be more restrictive and concern ourselves only with those plants or parts of plants which are regarded as vegetables from a culinary and nutritional point of view. This still covers a large and somewhat miscellaneous collection of plants, including leafy plants (cabbage and lettuce), roots and tubers (carrots, parsnips and potatoes) and even flowers (cauliflower, broccoli and artichoke). The members of one entire group of vegetables (tomatoes, peppers, aubergines and cucumbers) are, in fact, fruits.

Vegetables in the diet

Most vegetables are cooked before they are eaten; those which are not are sometimes referred to as 'salad vegetables'. The distinction is far from hard and fast however, as many vegetables which are normally cooked before eating are eaten raw in salads and, conversely, 'salad vegetables' may be cooked. Thus grated carrot is often eaten in salads and tomatoes in soups and sauces. Some vegetables are more-or-less inedible until they are cooked. Cooking softens them by dissolving pectins and hemicelluloses and gelatinizing starch (see p. 106).

Uncooked starchy vegetables are difficult to digest because the starch granules resist the action of the digestive enzymes.

Vegetables constitute such a diverse group that it is difficult to generalize about their nutrient content. Table 13.2 lists some of the common vegetables and from this it will be seen that, in general, green leafy vegetables and potatoes are good sources of vitamin C and legumes are good sources of protein. Like fruits, however, vegetables are mainly composed of water. Nevertheless, because large quantities of vegetables are eaten, they provide about half the vitamin C, 15–20 per cent of the vitamin A, thiamine, niacin and iron and about ten per cent of the protein and energy in the average British diet. They also provide about half the fibre.

Some common vegetables

GREEN VEGETABLES   Green leafy vegetables such as cabbage and Brussels sprouts are nutritionally important as sources of vitamin C, β-carotene (provitamin A, see p. 245), folic acid and iron. The dark green outer leaves contain more vitamin C and β-carotene than the paler inner leaves. Cabbage and Brussels sprouts contain 3–4 per cent fibre but lettuce has only about a third as much. Green vegetables rapidly lose vitamin C when they are kept, and if they are boiled up to half their vitamin C content can be leached out into the cooking water. Whenever possible, therefore, fresh raw green vegetables should be eaten in preference to those which have been cooked.

Broccoli and spinach are good sources of vitamins A and C. Cauliflower is also rich in vitamin C but contains little vitamin A except in the green outer leaves. Onions and leeks qualify as green vegetables because their leaves are green. The white parts we eat contain vitamin C but no vitamin A and are chiefly of value for their flavour. Celery consists of leaf stalk and, apart from dietary fibre, contains little of nutritional value.

ROOT VEGETABLES   Carrots, turnips, swedes and parsnips are the more important root vegetables. The potato is also commonly regarded as a root vegetable although it is in fact a tuber (see below). Root vegetables are good sources of fibre and carrots are an important source of vitamin A (as β-carotene). The colourless turnips, swedes and parsnips contain no β-carotene but they are much richer than carrots in vitamin C.

POTATOES   Potatoes are *tubers*, that is swollen tips of underground stems which store energy as starch to feed new stems which grow from the 'eyes'. Potatoes are easily grown and give good yields. They have played an important part in the British diet for over 200

Table 13.2 Average values for nutrient content of vegetables per 100 g of edible portion

Type of vegetable	Energy (kJ)	Protein (g)	Fat (g)	Carbohydrate (as monosaccharide) (g)	Water (g)	Calcium (mg)	Iron (mg)	Sodium (mg)	Vitamin A (retinol equiv.) (µg)	Thiamine (mg)	Riboflavin (mg)	Niacin (equiv.) (mg)	Vitamin C (mg)
Baked beans	345	4.8	0.6	15.1	74	48	1.4	550	12	0.08	0.06	1.3	0
Beans, runner (boiled)	83	1.9	0.2	2.7	91	22	0.7	1	67	0.03	0.07	0.8	5
Beetroot (boiled)	189	1.8	0	9.9	83	30	0.4	64	0	0.02	0.04	0.4	5
Brussels sprouts (boiled)	75	2.8	0	1.7	92	25	0.5	2	67	0.06	0.10	0.9	40
Cabbage (raw)	92	2.8	0	2.8	88	57	0.6	7	50	0.06	0.05	0.8	55
Cabbage (boiled)	66	1.7	0	2.3	93	38	0.4	4	50	0.03	0.03	0.5	20
Carrots (old)	98	0.7	0	5.4	90	48	0.6	95	2000	0.06	0.05	0.7	6
Cauliflower (cooked)	40	1.6	0	0.8	95	18	0.4	4	5	0.06	0.06	0.8	20
Celery	36	0.9	0	1.3	94	52	0.6	140	0	0.03	0.03	0.5	7
Cucumber	43	0.6	0.1	1.8	96	23	0.3	13	0	0.04	0.04	0.3	8
Lentils (cooked)	420	7.6	0.5	17.0	72	13	2.4	12	3	0.11	0.04	1.6	0
Lettuce	51	1.0	0.4	1.2	96	23	0.9	9	167	0.07	0.08	0.4	15
Onion	99	0.9	0	5.2	93	31	0.3	10	0	0.03	0.05	0.4	10
Parsnips (cooked)	238	1.3	0	13.5	83	36	0.5	4	0	0.07	0.06	0.9	10
Peas (frozen)	307	6.0	0.9	10.7	78	35	1.6	2	50	0.30	0.09	1.6	12
Peas (canned, processed)	366	6.9	0.7	18.9	70	33	1.8	380	10	0.10	0.04	1.4	0
Potatoes	315	2.0	0.2	17.1	79	8	0.4	8	0	0.2	0.02	1.5	8–19
Potatoes (boiled)	322	1.8	0.1	18.0	80	4	0.4	7	0	0.2	0.02	1.2	5–9
Potato crisps	2224	6.3	35.9	49.3	3	37	2.1	550	0	0.19	0.07	6.1	17
Potatoes (fried as chips)	983	3.6	10.2	34.0	44	14	0.84	41	0	0.2	0.02	1.5	6–14
Spinach (boiled)	128	5.1	0.5	1.4	85	136	4.0	120	1000	0.07	0.15	1.8	25
Tomatoes (fresh)	60	0.9	0	2.8	93	13	0.4	3	100	0.06	0.04	0.8	20

years. Apart from cheapness, ease of cultivation and culinary versatility, one of their advantages is the fact that they have a dormant period after harvesting. Thus they can be stored and used during the winter months when other vegetables are in short supply. Potatoes can be processed into a flour for use in baked goods or dehydrated to give a product which is easily rehydrated with hot water to produce 'instant' mashed potato. Most dehydrated potato contains added vitamin C and thiamine to replace that destroyed during processing.

Potatoes, like all vegetables, consist principally of water. Apart from water they are mainly made up of starch and for this reason they are often looked upon merely as a cheap source of energy. In fact, however, potatoes are of great importance in the diet as a major source of vitamin C, thiamine, folic acid and dietary fibre.

Table 13.2 shows that potatoes are by no means rich in vitamin C and about half of what is present is lost when potatoes are cooked (see Table 13.3). Nevertheless, they are an important source of vitamin C because we eat so many of them. Such a wide variation exists in the amount eaten, however, that statements about the contribution of potatoes to the *average* British diet (both here and elsewhere) must be interpreted with caution.

The figures given in Table 13.2 indicate that potato crisps and chips are richer sources of vitamin C than boiled potatoes. Their water content is considerably lower, however, (especially for crisps) and on a dry-weight basis their vitamin C content is also lower.

The vitamin C content of freshly dug new potatoes can be as high as 20 mg/100 g but it declines sharply to about half that value after three months storage. Loss of vitamin C continues, but much more slowly, during longer storage. Substantial losses of vitamin C occur when potatoes are cooked as shown in Table 13.3. Losses during cooking occur through oxidation and solution. If boiled potatoes are kept hot for a period of time before serving, or if air is beaten into them by mashing or creaming, further substantial losses of vitamin C may occur.

Only about four per cent of the protein in an average British diet is provided by potatoes but its biological value (see p. 171) is high and

Table 13.3 Retention of vitamin C in cooked potatoes

Method of cooking	Vitamin C content (percentage of raw value)
Boiling	55–65
Baking	70–80
Roasting	60–70
Frying (chips)	75–85

almost equals that of egg protein. People who eat large amounts of potatoes often have a diet which is far from ideal and in these circumstances the protein provided by potatoes is valuable. A daily intake of 1 kg of boiled potatoes or 0.5 kg of 'chips' will provide enough of all the essential amino acids except methionine.

Potatoes are a major source of thiamine and folic acid. They supply about 20 per cent of the thiamine in the average diet and are second only to green vegetables as a source of folic acid. Modern methods of analysis have shown that the folic acid content of cooked potatoes can be as high as 45 μg/100 g.

The iron content of potatoes is by no means high but a kilogram provides about half the RDA. Potatoes provide about five per cent of the iron in an average British diet and they are the third most important dietary source. The iron present in potatoes is well absorbed, possibly because vitamin C is also present and this may also promote the absorption of iron from other foods eaten at the same time.

Potatoes contain about 1.3 per cent dietary fibre and provide about 15 per cent of the average fibre intake. About half the fibre is classified as soluble fibre (see p. 153). It is sometimes suggested that potatoes cooked in their jackets contain substantially more fibre than those which have been peeled but the difference is actually insignificant.

Because potatoes are seldom included in 'calorie controlled' or 'slimming' diets they have acquired an undeserved reputation as a 'fattening food'. It is true that they are a *cheap* form of energy but they are no more 'fattening' than any other food and, as we have seen, they provide the diet with appreciable quantities of valuable nutrients. Although we eat so many potatoes they contribute only about five per cent to the energy content of the average British diet. In fact, because of their high water content their 'energy density' (i.e. the amount of energy available per unit weight) is quite small. If they are cooked or served with large additions of fat, as in crisps, chips or mashed potatoes with 'lashings' of added butter then, of course, the energy density of the product bears little relation to that of the potato itself.

When potatoes are exposed to light the poisonous alkaloid *solanine* may be formed at their surface. Fortunately they also become green when exposed to light and this colouration serves as a warning that solanine may be present. Green parts of potatoes and all shoots and eyes (which may also contain solanine) should be removed when potatoes are being prepared for cooking.

LEGUMES OR PULSES   Peas, beans and lentils, which grow as seeds inside a pod, are referred to collectively as legumes or pulses. Normally only the seeds are eaten but sometimes, as with runner beans, green beans and the aptly named 'mangetout' peas the pod also is eaten. In Britain the legumes, though nutritious, form only a minor part of the diet but in many parts of the world they are important foods. Pulses have a high protein content and are rich in lysine but poor in methionine. This is the opposite of cereal proteins and so a combination of a cereal based food with pulses provides an amino acid mixture of high biological value. The undoubted popularity of baked beans on toast is obviously based on sound nutritional principles!

Fresh or frozen peas and beans eaten in their pods supply vitamin C, thiamine, niacin and carotene. Many pulses are eaten after they have been dried and stored, however, and this destroys their vitamin C content. The vitamin C content of canned peas (which are often canned dried peas) is also zero.

Peas are the most popular green vegetable in Britain and well over half those sold are frozen peas. A large proportion of the green beans sold have also been frozen. Freezing often takes place within hours of harvesting and the vitamin C content of frozen peas and beans is often higher than that of their 'fresh' counterpart because of the time which the latter take to reach the market.

Over 200 types of bean are known and they are grown all over the world. Most of them are varieties of the species *Phaseolus vulgaris*. In Britain, kidney beans (green beans or runner beans) are popular and many other types are available. Baked beans are more popular in Britain than anywhere else. They are made from 'pea beans' or 'navy beans' by soaking them in water, baking them and canning them in a sweet tomato sauce. Baked beans, like all pulses, are a good source of protein and iron. Soya beans are not eaten as vegetables in Britain but they are, nevertheless, of great importance world-wide. The beans are rich in protein of high biological value and contain over 20 per cent fat. Soya bean oil, which is rich in PUFAs (see p. 47) is extensively used in the manufacture of margarine and as a cooking oil. The beans are also used for making textured vegetable protein (TVP; see p. 201).

Soya beans and red kidney beans contain an *antitrypsin factor* which interferes with the action of the protein-splitting enzyme trypsin in the small intestine. Some beans contain *lectins* which interfere with the absorption of nutrients in the small intestine and *haemagluttinins* which cause red blood cells to cling together. Fortunately, all these toxins are destroyed by heating and thoroughly cooked beans are not harmful.

Pulses are good sources of protein and are richer in B vitamins and dietary fibre than green vegetables and root vegetables. Fresh and frozen peas and beans eaten in their pods are also good sources of vitamin C.

Pulses sometimes cause digestive problems, or flatulence through intestinal gas production. They contain trisaccharides and tetrasaccharides which are not digested by enzymes in the small intestine. They pass unchanged to the large intestine, or bowel, where they are broken down by bacteria to smaller molecules including the gases carbon dioxide, methane and hydrogen.

Storage of vegetables

Most hard vegetables can be stored for long periods after they have been harvested. Potatoes, for example, deteriorate only slowly and they remain in good condition for many months. Some other vegetables, however, have a much shorter storage life. Broccoli, for example, is harvested while the plant is still developing and must be eaten before the flowers open.

The storage life of a particular type of vegetable depends mainly upon its quality at the time it was harvested but conditions, of course, are also important. Most vegetables mature gradually and, unlike fruits, they do not have a critical ripening period. They are often harvested before their growth is complete to ensure that tenderness and flavour are at a peak. If harvesting is delayed for too long vegetables may become tough or stringy through thickening of the cell walls. Their sugar content may also fall leading to a reduction in sweetness and flavour. Careful judgement is needed to ensure that vegetables are harvested at such a time that they are in peak condition when they are eaten or on arrival at a freezing or canning plant.

Once vegetables are harvested they begin to lose water by diffusion through the cell walls. This causes shrivelling or wilting as the cell membranes draw away from the walls and the vegetable becomes less firm.

When raw vegetables are eaten the contrast between fresh and 'not-so-fresh' is striking. A fresh vegetable offers an initial resistance to biting followed by release of copious juice. A vegetable which has begun to deteriorate, on the other hand, will be less firm and juicy. Leafy vegetables are particularly prone to wilt, because water loss occurs more readily owing to their large surface area.

For maximum storage life, leafy vegetables should be kept in cool and fairly moist conditions. Dry conditions obviously encourage water loss and wilting. The 'crisper' compartment of a refrigerator provides good conditions because the temperature is low and the humidity in the closed compartment is higher than on the shelves of the refrigerator.

Potatoes should not be kept in a refrigerator because they begin to convert their starch to sugar below about $8^\circ C$. They develop a sweet taste and chips or crisps made from them tend to be dark in colour. At temperatures below $0^\circ C$ cell rupture occurs and an inedible 'flabby' potato, which decays rapidly, is obtained on thawing.

In general, vegetables are not as easily damaged as fruits during storage and transportation. Nevertheless, they can be bruised and develop brown patches through the action of phenolase in the same way as fruits (see p. 272).

Stored vegetables gradually lose vitamin C as they age but the rate of loss may be slow if they are undamaged. Whole cabbage, for example, loses little or no vitamin C in a week. Vegetables which have an 'open' structure and wilt quickly, such as spinach and lettuce, lose vitamin C fairly quickly. Their storage life is not very long, however, so that they are usually eaten before serious loss of vitamin C has occurred. The other vitamins present in vegetables are not greatly affected by storage although some loss of thiamine may occur.

SENSORY QUALITIES OF FRUIT AND VEGETABLES

A preference for a particular type of fruit or vegetable is more likely to depend on its taste, smell and colour than on a knowledge of its nutritional qualities. Taste and smell both contribute to flavour and these qualities are so closely related that it is difficult to distinguish between them or to define them. All of them are of chemical origin inasmuch as they are caused by the presence of specific compounds in the fruit or vegetable. Beyond that, however, it is not always possible to say with confidence exactly why a particular fruit or vegetable should have the characteristic taste, smell and flavour we associate with it.

Taste and smell

The taste of a fruit is a subtle blend of sweetness and acidity (combined in some cases with astringency or bitterness) delicately complemented by the flavour of the particular fruit. Fruits are sweet because of the presence of sugars which are formed when a fruit ripens and if 'fruit acids' are also present they will produce a sharp taste. The relative amounts of sugar and acids present largely determines whether a particular fruit is sweet or sour.

The chemical structures of some acids which commonly occur in fruit are shown below. All of them contain a carboxyl group ($-COOH$). Oxalic acid, malic acid and tartaric acid are dicarboxylic acids because their molecules contain two carboxyl groups and, likewise, citric acid and isocitric acid, where there are three carboxyl groups, are tricarboxylic acids. With the exception of oxalic acid they are all hydroxy acids because their molecules contain hydroxyl ($-OH$) groups.

COOH	COOH	COOH	COOH	COOH
COOH	CHOH	CHOH	CH_2	CH_2
	CH_2	CHOH	COHCOOH	CHCOOH
	COOH	COOH	COOH	CHOHCOOH
Oxalic acid	Malic acid	Tartaric acid	Citric acid	Isocitric acid

The fruit acids are colourless, odourless water-soluble solids. Citric acid, as its name makes clear, is present in the juice of citrus fruits but it also occurs in other fruits. Malic acid also is present in most fruit juices and it is mainly responsible for the acidity of apple juice. Tartaric acid is the main acid in grape juice but citric and malic acids are also present. Blackberries owe their acidity mainly to isocitric acid. Oxalic acid occurs in unripe tomatoes and strawberries in very small amounts, but rhubarb stalks may contain as much as one part in 200. The presence of oxalic acid is undesirable because it combines with calcium present in other foods to form insoluble calcium oxalate and thus renders it unavailable to the body.

Bitterness and the closely related taste characteristic, astringency, are not related to pH or the presence of fruit acids and both sweet and sour fruits can be bitter or astringent. Bitterness (e.g., in Seville oranges) is caused by the presence of complex phenolic substances known as *flavanoids* or tannins. A low level of astringency contributes to the taste of many fruits and much higher levels are found in unripe fruit and some grapefruit.

Flavour is a more subtle property than taste, consisting as it does of a combination of taste and smell. Fruit owes its smell to the presence of a variety of volatile sweet-smelling organic compounds including acids, alcohols, esters (which are formed by reaction between acids and alcohols), aldehydes, ketones and hydrocarbons. A large number of such compounds may contribute to the flavour of a particular fruit and over 200 have been identified in ripe bananas. Some of them may be present in exceedingly small amounts but they can still be detected by the palate. It has been shown, for example, that the smell of ethyl 2-methylbutyrate can be detected in concentrations as low as one in 10^{10} (i.e. one part in 10 000 000 000). The simpler compounds are present in many ripe fruits and they provide a common background of ripeness and fruitiness but a single compound may be responsible for the characteristic flavour of a particular fruit. For example, the presence of benzaldehyde characterizes the flavour of cherries and almonds and ethyl 2-methylbutyrate that of ripe apples. When fruits ripen there is an increased production of volatile compounds and the proportions in which the various substances are present also changes.

Blends of esters with alcohols and other fragrant compounds are used in cooking and in manufactured foods as synthetic flavours or 'essences' as substitutes for genuine fruit flavours (see p. 382).

Vegetables do not, in general, have such pleasant tastes, smells and flavours as fruits. No-one has found it worthwhile to produce artificial cabbage or onion flavour for use in soft drinks or jellies, nor do we have vegetable-flavoured sweets. Nevertheless, vegetables do have tastes, smells and flavours which, though less prominent than fruit flavours, are equally distinctive.

Many of the less attractive (and perhaps more pronounced) vegetable smells are caused by sulphur compounds. Cabbage, Brussels sprouts and cauliflowers owe their smell to a group of sulphur compounds known as isothiocyanates or mustard oils. In raw and undamaged vegetables these offensive compounds are bound to sugar and are thereby rendered odourless. When plant tissues are damaged by cutting, bruising or chewing an enzyme catalyses the breakdown of the complex sulphur-containing compounds and the pungent isothiocyanates are released. The tastes and smells thus produced vary in intensity from the acrid odour of crushed mustard seed to the relatively mild smell of shredded cabbage.

When cabbage-type vegetables are cooked in boiling water the complex sulphur compounds break down and combine with other plant materials. New powerfully-smelling sulphur compounds are produced including the gas hydrogen sulphide.

Garlic, onions, leeks and chives also owe their similar but different smells and flavours to the presence of sulphur compounds. They contain a compound derived from the amino acid cysteine which is odour-free whilst inside the plant tissues. When the cells are broken by crushing, however, this odourless compound is converted enzymically to other sulphur compounds. Some of these have powerful odours while others are lachrymators, i.e. substances that cause a burning sensation in the eyes and make them water. The compound chiefly responsible for the well-known odour of garlic is diallyl sulphide, $(CH_2=CH-CH_2)_2S$.

Colour

There are three main groups of compounds which give colour to fruit and vegetables.

Chlorophylls – green colours
Carotenoids   – yellow, orange and red colours
Anthocyanins – red, purple and 'bluey' colours

CHLOROPHYLLS These are the green colouring matter of leafy vegetables and unripe fruit where their presence is essential for the conversion of carbon dioxide to simple carbohydrates by photosynthesis. Two types are present in fruit and vegetables – *chlorophyll a* which is bright green in colour and *chlorophyll b* which is less brightly coloured. The two versions differ slightly in structure but they both have large molecules as can be appreciated from the molecular formula of chlorophyll a which is $C_{55}H_{72}O_5N_4Mg$. Usually about three times as much chlorophyll a is present in plant tissues as chlorophyll b. From our point of view they are both equivalent and in what follows they will be referred to simply as chlorophyll.

The chlorophyll molecule has a large and complicated ring structure which in some respects resembles the molecule of haemoglobin, the red colouring matter of blood. The chlorophyll molecule, however, has a magnesium atom at its centre in place of the iron atom present in haemoglobin. It also has a long 'tail' containing 20 carbon atoms, derived from the alcohol phytol, $C_{20}H_{39}OH$, and this makes it fat-soluble.

The chlorophyll molecule is not particularly stable and both the central magnesium atom and the phytyl side chain are easily removed when fruit or vegetables are cooked or processed. The magnesium atom is displaced by heating in acid conditions and chlorophyll derivatives which are brownish in colour are produced. This is what happens when cabbage is overcooked. If sodium bicarbonate is added to the water in which 'greens' are cooked it 'preserves' their green colour by preventing or slowing down the loss of magnesium. This practice is not to be recommended, however, because it causes loss of vitamin C.

Displacement of the central magnesium atom from the chlorophyll molecule causes canned green vegetables to lose their natural colour. This may occur when they are canned or during subsequent storage and it is probably caused by release of organic acids from the plant tissues. The loss of colour cannot be prevented by adding sodium bicarbonate because when this is done the vegetables become soggy during processing and storage. To compensate for the loss of natural colour artificial dyestuffs are usually added to canned green vegetables (see p. 378).

The phytyl side chain may split off from a chlorophyll molecule during blanching, cooking or processing. The remainder of the molecule, which retains its green colour, is more soluble in water and colour loss may occur through leakage into the surrounding water.

CAROTENOIDS   Most yellow or orange (and some red) foods owe their colour to the presence of *carotenoids* which are the most widespread of all plant colouring matters. Carotenoids occur in all green plants alongside chlorophyll: they play an indirect role in photosynthesis by absorbing sunlight of certain wavelengths and making it available to chlorophyll.

There are many carotenoids but they fall into two categories – the *carotenes* which are hydrocarbons (i.e. contain only carbon and hydrogen) and the *xanthophylls*, which also contain oxygen. Carotenes are mainly found in orange or reddish coloured plants and xanthophylls in yellow plants.

Carotenoids have large molecules and they are highly unsaturated and fat-soluble. Their molecules contain numerous carbon—carbon double bonds and they are therefore susceptible to oxidation. β-carotene, for example, has the molecular formula $C_{40}H_{56}$ and its molecule contains 11 carbon—carbon double bonds. In plant tissues, however, they are in a protected environment and only small losses occur during storage or normal cooking operations.

Carotenes (but not xanthophylls) are converted to retinol (vitamin A; see Chapter 12) in the small intestine. They thus have a nutritional significance as well as their role in making food more visually attractive.

ANTHOCYANINS   The anthocyanins belong to a class of compounds known as *flavanoids* and they impart red, purple and blue colours to fruit, vegetables and wine. There are six anthocyanins which differ slightly in chemical structure but they all possess the same basic skeleton. In plant tissues they are combined with sugars and the number of different combinations, and hence different colours, is very large.

Anthocyanins are soluble in water and they easily leak out of fruit and vegetable tissues during cooking. They are sensitive to changes in acidity which cause them to change colour. Red colours predominate in acidic conditions (i.e. at low pH) but at higher pH the colour changes to yellow or blue. In most cooking operations, however, anthocyanins are quite stable and they retain their colour because the pH is kept low by plant acids. More severe conditions occur when fruit is canned and many coloured fruits are bleached or become discoloured. Artificial colouring matter is often added to canned fruit to compensate for loss of natural colour.

SUGGESTIONS FOR FURTHER READING

DUCKWORTH, R.B. (1966). *Fruit and Vegetables*. Pergamon Press, Oxford.
JOHNS, L. AND STEVENSON, V. (1979). *The Complete Book of Fruit*. Angus & Robertson, London.
LUH, B.S. AND WOODROOF, J.G. (1975). *Commercial Vegetable Processing*. Avi, USA.
YAMAGUCHI, M. (1984). *World Vegetables – Principles, Production and Nutritive Values*. Ellis Horwood, Chichester.

CHAPTER 14

Methods of cooking

Some foods are best eaten when freshly harvested without further preparation or cooking. Some vegetables, such as lettuce, tomatoes and radishes, and some fruit, such as strawberries, peaches and melons, come into this category. Their characteristic flavour and texture can best be appreciated when they are fresh and raw; cooking can only cause deterioration of these qualities. Most foods, however, are greatly improved by cooking which, if properly carried out, enhances the appearance, flavour, texture and digestibility of food. Cooking may also promote the safety and keeping qualities of food by killing moulds, yeasts and bacteria that are pathogenic or cause spoilage. Cooking, however, needs to be distinguished from the use of heat treatment to preserve food – a subject which is discussed in Chapter 16.

The preparation and cooking of food is both an art and a science. The art of cooking food has developed over centuries and in refining cooking techniques and recipes cooks have unwittingly developed the necessary scientific skills of careful experimentation and precise observation. The knowledge accumulated in this way has been passed down through many generations and represents the results of numerous experiments. Today, by applying scientific principles, we can understand the changes that occur during cooking and in this way we can improve the cooking process both in terms of the quality and the nutritional value of the food produced.

Cooking may be simply defined as the heat treatment of food carried out to improve its palatability, digestibility and safety. Cooking involves transfer of both *heat* and *mass*.

Traditional methods of cooking such as boiling and baking are all developments of the original method which used an open fire. The only cooking methods which are not derived from the open fire technique are microwave and electromagnetic induction methods of cooking.

METHODS OF HEAT TRANSFER   Food is a relatively poor conductor of heat and in traditional cooking methods the food is heated at its surface and heat is then transferred into the body of the food by conduction and/or convection. In methods using a high temperature source, such as grilling or toasting, heat is transferred to the surface of the food directly by radiation. More usually heat is transferred from the source of heat to the surface of the food using some intermediate medium such as water, steam, air or oil. Microwave cooking differs from traditional methods in that heat is generated *within* the food. Even so the heat so generated is then transferred to regions of lower temperature by means of conduction and convection.

Radiation involves the emission of heat from a high temperature source in the form of waves. Such waves, which are similar in nature to light waves, pass through air in straight lines to reach the surface of the food without any intermediate medium being involved. The energy of the radiation is absorbed by the surface of the food which, in consequence, heats up rapidly. Methods of cooking which utilize radiant heat, such as grilling and infrared cooking, are therefore liable to overcook the outside of a food while leaving the inside underdone. Where such a result is desirable, as in cooking steaks, grilling may be the preferred method of cooking.

While radiation involves *direct* heat transfer from heat source to food, conduction and convection are both *indirect* means of heat transfer. *Convection* is the transfer of heat as the result of the movement of a fluid, such as air or water, from a higher temperature region to one at a lower temperature. Thus when food is placed in a heated oven, the heated air in the oven rises and when it reaches the surface of the food some of the kinetic energy of the air molecules is transferred to the food surface as heat energy. The rising hot air displaces the cooler air from the top of the oven forcing it downwards so that it is reheated by the heat source. In such an oven there is therefore a circulation of heated air, the top of the oven being some $50°C$ hotter than the bottom. Convection heating in ovens depends upon the flow of hot air over the food, and this can be improved by increasing the rate of flow using a fan. This is the principle used in forced convection ovens which allow an even temperature to be achieved throughout the oven.

Conduction of heat occurs when a metal pan is placed on a heat source. Metals are good conductors of heat, meaning that they transfer heat energy rapidly and efficiently. In boiling, for example, the pan conducts heat from the heat source to the water in the pan. The heated water circulates, heating the surface of solid food by convection. Heat is then transferred to the interior of the food by conduction.

One other method of cooking by conduction involves *electromagnetic induction*. The principle of this method is that some metals, particularly aluminium alloys, can absorb electric currents of certain frequencies and generate from them eddy currents in the metal which produce a heating effect. The advantage of this technique is that heat for cooking is generated within the cooking vessel so eliminating the transfer of heat from its source to the pan used for cooking. Consequently this technique is both more efficient and safer than traditional methods; at present it is also more expensive.

The rate of heat transfer involved in cooking can be considered in two stages:

1. the rate at which heat is absorbed by the surface of the food;
2. the rate of conduction of heat to the centre of the food.

The art of cooking is to control these two rates so as to produce the desired result. The factors which affect the first rate are the rate of conduction if the heat source is in direct contact with the food, as in contact grills, or the rate of convection if the food is immersed in a fluid such as air or water, as in baking or boiling. The factors which affect the second rate are the temperature of the surface of the food, the thickness of the food, the rate of evaporation from its surface of the food and its thermal conductivity.

TRANSFER OF MASS   Apart from transfer of heat, cooking also involves transfer of mass, mainly the transfer of water through the food as cooking proceeds. During cooking, water travels from the centre of the food to the surface where it evaporates. Soluble nutrients and flavours also travel with the water, and in moist heat cooking methods may be lost from the surface of the food into the cooking water – a process which is known as *leaching*. Fat is also transferred during cooking and may either be added to the food, as during frying, or lost from the food as 'cooking drip' in grilling or roasting.

GENERAL EFFECTS OF COOKING   The transfer of heat and mass produces many of the changes of colour, flavour, volume, texture and digestibility that occur during cooking. The art of cooking is to promote the desirable changes while minimizing the undesirable ones. For example, in the baking of pastry the rate at which heat is absorbed by the surface of the pastry determines the extent of browning which is brought about partly by the caramelization of sugar and partly by non-enzymic browning due to the interaction of sugar and lysine. If the rate of heating is too low the product will have an uncooked whitish appearance; if it is too high its colour may be too brown (or even black!) and there will be severe destruction of lysine due to excessive non-enzymic browning.

In addition to the changes mentioned, there will also be changes in nutritional value during cooking. Some specific effects may be beneficial, for example, destruction by heat of the substance that inhibits the enzyme activity of trypsin in many raw legumes, such as groundnuts and soya beans. Other specific effects may be detrimental, such as non-enzymic browning mentioned above. Some cooks, in attempting to improve the colour of green vegetables by the addition of sodium bicarbonate, cause destruction of vitamin C. Table 14.1 summarizes the main types of nutrient loss that occur during the preparation and cooking of food.

CHANGES OCCURRING AFTER COOKING   In the home food is usually consumed immediately after cooking, but this may well not be the case

Table 14.1 Summary of nutrient losses during preparation and cooking of food

Food	Process involved	Loss caused by	Examples of nutrient lost
Animal			
Chopped or ground meat, fish	Freeze-thaw	Thaw drip	Protein, B vitamins – especially niacin
Meat joints	Microwave cooking, grilling, roasting	Cooking drip	Fat, fat-soluble vitamins
All	Braising or stewing	Leaching	Water-soluble vitamins
All	Lengthy heating or high temperature as in baking or roasting	Oxidation or breakdown of nutrients	Thiamine, essential amino acids and fatty acids
Plant			
Fruit, vegetables	Bruising, long storage	Spoilage	Vitamin C
Fruit, vegetables	Cutting, chopping	Damage to cellular structure	-Vitamin C
All	Cooking, washing, soaking in water	Leaching	Loss of water soluble vitamins – vitamin C, thiamine, folic acid, pyridoxine
All	Lengthy heating or high temperature as in baking or roasting	Oxidation or breakdown of nutrients	Vitamin C, riboflavin, essential fatty acids and amino acids
All	Discarded cooking water	Leaching	Loss of water-soluble vitamins – vitamin C, thiamine, riboflavin, zinc

in commercial and industrial catering. Studies of catering in schools, hospitals and restaurants in many countries have shown that cooked food was kept warm for 0–7.5 hours before it was consumed. In the UK a study of the Meals-on-Wheels Service showed that food was kept warm for 50 minutes to $3\frac{1}{2}$ hours and that the internal temperature of the food varied from 38–47°C.

Provided that food is stored above 65°C microorganisms are unable to grow; below this temperature there is the possibility of microbial growth. If food is stored at safe temperatures above 65°C, however, there will be fairly rapid destruction of heat sensitive nutrients, particularly vitamins. The study of the Meals-on-Wheels Service showed that 31–54 per cent vitamin C was lost between the end of cooking and the first meal served. In general vitamin C is the vitamin most rapidly destroyed when food is kept hot, thiamine is less rapidly destroyed and only small amounts of riboflavin and niacin are lost.

Moist heat methods

Moist heat methods of cooking employ relatively low temperatures and destruction of nutrients by heat is therefore not great. Cooking times at such low temperatures tend to be long, however, and this results in extensive loss of water-soluble nutrients into the liquid used for cooking. Vitamin C is the nutrient which is most easily destroyed in cooking and hence loss of this vitamin may be taken as an index of the severity of the cooking process. If little vitamin C is lost it may be taken that the cooking process is a mild one and that there will have been little loss of other nutrients. The losses of vitamin C that occur when vegetables are boiled are shown in Table 14.2 and the figures demonstrate that the loss can be very severe.

It needs to be emphasized that nutrient losses in cooking quoted in the literature should be treated with caution. The figures quoted for a given food may show considerable variation and this is because nutrient losses depend to a large extent on the way in which the

Table 14.2 Vitamin C content of raw and boiled vegetables

Vegetable	Vitamin C (mg/100 g)		Per cent loss of vitamin C
	Raw	Boiled	
Runner beans	20	5	75
Savoy cabbage	60	15	75
Brussels sprouts	90	40	56
Carrots	6	4	33
Fresh peas	25	15	40
Old potatoes	8–20	4–14	50–70

cooking is carried out and also on the physical state of the food. For example, the quality of the fresh food, the treatment that the food receives before it is cooked, the time of cooking, the amount of liquid used, the extent to which air is excluded and the time for which the food is kept hot before it is consumed all affect the extent to which nutrients are lost. Figures quoted for nutrient loss during cooking must therefore be considered in conjunction with the conditions of the experiment. Unless this is done data in the literature appear to be contradictory.

In moist heat cooking, nutrients may be lost in a variety of ways. The most important is the leaching of water-soluble nutrients, primarily vitamins and mineral elements, into the cooking water. Nutrients are also lost by the action of heat in the presence of air. Vitamin C, for example, is very sensitive to such oxidative loss. The action of oxidizing enzymes also causes loss of nutrients and again vitamin C is easily lost, being rapidly destroyed by such oxidases in the presence of. oxygen in the cooking water. at As enzymes are destroyed by heat – ascorbic acid oxidase is rapidly destroyed at $100°C$ – loss of nutrients by enzymes is greatly reduced if food, particularly vegetables, is put into boiling water rather than being put into cold water which is then heated.

BOILING   Boiling is a common method of moist heat cooking; it utilizes the fact that because water has a high specific heat capacity it is an efficient heat reservoir and therefore is a convenient medium in which to transfer heat to food. Its ready availability is another important factor. One disadvantage of water as a medium for heat transfer is that, because of its good solvent action, food cooked in it may lose a considerable proportion of its soluble matter. Vegetables, for example, are commonly cooked by boiling and this results in an inevitable loss of some mineral elements and vitamins, the loss of the latter being the more important from a nutritional point of view. Although little or no carotene is lost when vegetables are boiled, considerable amounts of thiamine and ascorbic acid are destroyed as both these vitamins are water-soluble and easily destroyed by heat. In general about one-third of the thiamine and two-thirds of the ascorbic acid are lost, although, as discussed below, the amount lost varies considerably with changing conditions.

Although some loss of water-soluble nutrients by leaching is inevitable during boiling, the extent of this loss is governed to some extent by the amount of water used. The loss of both mineral elements and water-soluble vitamins increases as the amount of water used increases. For example, in a series of experiments it was found that cabbage, which lost 60 per cent ascorbic acid when cooked in a small volume of water, lost 70 per cent when cooked in a larger amount.

Table 14.3 Relation between cooking time and retention of vitamin C

Cooking time (min)	Per cent retention of vitamin C during boiling			
	Brussels sprouts	Cabbage	Carrots	Potatoes
20	49	–	35	–
30	36	70–78	22	53–56
60	–	53–58	–	40–50
90	–	13	–	17

Losses of thiamine, which are important when boiling cereal foods, follow a similar pattern and, for example, it was found that rice which lost about 30 per cent thiamine when cooked in a small volume of water, lost 50 per cent when cooked in a larger amount.

The time for which foods are boiled also affects nutrient loss. There is, for instance, a drastic loss of ascorbic acid when prolonged cooking times are used, as can be seen from Table 14.3 which shows the way ascorbic acid loss increases with cooking time for a number of vegetables. Another factor affecting nutrient loss is the treatment that the food receives before it is boiled. For example, the greater the surface area of a food the greater is the loss of water-soluble nutrients into the cooking water. Crushing, chopping, slicing and shredding of food not only increase the surface area but also releases enzymes that cause further loss. In one study, over one-third of vitamin C was destroyed during initial shredding and washing of spring cabbage prior to cooking.

Table 14.4 shows the effect that the size of food has upon the loss of nutrients that occurs during the boiling of vegetables, such as carrots, swedes and sprouts. Peeling vegetables before cooking also increases loss of nutrients and it has been found that whereas whole unpeeled potatoes lost about one-third ascorbic acid when boiled, peeled potatoes lost an additional ten per cent.

STEAMING AND PRESSURE COOKING   Steaming involves using the steam produced from boiling water. As the contact between the food and water is less than in boiling there is smaller loss of soluble matter

Table 14.4 Effect of size on per cent of nutrient loss occurring in boiled vegetables

Nutrient	Large pieces	Small pieces
Vitamin C	22–23	32–50
Mineral salts	8–16	17–30
Proteins	2– 8	14–22
Sugars	10–21	19–35

but, as a longer time of cooking is required, the amount of ascorbic acid which is decomposed by heat is increased.

The rate of cooking in steaming may be increased by the use of steam under pressure, this being the principle of the pressure-cooker. As increase of pressure raises the temperature at which water boils, the cooking temperature is greater than $100°C$, for example, when a pressure-cooker is used at its highest working pressure of 1.05 kg/cm^2 the boiling point is $120°C$. It is found that the amount of ascorbic acid lost because of the increased temperature is more than compensated for by the amount conserved due to the shorter cooking time, though the difference is small. The relatively short cooking times involved in pressure cooking are illustrated by the fact that, using the highest pressure setting, potatoes cook in about five minutes and a beef casserole in 20 minutes.

STEWING   Stewing involves cooking food in hot water, the temperature of which is kept below its boiling point. The changes which occur in stewing are, therefore, similar in character to those which occur during boiling, although they occur at a slower rate. Stewing, as a slow method of cooking, results in considerable loss of soluble matter. For example, in stewing fish, a third of the mineral salts and extractives may be lost as well as water-soluble vitamins. Stewed fish, therefore, lacks flavour and has less nutritional value than when it is raw. However, if the liquor in which the food has been cooked is used for making a sauce or soup, then the passage of nutrients into the cooking water involves no loss of nutritional value and the liquor retains the flavour lost from the stewed food.

One of the advantages of stewing is that, because of the low temperature used, protein is only lightly coagulated and is, therefore, in its most digestible form. Another advantage is that it exerts a tenderizing effect on protein food as insoluble tough collagen is converted into soluble gelatin by prolonged contact with hot water. Stewing is, therefore, a particularly suitable method for cooking tough meat.

Many fruits, such as strawberries and melons, are best eaten raw but others, such as rhubarb and damsons, are undoubtedly improved by cooking. Fruit may be cooked by stewing in water to which sugar has been added. Dried fruits are soaked before cooking to allow maximum absorption of water by osmosis, and less sugar is added to the cooking water because, during soaking, sugar diffuses out of the fruit. During stewing, cellulose is softened, protein is lightly coagulated and soluble matter is lost to the cooking liquor. At the same time, where stewing is done in syrup, sugar is absorbed by the fruit. Because of the low temperature used and the presence of fruit acids, which maintain pH below seven, destruction of thiamine (which is

Table 14.5 Vitamin C content of raw and stewed fruit

Fruit	Vitamin C (mg/100 g)		Per cent loss of vitamin C
	Raw	Stewed with sugar	
Apples	15	11	27
Blackberries	20	14	30
Damsons	3	2	33
Peaches	8	trace	100
Pears	3	2	33
Plums	3	2	33
Rhubarb	10	7	30

only present in small amounts) and ascorbic acid is small, though, due to their solubility, they gradually diffuse into the surrounding liquor. The loss of vitamin C that occurs during the stewing of fruit is shown in Table 14.5.

Dry heat methods

Dry heat methods of cooking are characterized by the use of higher temperatures than in moist heat methods, and the use of air as the intermediate medium conveying heat from the heat source to the surface of the food. When food is placed in a hot oven about 80 per cent of heat reaching its surface is conveyed by convection and about 20 per cent by radiation. Heat is conveyed from the surface of the food to the interior by conduction, this being a relatively slow process due to the poor thermal conductivity of food.

The rate of heat transfer in dry heat cooking can be illustrated by considering the roasting of a joint of beef. If the meat is to be cooked so that it is 'rare', the centre of the meat should reach a temperature of about 63°C while if it is to be 'well done' the temperature should be in the range 80–88°C. If a joint of meat weighing 4 kg is placed in a hot oven, heat is conducted from the surface towards the centre of the joint. The temperature gradient through the meat after three hours cooking is shown in Fig. 14.1. This figure illustrates that heat is conducted through the meat slowly and that after three hours cooking the meat is in a 'rare' state. If the meat is removed from the oven after three hours heat continues to be transferred from the hot outer surface towards the cooler interior so that cooking continues for some time after the meat is removed from the heat source.

The higher temperatures used in dry heat compared with moist heat cooking means that loss of nutrients which are sensitive to heat is correspondingly greater. Apart from mineral salts, which are stable to heat, all nutrients are affected to some extent by dry heating. Fats

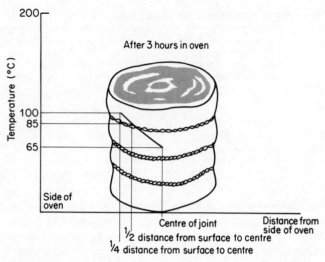

Figure 14.1   The temperature gradient through meat being cooked in an oven.

are stable to moderate heating and, although they darken, little breakdown occurs unless they are heated to high temperatures when they start to decompose with the formation of *acrolein*, which has an unpleasant acrid odour (see frying). Carbohydrates are affected by dry heat: starch is converted into pyrodextrins which are brown in colour and which contribute colour to toast and breadcrust; sucrose is converted into dark-coloured caramel in a complex multi-stage reaction that involves its initial breakdown into monosaccharides and its final polymerization into coloured substances.

Dry heat cooking destroys those vitamins which are unstable to heat, notably ascorbic acid, which as we have already noted is destroyed at quite low temperatures.

Table 14.6 shows the average percentage losses of B vitamins which occurs when meat and fish are cooked using dry heat. It is apparent from this table that 20 per cent of B vitamins are destroyed.

Proteins, as we saw in Chapter 8, are extremely sensitive to heat but their nutritive value is not significantly affected unless they are heated to a fairly high temperature, such as occurs in roasting. Loss of nutritive value depends not only on the cooking temperature but also on the time of cooking and the presence of other nutrients, particularly carbohydrates.

Amino acids are only destroyed at high temperatures such as are employed in roasting, and even then the loss of protein is small and confined to the surface of the food. A much greater loss of nutritive

Table 14.6 Percentage loss of B vitamins when meat and fish are cooked using dry heat

Vitamin	Meat Roasted and grilled	Fish Baked	Grilled
Thiamine	20	30	20
Riboflavin	20	20	20
Niacin	20	20	20
Pyridoxine	20	10	20
Pantothenic acid	20	20	20

value results from a change in protein structure which affects the linkages between amino acids in such a way that they become resistant to enzymic hydrolysis. Amino acids affected in this way – notably aspartic acid, glutamic acid and lysine – cannot be released by enzymes during digestion and are therefore unavailable to the body.

NON-ENZYMIC BROWNING   As already mentioned, when protein and carbohydrate exist together in the same food an additional loss of nutritive value may occur due to *non-enzymic browning* (also called the *Maillard reaction*). Reaction occurs between amino groups projecting from a protein chain or peptide or amino acid and the carbonyl group of a reducing substance such as glucose. The details of the reaction are not completely understood but it involves a number of steps, the first of which is an addition reaction between the amino and carbonyl groups, and the last is a polymerization to form a brown substance. Several amino acids undergo this reaction, notably lysine and methionine. The reaction results in the formation of substances which cannot be hydrolysed by enzymes, and the proteins affected are thus unavailable to the body.

Non-enzymic browning occurs particularly at high temperatures and at pH values of seven and above: a certain amount of moisture is also necessary. The reduction in nutritive value of protein due to non-enzymic browning has been studied intensively in terms of the loss of lysine during cooking. For example, it has been found that bread loses 10–15 per cent during toasting. Some lysine is also lost during the roasting of meat, though in home cooking the loss has been found to be small.

Apart from causing some loss in the nutritive value of proteins during dry heat cooking, non-enzymic browning is also responsible for producing some desirable changes in the flavour, colour and aroma of food during roasting, baking and toasting. For example, it improves the quality of bread during baking and toasting, and of nuts and coffee beans during roasting. It is also partly responsible for the flavour of such diverse products as meat extract, biscuits and breakfast cereals.

Frying

Frying is a convenient method of cooking where a high temperature and rapid cooking are desired. Fat is the medium used in frying to provide the necessary high temperature. It is chosen because of its high boiling point and because it can be heated almost to its boiling point without much decomposition occurring. There are two methods of frying, the most important being *deep frying*.

Deep frying, as the name suggests, is done in a deep pan, the food being lowered into the fat when it is very hot, normally between 175°C and 200°C. As soon as the food comes into contact with the hot fat there is a violent bubbling as the water on the surface of the food is vaporized. When potato chips are immersed in oil at 190°C the steam produced forms a layer around the chips in the form of a 'stationary layer' which greatly reduces the rate of heat transfer from the oil to the potato surface. However, once heat has passed through this barrier it is conducted rapidly from the surface of the food to its interior, because of the high temperature of the oil. Cooking proceeds so rapidly that loss of mineral salts and nitrogenous substances is reduced to a minimum. Cooking is complete when the outside of the food is crisp and usually golden-brown in colour.

The second method of frying is *shallow frying*, which is done in a shallow pan, the bottom of which is covered with fat. The main role of the fat is to prevent the food from adhering to the pan, the cooking being done mainly by direct conducted heat. In this method of cooking, heat is applied only to one surface of the food at a time, so that uneven cooking may result unless the food is turned regularly.

Fats used for frying must be pure because impurities are likely to decompose at the high temperatures employed, producing unpleasant flavours and odours. Vegetable oils, providing that they have been carefully refined, may be used for frying and such oils can often be heated to higher temperatures than the more conventional fats without decomposition occurring. When using a fat or oil for deep frying the temperature of the fat should be checked with a thermometer and it should not be allowed to rise above the value recommended. This is because if the temperature used is too high the *smoke point* of the fat may be reached, at which temperature blue smoke appears indicating incipient decomposition. If the temperature is raised above the smoke point the rate of decomposition increases rapidly. The smoke point for oils and fats is in the range 135–245°C, coconut oil being at the lower end of this range and pure groundnut oil at the top.

CAUSES OF FAT DETERIORATION DURING FRYING

1. *Too high a temperature.* If the frying temperature is high steam generated during frying causes some hydrolysis of fat and glycerol and free fatty acids are formed; the latter may undergo dehydration with the formation of *acrolein*. This is a simple unsaturated aldehyde having an unpleasant acrid odour. It is probably present in small quantities in the smoke from the fat:

$$CH_2OHCHOHCH_2OH \rightarrow CH_2=CHCHO + 2H_2O$$
Glycerol   Acrolein

2. *Access of air.* At high temperatures oxygen in air causes rapid oxidation, and consequent deterioration, of oils used for cooking. Such deterioration is accelerated by light. Provided, however, that pans used for deep frying have lids, the hot cooking oil is covered at its surface by a layer of steam released from the food being fried and this prevents access of oxygen to the surface of the oil.

3. *Contamination caused by food residues.* When food is cooked by deep frying particles of food become detached and should be removed from the oil before it is used again for frying. If such food residues are not removed repeated reheating will cause them to become charred and cause deterioration and darkening of the oil.

4. *Loss of natural antioxidants.* Natural oils and fats contain antioxidants that help to prevent oxidative changes which produce rancidity. When food is fried the concentration of antioxidants is reduced and, in continuous frying, the removal of food reduces the amount of cooking oil in the pan which further reduces the amount of antioxidant present. The loss of antioxidants can be reduced by 'topping' up the cooking oil as frying proceeds and by using fresh oil regularly.

5. *Effect of traces of copper.* Extremely small traces of copper at levels of 0.1 ppm can lead to the development of rancidity in cooking oil. It is therefore important to use pans made of stainless steel, which does not contain any copper.

It is evident from the above that repeated use of oil leads to deterioration, and consequently that oil used for cooking should be changed frequently.

Other factors which affect the choice of fat for frying are flavour and spattering properties. Spattering is due to the presence of water in the fat; the water vaporizes on heating and causes the fat to bubble and froth. When the bubbles burst, the fat is said to 'spatter'. Pure cooking fats do not contain water and so give smoother frying than butter and margarine. Efforts are usually made to reduce the spattering properties of margarine by the addition of lecithin, which also improves emulsification.

NUTRITIONAL CHANGES   When food is fried in oil some of the fat used as a heat transfer agent becomes a part of the cooked product and this clearly affects its nutritional value. For example, while the fat content of raw potatoes is negligible (0.2 g fat/per 100 g potatoes) the fat content of the potatoes when fried as chips increases to 7–15 g fat/100 g potatoes depending on the method used for frying. Such an increase in fat content is clearly of considerable nutritional significance, especially in view of the weight of opinion advocating a reduction in fat intake as an important health goal. Experiments to discover how the increase in fat content of food varies with the way in which frying is carried out have shown that the increase can be minimized by using a combination of steaming, dipping in hot oil and baking in an oven rather than traditional deep frying.

When food is added to very hot oil there is rapid evaporation of water and natural juices from the surface of the food leading to rapid dehydration of the food surface. This produces the crisp texture and attractive flavour associated with fried food. Losses of nutrients which occur during frying have not been extensively investigated but in general they appear to be similar to losses which occur during roasting. Vegetables suffer a greater loss of vitamins when fried than when they are boiled. The loss of ascorbic acid in potatoes which are fried has been investigated but the results show considerable variation according to the cooking conditions. Retention of ascorbic acid was found to be greatest when the potatoes were cooked rapidly in deep fat and lowest when they were cooked slowly in shallow fat. When meat is fried some loss of B vitamins occurs the results being similar to those given for roasting in Table 14.6.

Microwave cooking

In the methods of cooking considered so far heat is applied to food from an outside heat source. In microwave cooking heat is generated within the food and the dramatic reduction in cooking time which results is the main advantage of this method of cooking.

The essential component of a microwave oven is a magnetron which converts electrical energy into microwave energy. The magnetron receives electrical energy at very high voltage and converts it into microwaves with an extremely high frequency (2450 MHz). Such microwaves come into the same category as visible, infrared and radio waves, all of which are non-ionizing forms of radiation (unlike X-rays). Microwaves travel in straight lines and are reflected by metals. Microwaves pass into the oven, being evenly distributed with the help of a stirrer or paddles, and are reflected by the metal sides, base and roof of the oven onto the food to be cooked. Microwaves penetrate the food to a depth of about 4 cm and food molecules which absorb the high frequency microwaves start to oscillate at the same frequency as the microwaves, namely 2450 million cycles per second! This rapid oscillation of the food molecules generates friction between them which produces heat.

The degree to which microwaves are absorbed by substances depends upon the dielectric constant of the material. Materials with a high dielectric constant absorb microwaves to a greater extent than those with low dielectric constants. Water has a particularly high dielectric constant compared with solid materials and thus in microwave cooking it is the water in the food which mainly absorbs the microwave radiation. Hence foods with a high water content cook more rapidly than those containing less water.

Fat has a much lower dielectric constant than water and consequently fatty parts of food heat up much more slowly than those with a high water content. It follows that foods containing a mixture of watery and fatty tissues, such as bacon, cook unevenly in a microwave oven. The more homogeneous a food is, the more evenly it is cooked by microwaves.

Some materials, such as glass, plastic, china and earthenware have very low dielectric constants and absorb very little microwave radiation. In consequence they are heated to only a small extent by microwaves and so make suitable containers for food which is cooked in this way. (Metal containers cannot be used because they reflect microwaves.)

Microwaves penetrate food to a depth of 3–5 cm depending on the composition of the food. Thus, when small pieces of food are exposed to microwaves the radiation completely penetrates it and heat is generated throughout the food resulting in rapid cooking. Larger pieces of food are cooked more slowly because those parts of the food which are not penetrated by microwaves are heated by conduction from the outer layers which have been penetrated by the radiation. Similarly, foods having an irregular distribution of water do not heat up uniformly as the water is heated rapidly by microwaves and surrounding regions are then heated more slowly by conduction.

Food cooked by microwave heating differs in some respects from food cooked by other methods. For example, the cooking time required when using microwaves is so short that slow chemical changes, which are important in slow cooking methods, do not have time to occur. Thus food cooked by microwaves does not turn brown or develop crispness, but it does retain most volatile substances; consequently the food has a different flavour from usual and this may make it less acceptable.

The fact that a simple microwave oven is unable to brown and crisp food has led to the development of combination ovens in which microwave heating is complemented by convection heating or heating with halogen lamps. Some combination ovens also provide a grill. The use of such ovens produces cooked food with similar characteristics to that produced in conventional ovens.

It is difficult to ensure even distribution of microwaves within a microwave oven and this can result in 'hot spots' within the food. To obviate this difficulty food may be cooked on a rotating turntable. Alternatively a stirrer fan situated in the roof of the oven and having rotating blades, which deflect microwaves off the metal walls of the oven, may be used.

The particular merit of microwave cooking is the short cooking time required. For example, a fish fillet is cooked in only 30 seconds, a chop in one minute, a chicken in two minutes and a baked potato in four minutes. The rapidity with which microwaves heat food makes them very useful as a means of quickly reheating pre-cooked foods, a process which takes only a few seconds. In canteens, restaurants and hospitals cooked food often has to be kept hot for long periods, with consequent loss of flavour and nutritional value, before it is eaten. In these circumstances the use of microwaves enables cooked food to be reheated rapidly just before it is eaten, thus eliminating the need for keeping the food hot. The use of this technique is the basis of most fast food establishments.

It is apparent that microwave cooking is convenient for both the *fast cooking* and the *fast reheating* of food. It is also valuable for a third purpose, namely the *fast defrosting* of frozen food.

Frozen food is normally thawed by allowing it to stand at room temperature. However, this is a lengthy process because most frozen food has a high water content – at least 60 per cent – and because water has a high specific latent heat of fusion. This means that to convert ice at $0°C$ to water at $0°C$ requires a large amount of energy supplied as heat. Microwaves, however, penetrate, heat and thaw frozen food very rapidly so making rapid thawing a practical possibility. For example, a turkey weighing 4.5 kg which takes about 36 hours to thaw at room temperature takes only about $3\frac{1}{2}$ hours to thaw in a microwave oven. Microwave cooking can be used to heat pre-cooked frozen food very rapidly and this is an important advantage of this technique.

NUTRITIONAL CHANGES　In theory, microwave cooking should result in a reduction in nutrient loss compared with other methods because it involves a shorter cooking time, a lower temperature at the surface of the food and the use of little or no added water. In practice, the changes or losses of nutrients that occur in microwave cooking are comparable to those occurring in other methods. Research on animal proteins and minerals suggests that microwaves have little effect on those nutrients.

Table 14.7 Comparison of vitamin C content in mg/100 g of fresh vegetables when cooked by microwave or boiling (after Cross and Fung, 1982)

Vegetable	Microwave	Boiled
Asparagus	20	18
Broccoli	117	73
Cabbage	43	25
Cauliflower	85	48
Runner beans	6	5
Spinach	24	15
Turnips	26	14

The loss of water-soluble vitamins, particularly vitamin C, in microwave cooking has been extensively investigated. Early experiments showed little difference in the loss of vitamin C in fruit and vegetables when cooked by microwave and by conventional means. However, more recent research, some results of which are summarized in Table 14.7, shows that the vitamin C content of vegetables cooked by microwave is higher than when they are boiled. It is considered that this is due to shorter cooking times and the use of little added water rather than any effect of microwave heating.

Raising agents

The baking of a dough or batter used in making bread, cakes or buns involves the use of an aerator which causes the mixture to rise during baking to give a product of even texture and, in the case of bread and sponge cakes, of large volume and open cellular structure.

Sometimes enough air may be incorporated by mechanical mixing to produce sufficient aeration during baking. Usually, however, carbon dioxide is used as an additional aerating agent. In bread making, carbon dioxide is normally produced by fermentation as described earlier, but in making other types of baked confectionery carbon dioxide is produced by a chemical *raising agent* or *baking powder*.

The simplest raising agent is *sodium bicarbonate* (strictly called sodium hydrogen carbonate) or 'baking soda' which produces carbon dioxide when it is heated:

$$2NaHCO_3 \xrightarrow{\text{heat}} Na_2CO_3 + H_2O + CO_2$$

It will be seen from the equation that the sodium bicarbonate is converted into sodium carbonate. If sodium carbonate is present in appreciable quantities it imparts an alkaline taste and a yellow colour to the product. This unfortunate result is sometimes noticeable in home-made scones in which too much baking soda has been used. For this reason sodium bicarbonate is only used in making products like gingerbread and chocolate cake, which have a strong flavour and colour of their own.

A baking powder is a mixture of substances which. when mixed with water and heated, produces carbon dioxide. For most purposes a baking powder is a more suitable aerator than baking soda. It consists of three ingredients: sodium bicarbonate as the source of carbon dioxide, an acid or acid salt to liberate the gas from the bicarbonate, and some form of starch, often cornflour or ground rice, as an inert filler to absorb moisture. A typical baking powder might contain 20 per cent sodium bicarbonate, 40 per cent acidic material and 40 per cent filler.

The action of a baking powder is most simply illustrated by the reaction of a solution of hydrochloric acid with sodium bicarbonate. On mixing, carbon dioxide is given off in the cold without any unpleasant tasting residue being left behind:

$$NaHCO_3 + HCl \rightarrow NaCl + H_2O + CO_2$$

In practice this mixture is not used, because reaction is so rapid that much gas would be lost before baking started. Also the use of an acid in solution is inconvenient, and the amount of acid needed would have to be most carefully controlled so that no free acid remained in the product after baking.

If hydrochloric acid is replaced by tartaric acid the evolution of gas, which occurs when water is added to the baking powder, is rather slower. On reaction with sodium bicarbonate, tartaric acid is converted into the harmless salt, sodium tartrate. In modern baking powders the acid is replaced by the acid salt, potassium hydrogen tartrate, better known as cream of tartar. This salt is less soluble in cold water than the acid, so that when the baking powder is mixed with water very little reaction occurs; when the mixture is warmed, however, a copious stream of gas is produced:

CHOHCOOH
| + NaHCO$_3$ $\xrightarrow{\text{heat}}$ CHOHCOONa
CHOHCOOK | + CO$_2$ + H$_2$O
                                        CHOHCOOK

Cream of tartar   Sodium potassium
                                        tartrate

A baking powder containing cream of tartar keeps better than one containing tartaric acid, because exposure to moisture has less effect. Also it is more convenient to use because carbon dioxide is not evolved in large quantities until the dough reaches the oven. As both the acid and the acid salt cost approximately the same amount, cream of tartar is normally preferred.

Several other acid salts may be used in place of cream of tartar. *Calcium hydrogen phosphate*, $CaH_4(PO_4)_2$, often called acid calcium phosphate (ACP), has the virtue of cheapness but, like tartaric acid, it reacts slowly with sodium bicarbonate in the cold when moisture is present; hence baking powders containing it have poor keeping qualities. *Disodium dihydrogen pyrophosphate*, $Na_2H_2P_2O_7$, usually referred to as acid sodium pyrophosphate (ASP), is preferred because of its superior keeping qualities. It is the acid salt of pyrophosphoric acid, $H_4P_2O_7$. Sometimes the two salts are used together. *Sodium aluminium sulphate* may also be used to replace cream of tartar.

Glucono-delta-lactone (GDL) is also used as the acid component in baking powder, particularly when making chemically leavened bread. It slowly hydrolyses in water or in a dough at room temperature producing *gluconic acid* and the rate of hydrolysis, and hence the rate of production of carbon dioxide, is markedly increased at higher temperatures.

The effectiveness of an acidic component of baking powder is measured in terms of its 'neutralizing value' or 'strength', and is defined as the parts of sodium bicarbonate neutralized by 100 parts of the acidic component. On this basis the strength of acidic substances used is as follows: ACP 80; ASP 74; cream of tartar 45; and GDL 45.

Instead of adding a calculated amount of baking powder to flour it is sometimes more convenient to use a self-raising flour. This is flour to which sodium bicarbonate and an acid substance are added in such proportions that on reaction the correct amount of carbon dioxide is produced to aerate the flour. One of the advantages of self-raising flour is that because the baking powder and flour are present in the correct proportions, there is no danger of too much or too little aeration, and unpleasant tastes are avoided.

The acid substances added to self-raising flour are the same as those used in baking powders. Both acid calcium phosphate and acid sodium pyrophosphate are used to a large extent, both singly and together. The large excess of flour present absorbs any moisture and so prevents the deterioration of the acid calcium phosphate.

The use of baking powder or self-raising flour causes some loss of nutritive value, notably of thiamine, during baking. Thiamine is stable at low pH values but at pH values of six or above it is rapidly destroyed by heat. Baked goods involving the use of self-raising flour or baking powder usually have a pH of about 6.8–7.3, and at this pH thiamine is rapidly destroyed during baking. Losses of thiamine may be very high in such baked goods, and during the baking of cakes, for example, all the thiamine may be destroyed.

SUGGESTIONS FOR FURTHER READING

GRISWOLD, R.M. *et al.* (1980). *The Experimental Study of Foods*, 2nd edition. Constable & Co., London.

MCGEE, H. (1986). *On Food and Cooking*. Allen & Unwin, London (An interesting and unusual book on the physics and chemistry of cooking.)

WEBB, L. (1977). *Microwave, the Cooking Revolution*. Forbes Publications, (An elementary but practical and readable review.)

CHAPTER 15

Diet and health

There is no such thing as a perfect or complete food, which means that there is no single food that provides sufficient of all the essential nutrients to keep us healthy. It follows that we need to eat a variety of foods to nourish us; also because different foods have widely differing nutritional contents we need to select the foods we eat in such a way that they provide us with a satisfactory or healthy diet. The information given in other parts of this book will not be of great practical value unless it can be applied so as to establish the nature of satisfactory patterns of eating. A discussion of diet therefore forms a practical and logical conclusion to the principles of nutrition discussed earlier.

MINIMUM REQUIREMENT Before considering diet itself it will be useful to define what is meant by nutritional need or requirement, for unless this is known it is impossible to plan healthy diets. An individual requires a certain amount of every nutrient each day. This amount must either come from the diet or from stores of that nutrient within the body. If it is the latter the store will be used up eventually and will need to be replaced. The amount of nutrient used daily in order to maintain good health is the physiological or minimum requirement.

While it is easy to define minimum requirement it is much more difficult to measure it. Different methods are employed for different nutrients. For example, for nutrients that are excreted from the body, such as protein, balance studies are used in which decreasing amounts of the nutrient are given until a further decrease in intake does not cause a decrease in excretion. This indicates that the minimum requirement has been given and that for this nutrient the body is in balance. For nutrients that are not excreted other techniques must be used, some of which are complex and time-consuming. For certain nutrients experimental methods to determine minimum requirement have not proved possible and so values are determined indirectly by observing actual intakes in healthy populations. Such indirect methods have been used for vitamin D, calcium and nutrients such as vitamins A and B_{12} of which there are large reserves in the body.

There are some nutrients for which a minimum requirement has not been determined because they are so widely distributed in food that it would be difficult to have a diet containing less than the minimum requirement. The minerals sodium and magnesium, many trace elements and vitamins E and K come into this category.

RECOMMENDED AMOUNTS   The minimum requirement of a nutrient may vary according to a person's age, size and level of activity. In addition it may not always be true that the minimum amount is also the optimal amount; it may be, for instance, that an intake of a nutrient greater than the minimum may be advantageous in combating disease. Also, the difficulty of determining minimum requirement means that the values obtained may not be very accurate. For all these reasons recommended amounts of nutrients rather than minimum requirements are used by national and international bodies in formulating dietary needs.

The recommended amount of a nutrient is the amount of the nutrient which should be provided per head in a group of people if the needs of practically all members of the group are to be met. The recommended amount does not refer to an individual but is an *average* amount that will be adequate for most people. It is usually calculated by adding 20 per cent to the minimum requirement figure and should therefore meet the needs of the great majority of people.

The recommended amount for energy differs from that for nutrients in that it is the same as the average minimum requirement. This is because it is undesirable for people to receive more than their requirement of energy as this can lead to obesity.

In the UK the recommended daily amounts of energy and nutrients are those formulated by the Department of Health and Social Security (DHSS) in 1979 and reproduced in Appendix I on page 396. The recommended amount of nutrients given in Appendix I are for *healthy* people – they make no allowance for additional needs resulting from any type of disease. Moreover, the amount recommended for any particular nutrient is based on the assumption that the needs for all other nutrients and for energy are satisfied. When the recommended amount of nutrients is used to plan diets that are nutritionally adequate, or for interpreting the results of dietary surveys, allowance must be made for any sort of wastage because the recommendations refer to food that is actually eaten. It is also important to appreciate that, because all nutrients can be stored in the body for at least a few days, it is not essential to consume the recommended amount every day.

The principles used in formulating the recommended amount of individual nutrients are discussed in the chapter concerned with that nutrient. Apart from the recommended amount given in Appendix 1, a number of national and international bodies have produced sets of recommended nutrient allowances, among the most important of which are those of the National Research Council (NRC) of the USA and the international standard compiled by the Food and Agricultural Organization (FAO) of the United Nations.

The nature of diet

The nature of our diet is important. As explained in Chapter 3, until recently the aim was to have a diet that was balanced, that is one which provided a mixture of foods which included sufficient of all the essential nutrients for the prevention of deficiency diseases. Such balanced diets could most easily be achieved by eating plenty of animal foods, such as meat, fish, cheese, butter, eggs and milk which have a high nutrient content, particularly of protein and fat, together with vegetables and fruit as a source of minerals and vitamins.

In Western countries, however, deficiency diseases are now rare and 'diseases of affluence' such as coronary heart disease and cancer have become commonplace. As explained in Chapter 3, such modern diseases are widely believed to be associated with diet. It is now considered that a more healthy diet can be achieved, not by a better balancing of nutrients, but by a careful selection of foods that will promote health. Thus animal foods should be eaten in moderation so as to reduce intake of fat, particularly saturated fat, while cereal foods, particularly unrefined cereal foods such as wholemeal bread, should be eaten plentifully so as to provide sufficient intake of fibre.

The pursuit of 'healthy diets' is promoted in many countries by governmental and other official bodies who issue guidelines on healthy eating. In Britain understanding of the term 'healthy diet' was promoted by the publication of the NACNE (National Advisory Committee on Nutrition Education) report of 1983 which presented a 'discussion paper on proposals for nutritional guidelines for health education in Britain'. This report, which surveyed the evidence and information contained in eight earlier reports, is a foundational document in seeking to establish the nature of a healthy diet. It establishes a new basis for evaluating diets and presents practical conclusions about changes that are needed in the British diet to make it 'more healthy'. These practical conclusions are considered at the end of this chapter.

NUTRIENT DENSITY   In considering diet it is useful to be able to compare nutrient values of different foods. The simplest way of doing this is to consider the relative amounts of a particular nutrient expressed per hundred grams of food. This has already been done many times in this book. Another way is to describe the amount of nutrient present in relation to a unit of energy, that is the *nutrient density* of the food. An example will make this clearer.

> Cheddar cheese has an energy value of 1682 kJ/100 g, a protein content of 26 g/100 g. Milk has an energy value of 272 kJ/100 g, a protein content of 3.2 g/100 g. The protein density of these foods is the amount of protein they contain per 1000 kJ.

> The protein density of Cheddar cheese is $26 \times \dfrac{1000}{1682} = 15.5$

> The protein density of milk is $3.2 \times \dfrac{1000}{272} = 11.8$

Such a comparison makes it very clear that the protein density of Cheddar cheese is much greater than that of milk.

Nutrient densities can be calculated for any food for all the nutrients that it contains. Some foods, such as processed foods, tend to have very low nutrient densities. This may be because nutrients are removed during the refining process as happens in the production of flour of low extraction rate. During the milling process the outer layers of the wheat grain, which are relatively rich in thiamine, calcium and iron, are discarded, so that flour of low extraction rate is relatively high in starch (and therefore energy) and relatively low in other nutrients. It follows that as the extraction rate is lowered the nutrient density of the flour falls. The nutrient density of processed foods may also be low because fat or sugar are added during processing thus increasing the proportion of energy-providing nutrients. For example, fruit yoghurt has a lower nutrient density than natural yoghurt because of the sugar added during the production of fruit yoghurt.

The use of the nutrient density concept is well illustrated in considering the diet of young children. Such children have a high requirement of energy and nutrients relative to their size in order to sustain a rapid growth rate. However, because they have small stomachs they cannot consume large quantities of food and so they require to eat foods with a high nutrient density. If they eat a bulky high fibre diet with a low energy content they will not receive sufficient nutrients to sustain growth.

NUTRIENT CONTENT AND CONTRIBUTION TO DIET   It is very important to make a clear distinction between the nutrient content or density of a food and the nutrient contribution which that food makes to our diet. Some examples will illustrate the point. Dried peas have a high energy value and are rich in protein (over 20 per cent) and some vitamins, notably thiamine. Yet the quantity of dried peas consumed in an average diet is so small that their contribution to our nutritional needs is very low indeed. Moreover, it is misleading to consider the nutrient content of dried foods if, before being eaten, they are soaked in water. When dried peas are soaked in water and cooked their water content increases by about 60 per cent with consequent decrease in their nutrient density. The contribution of all pulses (peas, beans and lentils) to an average diet is extremely small – less than one per cent of the total energy and protein and less than two per cent of the thiamine.

Compared with peas the nutrient content of potatoes is small. Their energy value is only one-third that of dried peas and they contain little protein (about two per cent) or vitamin C (10–20 mg/100 g). Yet because considerable quantities are eaten – on average about 1.25 kg/week – they contribute about five times more energy and protein to the diet than pulses. They also contribute over one-fifth of the vitamin C intake. Similarly, no-one would claim potatoes as a food rich in iron. They contain only one-twentieth as much iron as liver. Yet because we eat so much potato and so little liver, potatoes contribute over twice as much iron to the average diet as liver.

VARIETY IN DIET   A study of food habits of different races in different parts of the world reveals the fact that although nutritional needs remain essentially the same, the ways in which those needs may be met are unlimited. Diets may vary in an infinite number of ways and yet each diet may supply adequate amounts of the essential nutrients. Many variable factors, such as standard of living, custom, local abundance of certain foods and religious taboos all influence diet.

In highly industrialized countries such as Britain a wide variety of foods is available, whereas in less developed parts of the world the choice of foodstuffs may be more restricted. Such restricted diets, however, may be satisfactory and supply adequate amounts of all the nutrients. For example, a Central African tribe called the Masai exist on a diet which is largely composed of milk, meat and blood. These items are supplemented by concoctions made from barks of trees and certain roots which are drunk as beverages. This may seem a spartan and even repulsive diet to our more sophisticated taste, yet it keeps the tribe in good health and no deficiency diseases have been noted.

Another example from Britain itself illustrates the same point and shows that there is a remarkable difference in food habits even in such a small country. A survey carried out on the island of Lewis in the Hebrides some years ago revealed that the main articles of the diet were milk, fish, oatmeal, potatoes and turnips. To a town-dweller this must seem an unbearably frugal diet, but its monotonous character was dictated by necessity not by choice. The poor soil and lack of sun allowed only such undemanding crops as oats, potatoes and turnips to be grown successfully, while the lack of good grazing no doubt restricted livestock to a few cows which were, therefore, used mainly for milk. The resources of the sea were easily tapped to provide a varied supply of fish. In spite of its lack of variety the satisfactory nutrient content of this diet was shown by the fact that the inhabitants, both children and adults, were completely healthy.

During the 1939–45 war, when great efforts were being made by Britain to reduce food imports, a simple diet which would have been adequate nutritionally and would have required a minimum of imported food, was worked out. The main articles of this diet were milk, wholemeal bread and green vegetables. Scientifically such a diet would have been quite satisfactory but it was never adopted because it was thought that it would have been too unattractive for a modern industrialized nation.

These examples are not intended to show that diets restricted to a few foods only have some special virtue, but they do make it clear that such diets may be nutritionally satisfactory, provided that the foods selected are chosen with care.

Any diet is based on certain *staple foods* which form the bulk of the diet. The nature of these staple foods varies in different parts of the world. The Masai diet already quoted contains three staple foods: milk, meat and blood. The wartime diet,proposed for Britain in 1940 was based on three staple foods: milk, wholemeal bread and green vegetables. More generally it is found that staple foods are either cereals, such as wheat and rice, or starchy roots and tubers, such as potatoes and cassava. These staple foods provide a considerable proportion of the total energy and nutrients of the diet. In undeveloped countries the staple food is particularly important and when supply of that food is restricted or cut off because of crop failure, famine and death can quickly result. In Western countries staple foods are less important because a much greater variety of foods is available. Nevertheless, staple foods remain important and in Britain today wheat and potatoes are the staple foods in the diet as they have been for many years.

Although staple foods are the basis for any diet, the complete diet contains a number of other foods which contribute both variety and the supply of essential nutrients. Although, as we have seen above, satisfactory healthy diets may be achieved using only a few foods, it is normally desirable to eat a wide variety of foods. However, variety in diet does not of itself guarantee that the diet is a healthy one so that even in this situation food choice remains important. The nutritional guidelines discussed later in the chapter provide information which enable sensible choices to be made in the selection of food.

The British diet

The RDAs shown in Appendix I give the daily amount of energy and nutrients recommended for the maintenance of health, but the question of how these nutrients can best be supplied still remains. The most obvious way of forging this important link between recommended nutrient intakes and their practical application in terms of diet is to carry out dietary surveys and to relate diets which are nutritionally satisfactory to the analysis of nutrients contained in them. In Britain such surveys are carried out each year, diets being analysed not only in terms of nutrient content but also in terms of geographical and sociological differences. The results are summarized in the Annual Report of the National Food Survey Committee which provides valuable information on the way diet is changing in the country as a whole, in different regions of the country and in different classes of society. The data contained in Tables 15.1 and 15.2 are obtained from this source. Such data refers only to *household* consumption and so does not include food eaten outside the home.

In considering the average British diet it is convenient to group the wide variety of foodstuffs consumed into six main categories, namely; (a) milk, cheese and eggs; (b) meat and fish; (c) cereals; (d) fruits and vegetables; (e) fats; and (f) sugar and preserves. It is true that not all the items of our diet are included within these groups but excluded items, such as beverages, make a negligible contribution to the diet as a whole. The contribution of these food groups to the energy, protein, fat, calcium and iron in an average British diet is shown in Table 15.1, and their contribution to vitamins in Table 15.2.

Milk, cream, cheese and eggs. The importance of this group of foods is evident from Table 15.1. These foods are usually considered as valuable sources of high-quality protein, and after meat and fish they are the most important sources of protein in the British diet. Milk in particular is a notable source of dietary protein and in spite of its

Table 15.1 Per cent contributions made by important food groups to an average British diet

Food group	Energy	Protein	Fat	Calcium	Iron
Milk, cream, cheese, eggs	16	27	21	58	5
Meat, fish	17	36	26	5	23
Cereals	30	25	11	25	42
Fruits, vegetables	12	11	3	9	22
Fats	15	< 1	36	< 1	< 1
Sugar, preserves	8	–	–	< 1	1

Table 15.2   Per cent vitamin contributions to an average British diet

Food	Vitamin A	Thiamine	Riboflavin	Niacin	Vitamin C	Vitamin D
Milk, cream,						
cheese, eggs	21	12	44	3	5	21
Meat, fish	37	15	19	44	2	15
Cereals	1	50	20	19	1	11
Fruits,						
vegetables	23	20	10	23	90	–
Fats	18	–	–	–	–	52
Sugar, preserves	–	–	–	–	1	–

low protein density it provides 14 per cent of the total intake of protein as a result of the relatively large amounts consumed.

This group of foods supplies one-fifth of the fat in the British diet and also 16 per cent of the energy intake. Milk alone provides ten per cent of the total energy intake because, in spite of its high water content (88 per cent) and low nutrient density, consumption is high, as mentioned above.

Milk, cream and cheese are outstanding sources of calcium accounting for no less than 58 per cent of the total intake. Thus these foods are especially important during growth when the need for calcium is high. Milk alone provides 36 per cent of the calcium in an average diet.

Milk, cheese and eggs provide 44 per cent of the riboflavin in an average diet – more than any other food group. They are also valuable sources of both vitamin A and vitamin D, providing 21 per cent of the total intake of each.

Meat and fish.   Meat and fish make a major contribution to all the nutrients shown in the tables except calcium and vitamin C. Meat is the outstanding source of protein in the British diet, providing 31 per cent of the total intake. It also provides 25 per cent of our fat intake though much of this fat, particularly in lamb and beef, is saturated. Meat is a notable source of iron (21 per cent) and of B vitamins, notably niacin (34 per cent). Although meat appears to be an outstanding source of vitamin A (37 per cent), it needs to be appreciated that nearly all of this comes from liver.

Fish has a similar nutrient content to meat, but it contributes little to the diet owing to the small quantities eaten. Fish only makes a significant contribution to the intake of vitamin D (14 per cent) due to the high vitamin D content of fatty fish, such as herring and tuna.

Cereals.   Cereals include wheat and wheat products which are one of the staple foods of the British diet. Thus cereals form a major food group and make substantial contributions to every class of nutrient except fat. They contribute half the carbohydrate in the diet and provide more energy (30 per cent) than any other food group. They are also an important source of B vitamins, particularly of thiamine of which they supply 50 per cent of total intake.

Of all cereal foods in the British diet, bread is by far the most important. Despite a decrease in consumption from 200 g per day in the 1950s to 125 g per day at present, it still makes a major contribution to the diet. Although the overall consumption of bread has declined, consumption of brown bread is increasing and now accounts for about 20 per cent of all bread eaten. Bread of all types provides 13 per cent of the total energy in the diet. As explained in Chapter 7, certain nutrients – calcium, iron, thiamine and niacin – are added to flour (except wholemeal) to make up for nutrient losses that occur during milling. This explains why bread is such an important source of these nutrients; it provides 14 per cent calcium, 21 per cent iron, 25 per cent thiamine and five per cent niacin in an average diet.

Although cereals are not notably rich in riboflavin or niacin, they provide 20 and 19 per cent respectively of the total intake of these vitamins. This is because some cereal products, such as breakfast cereals, are fortified with these vitamins and because some cereal products are made with milk and eggs which provide them.

Fruits and vegetables.   Most fruits and vegetables are characterized by having a high water content, usually in the range 70–90 per cent, and consequently low nutrient density. Fruit adds almost nothing of nutritional value to the diet except vitamin C, of which they provide 40 per cent. Of this, ten per cent is provided by citrus fruit and 20 per cent by other fruit and fruit products, mainly fruit juices, most of which are rich in vitamin C.

Vegetables vary considerably in their nutrient composition and are eaten in widely varying amounts. Potatoes, as we have already mentioned, are a staple food of the British diet and consequently eaten in large amounts and throughout the year. Salad vegetables and Brussels sprouts, on the other hand, are seasonal vegetables and eaten in much smaller amounts. Vegetables provide 50 per cent of the total vitamin C intake, 22 per cent vitamin A (mainly from carrots) and useful amounts of thiamine (17 per cent), riboflavin (eight per cent) and niacin (15 per cent). Potatoes and pulses (peas and beans) supply most of these three B vitamins. Potatoes are a most important source of vitamin C and contribute 22 per cent of the total intake. They also provide most of the carbohydrate and protein contributed by vegetables to the diet.

Fats. These include all the culinary fats in the British diet, i.e. butter, margarine, low-fat and similar spreads, oils and cooking fats. These fats, whether used as table spreads or for cooking, provide 36 per cent of the total fat in the diet. In view of the current interest not only in the amount of fat consumed but also in its nature it is worth pointing out that the ratio of polyunsaturated fatty acids to saturated fatty acids in the diet (the P:S ratio) is 0.3. Fats contribute 15 per cent of the energy content of an average diet and 18 and 52 per cent of the fat-soluble vitamins A and D respectively.

Sugar and preserves. It is important to recognize that the figures quoted in Tables 15.1 and 15.2 relate to foods brought into the household, i.e. they do not include sweets and other sugary foods bought and consumed outside the home. Sugar consumption in Britain is about 38 kg per person per year, a figure which is high compared to consumption in other countries. Sugar and preserves brought into the household contribute only eight per cent of the total energy intake. The amount contributed by other sweets and confectionery is unknown, but for some groups of people, for example children, it may be significant.

Trends in the British diet

Social changes that have taken place over recent years have altered considerably the pattern of British eating habits. Formal family meals have been replaced to some extent by more informal eating styles. Cooked school meals planned to give a balance of nutrients have been replaced by individual choice and cafeteria-style eating or packed meals so that even young people have become accustomed to making their own choices about what they will eat. Workers are increasingly likely to have a midday snack rather than a complete meal. In the home breakfast is often not eaten, while TV snacks in the evening may replace a family meal. All these social changes tend to reinforce personal choice and increase the likelihood that processed foods, which are easily and quickly prepared, will be eaten instead of more conventional meals prepared from fresh foods.

The changes in eating patterns that have taken place in Britain over the ten year period 1975–85 can be appreciated by noting how consumption of particular foods has changed. Table 15.3 highlights the main changes. Considering staple foods, consumption of potatoes declined by ten per cent while the 50 per cent decline in the consumption of white bread was matched by a 50 per cent increase in the amount of brown bread eaten.

The pattern of fat consumption changed in favour of margarine, non-butter spreads and cooking oil (all up 50 per cent) at the expense

Table 15.3   Changes in British eating habits (1975–1985)

Change	Foods affected
Up 50%	Brown bread, frozen vegetables, fruit juice, cooking oil, margarine and non-butter spreads
Up 25%	Frozen fish, poultry (fresh and frozen)
Up 10%	Cereals, pickles and sauces
No change	Fresh fruit, meat products, processed vegetables (excluding frozen), lard, cheese, instant coffee, salt
Down 10%	Fresh green vegetables, fresh potatoes, bacon and ham, sugar
Down 25%	Fresh fish, flour, cakes and biscuits, carcass meat, milk, eggs, sugar, jam, tea, canned soup
Down 50%	White bread, butter, canned fruit

of butter (down 50 per cent) while consumption of sugar declined by ten per cent. This apparent reduction in sugar content may be misleading, however, as it refers to sugar bought as such and added to food directly in the home or used in cooking. It does not include the sugar added to processed foods. If this is taken into account it is probable that total sugar consumption over this period has been fairly constant.

Table 15.3 highlights the decline in the consumption of fresh and canned foods and the increasing use of frozen foods. It also notes the increasing popularity of fruit juice, consumption of which has increased so much that it is now a major contributor to our intake of vitamin C.

One item of diet not so far considered is alcohol. Alcohol consumption in Britain has been rising since about 1950 and alcoholic beverages currently supply 4–9 per cent of the total energy intake.

Dietary needs of special groups

The preceding discussion has been concerned with the average diet in Britain as assessed by annual surveys. While this is valuable because it allows general trends to be recognized, it does not help in evaluating the dietary needs of particular groups of people who have special needs. Such groups include pregnant and nursing women, babies and young children, elderly people and immigrants.

There are also other groups of people with special dietary needs, such as those who are ill, those who for reasons of health or fashion are slimming and those who for reasons of conscience or preference are vegetarians. Three examples of the nutritional needs of special groups are given below.

BABIES' DIETS *Breast feeding*. Mother's milk is the ideal food for a baby and all mothers should be encouraged to breast feed. It is desirable that breast feeding continues throughout the first year but especially for the first six months of a baby's life.

The advantages of breast feeding are well documented. Mother's milk provides the correct balance of nutrients for the baby's needs. The infant's requirements of energy and protein (Table 15.4) over the first six months of life are provided naturally and easily by breast feeding, and no nutrient supplements should be required as long as the mother is receiving an adequate diet. Nursing mothers are advised to take vitamin D supplements to increase their intake to the RDA of 10 μg.

Another advantage of breast feeding is that the milk is available at the right temperature and in the right quantity. Also with breast feeding the risk of infection is much less than in bottle feeding; the young baby is protected by antibodies and other substances in the mother's milk at a time when its own protective defences are not properly developed. Breast feeding reduces the risk of diarrhoea from contaminated milk because the milk passes directly from mother to baby without any external contact. Non-nutritional advantages of breast feeding include the fostering of a close physical relationship between mother and baby and a beneficial effect on the health of the mother. Finally, it is worth mentioning that babies are less likely to become obese when breast fed than when they are bottle fed because in the latter method there is a tendency to make feeds too strong and to add energy-rich cereal foods to the milk.

Bottle feeding. For many mothers there may often be a good reason why they cannot or should not breast feed, in which case they will bottle feed their baby using a commercial baby milk. Cow's milk has a composition very different from that of human milk and on its own is an incomplete food for babies; hence many attempts have been made to modify it so as to make it equivalent to breast milk. Although such

Table 15.4 Recommended daily amounts (RDA) of food energy and protein for infants

Age range in months	Body weight (kg)		RDA food energy (kJ)		RDA protein (g)	
	Boys	Girls	Boys	Girls	Boys	Girls
0–3	4.6	4.4	2200	2100	13	12.5
3–6	7.1	6.6	3000	2800	18	17
6–9	8.8	8.2	3700	3400	22	20
9–12	9.8	9.0	4100	3800	24.5	23

attempts have not been completely successful many commercial products are available that are satisfactory.

Commercial baby milks are normally in concentrated form – either dried or evaporated – and are reconstituted by the addition of water. Cow's milk is modified in a number of ways so that it more closely resembles human milk. The main objects are to reduce the mineral and protein content and increase the lactose content. In addition, such products are usually fortified with vitamin D as neither mother's milk nor cow's milk contain sufficient for the baby's needs. In some products the animal fat of cow's milk is replaced by vegetable oils. Ordinary skimmed milk is not suitable for babies because it contains less vitamin A and has a lower energy content than cow's milk.

Solid food. Although weaning, that is the transition from milk to solid food, will take place at different ages, few babies should be given solid food before the age of three months. Most babies, however, should be having some solid food by the age of six months.

There is no sound nutritional reason for introducing solid foods before three months and indeed there are positive dangers related to earlier weaning. For example, early addition of solids with high energy-density to feeds can produce obesity in babies and the early use of cereal foods containing wheat gluten can predispose the baby to *coeliac disease* (see Chapter 3).

When solid foods are introduced babies can be given most of the foods eaten by the rest of the family (apart from strongly spiced ones) provided that they are minced or sieved. Such ordinary food is to be preferred to commercial baby foods which are often sweetened and salted. When babies start to drink ordinary cow's milk they should start receiving vitamin supplements, particularly of vitamin D if there is too little exposure to sunlight to synthesize enough for the baby's needs. Cod-liver oil will supply vitamins A and D and orange juice will supply vitamin C. Alternatively, all three vitamins may be supplied as a vitamin supplement given in the form of drops. Such supplements should be continued until the age of two years and preferably up to five years.

In spite of the availability of suitable vitamin supplements the diets of some babies and young children are lacking in these vitamins and in a very small minority this deficiency is sufficiently serious to produce rickets and scurvy. It should be emphasized, however, that the number of children suffering from vitamin deficiency diseases in Britain is extremely small. With increasing understanding of the dietary needs of the young, better dissemination of such knowledge to mothers and better availability of vitamin supplements and fortified foods, rickets, which at the beginning of the century was a common disease among children, is now rare among British children, although unfortunately it is not uncommon among the children of Asian immigrants.

Apart from vitamins the main nutrient likely to be lacking in an infant's diet is iron. At birth babies have a reserve supply of iron which lasts for several months and this, together with the iron received from human or cow's milk, supplies their needs for some 4–5 months. After this time, however, they need to be given foods containing iron – sieved green vegetables, minced meat and eggs all being suitable. Although severe iron deficiency resulting in anaemia is not common in British children it is frequently found among immigrant children from countries, such as the West Indies, in which the traditional infants' diet is lacking in iron.

DIET OF THE ELDERLY   Elderly people may suffer from an inadequate diet for a variety of reasons, including loneliness, poverty, reduced enjoyment of food due to loss of taste and smell, mental and physical lethargy, or illness and inability to chew and digest food properly. In addition, elderly people often lack an understanding of the principles of nutrition. For these and other reasons elderly people more often suffer from malnutrition than the rest of the population. As the proportion of elderly people in Britain is increasing, the task of ensuring that they receive an adequate diet is of growing importance.

Although the dietary recommendations deal to some extent with elderly people, their nutritional needs require further investigation. It is known that energy needs decrease with age because of the reduction in physical activity and the recommendations concerning the levels of food energy intake appropriate to different age groups take this into account (see Appendix I).

Several studies have recently been carried out on the diet of the elderly. These conclude that the number of malnourished people is probably small and does not constitute a serious problem; it is believed that it is disease rather than malnutrition which is the primary problem. Some seven per cent of elderly people are malnourished, the nutrients most likely to be deficient being vitamins C and D and folic acid. Those over 80 years of age are more at risk than younger people. Obesity in elderly people, however, is a more important cause of nutritional disorder than lack of nutrients in the diet.

There are some particular hazards for older people. For example, old people who have difficulty in peeling fruit or cooking potatoes may lack sufficient vitamin C, while the housebound will have little or no chance of being in the sunshine and consequently may lack vitamin D. Vitamin D supplement may well be beneficial for older people, particularly in winter. Finally, old people suffer from loss of calcium from bone – a disease called *osteoporosis*. Although this condition cannot be prevented or remedied by diet, foods which are rich in calcium such as milk or cheese should be included in diets for the elderly.

SLIMMING DIETS The nature and effects of obesity and its prevalence in the UK and other Western countries where it is a major health problem have already been considered in Chapter 3. Here we shall consider ways in which diet can be used to reduce weight.

Some people slim for the simple reason that it is fashionable to be slim, but it is becoming increasingly recognized that many people are too fat and that this is objectionable not only for aesthetic reasons but for health reasons as well. People who are overweight impose a strain on the heart and other organs and they are more likely to suffer from mechanical disabilities, owing to the strain put upon joints and ligaments, than slim people. Moreover, fat people are predisposed to metabolic disorders. Statistics show that for men who are ten per cent overweight life expectancy is reduced by 13 per cent, while for those who are 30 per cent overweight it is reduced by 42 per cent.

It is estimated that 30 per cent of men and 65 per cent of women in the UK are following some sort of diet to control or reduce weight. Most people who try to slim do so by modifying their eating habits in some way and traditional methods of slimming mainly concern themselves with reducing energy intake by cutting down the amount of food eaten. This is because people become fat when the energy intake derived from food is greater than the total energy used by the body. The food which is surplus to the body's energy requirements is stored as fat in the fat depots of the body. In this connection it is important to appreciate that it is not only fatty foods which contribute to fat reserves; excess carbohydrate and protein also contribute.

It is evident from the above that fat reserves may be depleted by reducing the energy intake to below that used by the body. Alternatively, the energy intake can be maintained and the energy used increased by greater physical activity. Unfortunately, a great deal of exercise is needed to have much effect − about two hours strenuous exercise is needed to dispose of a good meal − and as increased exercise leads to increased appetite, loss of weight achieved through exercise is likely to be counteracted by subsequent increase in weight through extra eating and drinking.

In surveying different types of slimming diet some distinct trends can be recognized. In the 1960s and 1970s low-carbohydrate diets and calorie-counting diets were recommended while in the 1980s low-fat or high-fibre diets have been in vogue.

The essence of the low-carbohydrate diet is simple: if intake of carbohydrate foods is restricted there will be a considerable reduction in energy intake. It is argued that fat consumption will also be reduced. For example, if less bread is eaten then less butter will also be eaten. Thus, although the low-carbohydrate diet allows foods rich in fat and/or protein to be eaten without restriction, the hope is that reduction in bread consumption will also reduce consumption of fat associated with it.

Although the low-carbohydrate diet has in the past enjoyed great popularity it has helped create some of the nonsense which surrounds the subject of slimming. Carbohydrates have gained a bad reputation because they are considered to be particularly 'fattening' which is quite untrue. In this blanket condemnation of carbohydrates sugar and starch are both included. This view is in direct conflict with current nutritional opinion which attaches considerable importance to maintaining our intake of starch (and dietary fibre) while reducing sugar consumption.

Calorie-counting diets, once so popular, are now much less used. In such diets the slimmer can eat any foods provided that a selected energy intake is not exceeded. Such diets have the advantage that they permit complete freedom of choice but they have the major disadvantage that all food eaten must be weighed and energy values calculated. These are tedious activities and provide a strong disincentive for adopting this type of diet as a 'normal' way of eating.

Low-fat diets exclude or severely limit foods which are high in fat but allow other foods to be eaten freely. Fats have more than twice the energy value of carbohydrates or proteins so that a low-fat diet is likely to be a low-energy diet. This type of diet is currently recommended by many doctors because lowering fat intake, and particularly lowering intake of saturated fat, is considered to be healthy and may be a way of reducing the risk of coronary heart disease.

High-fibre diets are now very popular and aim at establishing a diet that has a low energy value but a high fibre content. Foods rich in fibre have the advantage that they have a low energy value while, because of their capacity to 'hold' considerable amounts of water, they provide bulk which gives a feeling of fullness. Moreover, as already mentioned, it is believed that an increased fibre intake is desirable in itself, particularly an increased intake of cereal fibre. Another advantage of a high-fibre diet is that it promotes the use of whole foods in place of refined convenience foods.

Dietary aids to slimming.   Many commercial products are available which are intended to aid slimming. The most drastic of these are complete diets in the sense that they replace normal foods completely. 'Very low calorie diets' have been developed as complete diets intended to produce very rapid loss in weight. They provide less than 2.5 MJ per day and a typical product would provide as little as 1.3 MJ per day. They also provide enough of all nutrients to satisfy recommended intakes. Such products are usually based on milk powder to which minerals and vitamins are added; they are often produced in the form of drinks which completely replace meals.

Very low calorie diets have become popular because they lead to rapid weight reduction; as much as 1.5 kg per week may be lost. Nevertheless, such diets need to be treated with caution because they are so drastic that they may produce unpleasant side-effects such as nausea and diarrhoea. They are most useful for very obese people who probably lose proportionately less body protein to fat than only moderately obese people when energy intake is severely restricted. Such diets are considered to be reasonably safe for up to four weeks, but should not be extended further without medical advice.

One obvious disadvantage of very low calorie diets and similar artificial eating regimes is that they do nothing to change eating habits or restrain the appetite. Once weight loss has been achieved through such a diet there is a strong likelihood that the slimmer will revert to previous eating habits and that the weight lost will soon be replaced. Such diets can only be commended for the extremely obese who require a brief form of 'dietary shock treatment' to initiate rapid weight loss. They should then transfer to a diet such as the low-fat or high-fibre diets already discussed, which will allow them to form new and permanent eating habits.

Apart from very low calorie diets there are a number of products available which are intended to aid slimming without replacing normal meals. The main types of such products are shown in Table 15.5.

One type of appetite-reducer contains glucose and is eaten, usually in tablet form, shortly before having a meal. There is some evidence to suggest that taking glucose in this way does reduce appetite, though the amount of glucose contained in such products is so small that it is

Table 15.5 The popularity and effectiveness of slimming aids (from Consumers' Association Survey)

Type	Per cent of slimmers using them	Verdict
Appetite reducers	17% for methyl cellulose type 22% for glucose type	Very few found them effective
Meal replacement (calorie counted meals)	50% had tried them	35% thought them helpful. Meals in drink form and in tins were best liked.
Low energy foods (a) sugar substitutes	90% had tried them 50% for pellet forms 25% for liquid forms	Majority found them helpful.
(b) low calorie drinks	67% had tried them	80% thought them helpful.
(c) low fat spreads	Most had tried them	Over 66% found them helpful.

unlikely that it could have an appreciable effect in reducing appetite. Those who have tried them (Table 15.5) confirm this.

Another type of appetite-reducing tablet contains *methyl cellulose* which when consumed absorbs water and swells in the stomach, giving a feeling of fullness. Such a product provides bulk but has no energy or nutrient value whatever. The amount of methyl cellulose contained in commercial products is so small, however, that they are unlikely to be of much practical benefit.

Meal replacement products are used to replace some (but not all) normal meals as part of an energy-controlled diet. They may be in the form of biscuits or confectionery bars or a powder which is added to milk. Such products contain a variety of nutrients as well as ingredients designed to make the slimmer feel full. Carrageenan, methyl cellulose and bran are all used for this purpose, the former being an interesting substance which is extracted from red seaweeds and which acts as a thickening agent by binding water. Thus the presence of carrageenan gives body to what would otherwise be a 'watery' drink and psychologically this gives slimmers the feeling that they have had a satisfying meal rather than a mere drink. Such products offer a convenient but expensive approach to slimming and suffer from the disadvantage that they do not help to establish better eating habits.

The most popular dietary slimming aids are low-energy versions of ordinary foods. The objective is to make products which resemble the original as closely as possible but which have a much reduced energy content. This is usually achieved by replacing energy-rich substances such as fat and sugar by energy-free substances such as air, water, bran and saccharin. Such products include sugar substitutes, low-energy drinks and soups, low-energy salad dressings and fats, starch-reduced breads and rolls, low-energy crisp breads and high-fibre bars and biscuits.

Guidelines for a healthier diet

The average diet in Britain today is nutritionally adequate in the sense that it provides the RDAs of nutrients recommended (Appendix 1). In spite of this, diseases of affluence such as coronary heart disease (CHD) and cancer, which are believed to be in some way related to diet, have become prevalent. Both CHD and cancer are major causes of death in industrialized countries and in Britain CHD is the leading cause of death. The nature of the link between these diseases of affluence and diet has been explored in Chapter 3 while the possible role that specific nutrients may have in these diseases has been explained in the chapters concerned with those nutrients.

The fact that diseases of affluence are considered to be in some measure diet-related has led to much concern about the nature of a healthy diet, and as mentioned earlier, the NACNE report of 1983 is an important attempt to establish guidelines as to what might constitute a healthy diet for the UK. The main recommendations of the NACNE report are given in Table 15.6.

The NACNE report gives recommendations both for the short term (the 1980s) and the long term (up to the year 2000), and presents its

Table 15.6 Nutritional guidelines proposed by NACNE (1983)

Dietary component	Current estimated intake	Proposal Long-term	Short-term
Energy intake	–	Recommended adjustment of the types of food eaten and an increase in exercise output so that adult body weight is maintained within the optimal limits of weight for height.	
Fat intake	38% of total energy	30% of total energy	34% of total energy
Saturated fatty acid intake	18% of total energy	10% of total energy	15% of total energy
Polyunsaturated fatty acid intake	–	No specific recommendation(a)	
Cholesterol intake	–	No recommendation	
Sucrose intake	38 kg per head per year	20 kg per head per year	34 kg per head per year
Fibre intake	10 g per head per day	30 g per head per day	25 g per head per day
Salt intake	8.1–12 g per head per day	Recommend reduction by 3 g per head per day	Recommend reduction by 1 g per head per day
Alcohol intake	4–9% of total energy	4% of total energy	5% of total energy
Protein intake	11% of total energy	No recommendation	

(a) In practice there is likely to be a greater consumption of both polyunsaturated and monounsaturated fatty acids, and a tendency for the ratio of polyunsaturated to saturated fatty acids to increase.

proposals in quantitative form where this is possible. Rather than giving vague advice about increasing or reducing the intake of certain nutrients it gives precise goals. It does so in the hope that these will provide a reference point which all those involved in providing nutritional advice and information can work towards. The distinction between short-term and long-term goals takes into account the relatively long time required for people to make radical changes to their eating patterns. It gives time for educationalists (hopefully aided by the media) to help the public understand the need for the changes proposed, time for food manufacturers and retailers to produce and promote 'healthier' diets and time for government to introduce any necessary new legislation.

The NACNE report recommends that for energy, intake should match output. It does not suggest that the overall intake of food should be reduced, though it does recognize the importance of the obesity problem. It strongly commends the practice of exercise from youth to old age both to promote energy output and health in general.

Fat in the diet, discussed in Chapter 5, is considered by many to be associated with CHD and the NACNE report recommends that total fat intake should fall to 34 per cent of the total energy intake in the short term and to 30 per cent in the long term. Such a reduction in fat consumption would also help in overcoming obesity. It is recommended that intake of saturated fat should fall to ten per cent of total energy intake in the long term. However, such a change would involve considerable change in eating patterns involving the eating of much less butter and other dairy products and meat, and it was considered that this would not be achievable in the short term. For the short term, therefore, it is recommended that intake of saturated fat should fall to 15 per cent of total energy intake and that some saturated fat should be replaced by polyunsaturated fat. This would mean an increase in the consumption of foods such as soft margarine which are rich in PUFA. Such a change would increase the ratio of polyunsaturated to saturated fats (P:S ratio) from 0.3 at present to 0.32.

The fact that no recommendation is made for cholesterol reflects the view that there is little evidence to implicate dietary cholesterol as a risk factor in CHD or other diseases of affluence.

The NACNE proposals for sugar and fibre may be considered together. It is recommended that sugar intake should be fairly drastically reduced until in the long term it is only about half the present consumption but that fibre intake should be substantially increased from 20 g per head per day at present to 30 g per head per day in the long term. Sugar consumption, which includes both sugar bought as such and added to food in the home as well as that used in manufactured foods, is exceptionally high in Britain. This is considered undesirable for several reasons including the fact that excess of sugary foods in the diet may reduce intake of other more nutritious foods and that sugar contributes to both obesity and dental decay.

Reduction of sugar intake would involve a reduction in the consumption of refined foods, whereas the recommended increase in fibre consumption would mean an increase in the consumption of unrefined or whole foods. Such a change represents a highly desirable step towards adopting a healthier diet. It means a considerable change in eating habits and eating less manufactured foods such as cakes, biscuits, sweets and confectionery as well as other manufactured foods in which sugar is a main ingredient. It means eating instead more whole foods such as wholemeal bread and wheat products made from wholemeal flour rather than white flour. It also means eating more cereals in an unrefined form such as brown rice as well as more fruit and vegetables.

The NACNE report recommends a reduction in our intake of salt. This is because salt intake is some 20 times higher than that needed for health, and because a high salt intake is undesirable for people with high blood pressure which is one of the main risk factors in CHD. Such a reduction in salt consumption could only be achieved with the cooperation of food manufacturers as three-quarters of our intake comes from manufactured food. Further reduction could be achieved by adding less salt during cooking and by not adding further salt at the table.

It is recommended in the NACNE report that in the long term average alcohol consumption should not exceed four per cent of the total energy intake. This is equivalent to nearly 20 g alcohol per day which means that average consumption should not be more than one pint of beer per day. Alcohol consumption varies greatly between individuals which explains why present consumption in Table 15.6 is given as a band of 4–9 per cent of total energy. Alcohol consumption has been rising in recent years and this is considered undesirable for health. NACNE recommend that no individual should drink more than four pints of beer per day.

Finally, the NACNE report makes no recommendations about increasing or decreasing protein intake. This is because if the other recommendations of the report are followed the diet will contain plenty of unrefined foods, especially cereal foods, and this will ensure that protein intake is adequate.

Since the NACNE report was published the DHSS has produced a report from the Committee on Medical Aspects of Food Policy (COMA) on the relationship between diet and CHD. The main recommendations of this report are given in Table 15.7. A comparison of Tables 15.6 and 15.7 shows some apparent differences, and the COMA proposals for fat appear to be less restrictive than the long-term NACNE proposals. However, in fact the two sets of

Table 15.7 Nutritional guidelines proposed by COMA (1984)

Ingredient of diet	Recommendation
Total fat	35% of total energy
Saturated fatty acids	15% of total energy
Polyunsaturated fatty acids	6.8% of energy
P/S ratio (Polyunsaturated/saturated fat)	0.45
Cholesterol	No specific recommendation
Simple sugars	No further increase
Alcohol	Excessive intake avoided
Salt	No further increase; consider ways of decreasing it
Fibre-rich carbohydrates	Increase intake

proposals are very similar because the NACNE recommendations include alcohol when assessing total energy intake, so that the *percentage* of total energy derived from fat is lower.

The COMA proposals are less comprehensive than the NACNE proposals and make no recommendations for either sugar or fibre in the diet, except to suggest that sugar consumption should not increase. The COMA recommendations for fat are not intended to apply to babies or young children under five who normally receive a major proportion of their energy intake from cow's milk.

The NACNE proposals give a general set of nutritional guidelines which offer a basis for an ongoing debate and the publication of further guidelines. Some nutritionists are concerned that nutritional guidelines, which present recommendations for the average person, cannot allow for the variations in people's needs, whether physical or psychological. For example, the NACNE recommendations result in a bulky diet containing more fibre and less fat than the traditional diet. While this may be desirable for most people it may not be appropriate for young children and people with small appetites who need to eat foods of high nutrient density if their nutritional needs are to be met.

Even if nutritional guidelines do not produce recommendations that can be applied to every individual, they do offer a way of improving the diet of most people without resorting to detailed dietary recommendations involving specific foods. Thus the goals proposed can be met in many different ways, so that individual choice of what specific foods to include in the diet is retained.

While the nature of specific guidelines may change, a broad consensus has now emerged about the changes that need to be made in Western diets to make them more healthy. The extent to which such diets may provide improved protection against diseases of affluence is unknown, though the adoption of such broad guidelines as the basis of the *average* diet can certainly do no harm. The consensus about the nature of these changes can be summarized as follows:

1. Eat less energy-rich food and so avoid being overweight.
2. Eat less fat, especially saturated fat.
3. Eat less sugar.
4. Eat less salt.
5. Drink less alcohol.
6. Eat more wholefoods, such as bread (preferably wholemeal), cereals, fruit and vegetables.

SUGGESTIONS FOR FURTHER READING

BRITISH CARDIAC SOCIETY (1987). *Coronary Disease Prevention.* BSC, London.

BRITISH DIETETIC ASSOCIATION (1987). *Children's Diets and Change.* BDA, London.

BRTTTSH DIETETIC ASSOCIATION (1985). *The Great British Diet.* Century Hutchinson, London.

BRITISH NUTRITION FONDATION (1985). *Nutrition During Pregnancy*, BNF Briefing Baper No. 6. BNF, London.

BRITISH NUTRITION FOUNDATION (1985). *Nutrition in Catering* , BNF Briefing Paper No. 4, BNF, London.

BRITISH NUTRITION FOUNDATION (1986). *Nutrition and the Elderly*, BNF Briefing Paper No. 9. BNF, London.

DEPARTMENT OF HEALTH AND SOCIAL SECURITY (1984). Committee on Medical Aspects of Food Policy (COMA), *Diet and Cardiovascular Disease.* HMSO, London.

DEPARTMENT OF HEALTH AND SOCIAL SECURITY (1988). Committee on Medical Aspects of Food Policy (COMA), *Present-day Practice in Infant Feeding*, Third Report. HMSO, London.

DEPARTMENT OF HEALTH AND SOCIAL SECURITY (1979). *Recommended Daily Amounts of Food Energy and Nutrients for the UK.* HMSO, London.

DEPARTMENT OF HEALTH AND SOCIAL SECURITY (1980). *Rickets and Osteomalacia.* HMSO, London.

DEPARTMENT OF HEALTH AND SOCIAL SECURITY (1987). Committee on Medical Aspects of Food Policy (COMA), *The Use of Very Low Calorie Diets in Obesity.* HMSO, London.

DRUMMOND, SIR J.C. AND WILBRAHAM, A. (1958). *The Englishman's Food: a History of Five Centuries of English Diet.* Jonathan Cape, London.

FOOD AND AGRICULTURE ORGANIZATION (1980). *Handbook on Human Nutritional Requirements.* HMSO, London.

FOOD AND NUTRITION BOARD, USA (1980). *Recommended Daily Allowances*, 9th edition. National Academy of Sciences, USA.

FOOD AND NUTRITION BOARD, USA (1980). *Towards Healthful Diets.* National Research Council, USA.

FRANCIS, D.E.M. (1986). *Nutrition for Children*, Blackwell, Oxford.

JOINT ADVISORY COMMITTEE ON NUTRITION EDUCATION (1985). *Eating for a Healthier Heart.* Health Education Council, London.

KENN, J.R. (Ed.) (1985). *Vitamin Deficiency in the Elderly*, Blackwell, Oxford.

MINISTRY OF AGRICULTURE, FISHERIES AND FOOD (annually). *National Food Survey Committee Reports: Household Food Consumption and Expenditure*. HMSO. London.

NATIONAL ADVISORY COMMITTEE ON EDUCATION (NACNE) (1983). *Proposals for Nutritional Guidelines for Health Education in Britain*. Health Education Council, London.

NATIONAL DAIRY COUNCIL (1988). *Nutrition and Children Aged One to Five*, Fact File 2. NDC, London.

ODDY, D. AND MILLER, D. (Eds) (1976). *The Making of the Modern British Diet*, Croom Helm, Beckenham.

ROYAL COLLEGE OF PHYSICIANS (1983). Report on *Obesity*. Royal College of Physicians, London.

SLATTERY J. (1986). *Diet and Health: Food Industry Initiatives*, University of Bradford, Food Policy Research Briefing Paper. Bradford.

WENLOCK, R.W. *et al.* (Eds) (1986). *The Diets of British School Children*. HMSO, London.

WHEELOCK, V. (1986). *The Food Revolution*. Chalcombe Publications, Marlow.

CHAPTER 16

Food spoilage and preservation

FOOD SPOILAGE

Most of our food is perishable and, in the natural course of events, it becomes inedible fairly quickly. Food spoilage occurs mainly as a result of chemical reactions involved in the processes of ageing and decay, through the action of microorganisms, or through a combination of both. A knowledge of why and how food spoilage occurs enables steps to be taken to prevent or minimize it. This, in essence, is the subject matter of this chapter.

Food which is not fresh is not necessarily harmful but it is usually less attractive to the consumer than fresher food. It may also be less nutritious and inevitably commands a lower price. Most unprocessed foods deteriorate fairly rapidly if kept at normal temperatures but some dry foods, such as cereal grains, will remain wholesome for long periods. Processed foods originate from a desire to make use of seasonal surpluses by prolonging the period for which they remain wholesome so that they can be used throughout the year. Butter, cheese, bacon and dried fruit are examples of early processed foods.

In addition to chemical spoilage and attack by microorganisms, drying, staling, contamination with dirt or chemicals and damage by animals or insect pests all play their parts in food spoilage. In many cases spoilage of this type can be avoided if care is taken in the transport and storage of food. Even though we now know a great deal about how food spoilage takes place vast losses still occur and it has been estimated that between 10 and 20 per cent more food would be made available if such wastage could be completely prevented.

Chemical spoilage

Almost all our food is produced by living organisms, whether they be animals or plants, and it is mainly composed, as we have seen, of organic compounds. In the living plant or animal these compounds are involved in a variety of complex and carefully controlled chemical reactions which, in the main, depend upon the presence of enzymes. When a plant is harvested or an animal is slaughtered many of these reactions cease. The enzymes present, however, will still be active and are able to continue catalysing reactions which can adversely affect the quality of the food.

When fruit is picked growth stops but it is still alive and ripening is able to continue. Once ripe, however, it will deteriorate rapidly, owing to the combined actions of enzymes and microorganisms, unless precautions are taken.

Vegetables, like fruit, remain alive after harvesting and they are prone to deterioration in the same way through the actions of enzymes and microorganisms and through loss of water. The ways in which fruit and vegetables deteriorate when stored are described in Chapter 13.

Changes occur in meat after an animal has been slaughtered and the initial changes are beneficial. Freshly slaughtered meat is liable to be tough and lacking in flavour. It is not usually eaten until certain posthumous changes have occurred which make it more tender and flavoursome. This process of 'hanging' or 'conditioning', as it is known, is discussed in Chapter 9 (see p. 185).

If meat is kept for too long at room temperature it becomes 'soggy' and unwholesome, partly owing to the breakdown of its proteins by proteolytic enzymes. Putrefaction will eventually set in with production of slime and foul odours caused by *Pseudomonas* bacilli: the meat will be offensive and inedible. For reasons that are not entirely clear the distinctive flavour of mild putrefaction is held to be desirable when venison, hare, pheasant and other game is eaten. Because of this these delicacies are hung for much longer than other meats.

In addition to spoilage caused by protein breakdown, meat may also suffer through oxidation of fats which are always present. Unsaturated fats are more likely to become rancid through oxidation and for this reason poultry, pork, lamb and veal cannot be kept as long as beef because they have a higher proportion of unsaturated fats. Oxidized fats are one of the main causes of 'off' flavours in cooked meats.

What has been said above about spoilage of meat applies also to fish but with greater force. The reason is that fish are cold-blooded creatures and their enzymes are able to operate efficiently at lower temperatures than those found in land-based animals. They continue to operate at freezing point and this is why fish 'goes off' more quickly than meat, even if it is kept in a refrigerator.

Microbial food spoilage

Microbes, or microorganisms, are extremely small living things. They range in size from certain algae just big enough to be seen by the

naked eye (about 100 μm) to viruses which are too small (about
0.1 μm) to be seen by a normal microscope. They can, however, be
seen by using an electron microscope.

Microorganisms need water and nutrients before they can multiply.
They cannot multiply on clean dry surfaces. Some of them, called
aerobes , also need oxygen from the air; others called *anaerobes* can
do without it.

Microorganisms like much the same type of food as we do and
moist food kept in a warm place is likely to be attacked by microorganisms which will feed on it and grow on its surface. Microorganisms do not multiply at low temperatures and they are killed by
high temperatures.

In addition to causing food spoilage some microorganisms, or the
toxins they produce, are harmful to human beings, and if food
contaminated by them is eaten *food poisoning* may result (see
Chapter 17).

Food which has been attacked by microorganisms may look
offensive or have a peculiar smell. In many instances, however, it is
not possible to tell by looking at a sample of food, or by tasting it,
whether it has been attacked. In fact, food may be heavily infected
and still appear to be wholesome and such foods are more likely to
cause food poisoning than those which have obviously deteriorated. It
must be emphasized, however, that the presence of microorganisms is
not always harmful. Indeed, many of the most highly prized food
flavours are a consequence of microbial activity. For example, blue
cheeses such as Roquefort, Stilton and Gorgonzola owe their characteristic flavours to the presence of the mould *Penicillium roqueforti*.

The microorganisms principally responsible for food spoilage are
moulds, bacteria and yeasts.

MOULDS   Moulds are a form of fungi. Unlike yeasts and bacteria
they are multicellular organisms. They grow as fine threads or
filaments which extend in length and eventually form a complex
branched network or mat called *mycelium*. At this stage mould
growth on foods is easily visible as a 'fluff'. Moulds also produce
spores, or seeds, and these can be carried considerable distances by air
currents and in this way infect other foods. Mould spores are almost
always present in the atmosphere and large numbers may be literally
'floating around' in food premises.

Most moulds require oxygen for development and this is why they
are usually found only on the surface of foods. Meat, cheese and
sweet foods are especially likely to be attacked by moulds. In alkaline
or very acid foods (pH below 2) mould growth is usually inhibited
although some moulds will grow even under these conditions. Moulds
grow best at a pH of 4–6 and a temperature of about 30°C; as the

temperature decreases so does the rate of growth, although slow growth can continue at the temperature of a domestic refrigerator. Mould growth does not occur above normal body temperature but it is very difficult to kill moulds and their spores by heat treatment. To ensure complete destruction of all moulds and their spores sterilization under pressure is necessary (i.e. at above 100°C). Alternatively, the food may be heated to 70–80°C on two or more successive days so that any spores germinating between the heat treatments will be destroyed.

Certain moulds produce poisonous substances known as *mycotoxins* which can be harmful if eaten. The mould *Aspergillus flavus* which can grow on groundnuts (i.e. peanuts) and cereals produces *aflatoxin* which causes serious illness if eaten. Moulds grow on other foods, including cheese and bread, and these are usually regarded as being unattractive but not harmful. It is possible, however, that they may contain dangerous mycotoxins and so it is safer not to eat them.

BACTERIA   Bacteria are simple single-cell organisms. They are minute living particles either spherical (cocci), rod-shaped (bacilli) or spiral (spirella). It is difficult to get any idea of the size of bacteria but 10^{13} of them (i.e. about 2000 times the population of the world) would weigh only about a gram. Bacteria and their spores are widely distributed. They are present in the soil, in the air and in and on human and animal bodies. Uncooked food of all descriptions will almost certainly be contaminated with bacteria. In fact, they occur so universally that it is exceedingly difficult to get away from them!

Bacteria grow by absorbing simple substances from their environment and when they reach a certain size the parent organism splits to form two new ones. In favourable circumstances this fission may occur every 20 minutes or so and in 12 hours one bacterium can provide a colony of some 10^{10} bacteria.

When bacteria multiply in or on food their presence becomes obvious when they are present to the extent of $10^{6}–10^{7}$ per gram of foodstuff. Bacteria grow most readily in neutral conditions and growth is usually inhibited by acids. Some bacteria, however, will tolerate fairly low pH. For example, *Lactobacilli*, which cause souring of milk with the production of lactic acid, and *Acetobacter*, which convert ethyl alcohol to acetic acid, flourish in acid conditions. Some bacteria will only grow in the presence of oxygen whereas others, known as *anaerobes*, will only grow in its absence.

Bacteria grow best within a given temperature range and in their vegetative stage (i.e. when they are actually growing) all of them can be killed by exposure to a temperature near to 100°C and at this temperature they are destroyed instantly. Bacteria, like moulds, can form spores. The heat resistance of bacterial spores varies from species to species and with the pH of the surrounding medium. This is considered in more detail on page 350.

The commonest food spoilage organisms are *mesophilic bacteria* which originate in warm-blooded animals but are also found in soil, water and sewage. Mesophilic bacteria grow best at about normal body temperature between 30 and 40°C and do not grow below 5°C. *Psychrophilic bacteria*, which have their origin in air, soil and water, grow best at a somewhat lower temperature – about 20°C – although some of them are quite happy at considerably lower temperatures. Such psychrophilic bacteria can grow quite easily at the temperature of a domestic refrigerator. A small group of bacteria can grow at temperatures up to 60°C and these are known as *thermophilic bacteria*. The spores of thermophilic bacteria can be very heat resistant.

Bacteria are not necessarily harmful to man and the species *Escherichia coli* exists in vast numbers in the human gut. Some bacteria present in the gut are positively beneficial because they synthesize some absorbable B group vitamins and vitamin K from the contents of the large intestine. Bacteria which do no harm to man are, nevertheless, capable of spoiling food.

YEASTS Yeasts are microscopic unicellular fungi but, unlike moulds, they reproduce themselves by budding, i.e. by the formation of a small off-shoot or bud which becomes detached from the parent yeast cell when it reaches a certain size and assumes an independent existence. Yeasts can also form spores but these are far less heat-resistant than mould spores and bacterial spores. Yeasts occur in the soil and on the surface of fruits. For example, the presence of yeasts on the skin of grapes is the reason why grape juice ferments to become wine. Yeasts can grow in quite varied conditions but the majority prefer acid foods (pH 4–4.5) with a reasonable moisture content. Most yeasts grow best in the presence of oxygen between 25 and 30°C. Some yeasts can grow at 0°C and below. Yeasts and yeast spores are easily killed by heating to 100°C.

Yeasts are used for making bread (and other fermented goods), beer and vinegar. They cause spoilage of many foods including fruit, fruit juices, jam, wines and meat. Although they may spoil food yeasts are not *pathogenic*, i.e. they do not cause diseases such as food poisoning.

Microorganisms attack food because they require energy and raw materials to support their metabolic processes and explosively rapid growth. In other words, like us, they need food – only more so. Microorganisms break down the complex molecules of the foods on which they are growing and convert them into smaller absorbable molecules. If microbes are allowed to grow unhindered their presence on food often (but not always) becomes apparent to the eye and, especially in the case of meat, to the nose.

Microbial growth usually ceases when the food has become really foul because the microorganisms themselves can no longer tolerate the conditions they have created. Organic acids are among the substances produced during microbial breakdown of food and these accumulate and eventually suppress further microbial growth. Given time, another species of microorganism may appear which is able to tolerate the more acidic conditions and attack on the food will then continue.

Vegetables and fruit have dry, relatively non-porous skins and their cell juices are mildly acidic. They are thus more likely to suffer from the growth of moulds and yeasts than from attack by bacteria. Yeast and mould spores are always present in the air but intact fruit and vegetables are not at great risk. However, if they are over-ripe or damaged, so that cell fluids leak onto the surface, mould or yeast growth is highly likely. If other conditions are favourable the food will deteriorate quickly. It is well known that 'one rotten apple in a barrel' can have a devastating effect. The reason is, of course, that the rotting specimen generates many millions of voracious microbes which quickly attack adjacent wholesome apples.

Fruit and vegetables will remain in good condition for the maximum possible time if they are clean, kept cool and handled with care.

Meat spoilage is mainly caused by bacteria and moulds although meat is not immune from attack by yeasts. Healthy carcass meat should be free from bacteria. In practice, the surface is usually contaminated from the hide and intestines when the animal is slaughtered and when the carcass is cut up. Poultry is particularly prone to bacterial contamination and the skin and interior surfaces usually harbour great numbers of bacteria.

When microbes grow on the surface of meat they break down the protein molecules and grow to form a film of bacterial slime. Carbon dioxide, hydrogen and ammonia are formed and the surface layer of meat becomes greyish-brown in colour owing to conversion of *myoglobin* to *metmyoglobin*. As putrefaction continues hydrogen sulphide, mercaptans and amines, all of which are foul-smelling, are formed and collectively demonstrate the inedible state of the meat.

It is very difficult to ensure that meat is completely free from bacteria and it almost always has some surface contamination. The bad effects of this can be minimized by storing meat at a low temperature (below $5°C$) to prevent bacterial growth. Before cooking it should always be wiped with a *clean* damp cloth to remove bacterial slime.

FOOD PRESERVATION

Microorganisms are present in the air, in dust, soil, sewage and on the hands and other parts of the body. They are so widely distributed that their presence in or on food is inevitable unless special steps are taken to kill them. If food is to be kept in good condition for any length of time it is essential that the growth of microorganisms be prevented. This can be done either by killing them and then storing the food in conditions where further infection is impossible or by creating an environment which slows down or stops their growth.

The ability to preserve food in good condition for long periods is an undoubted boon. The amount of food wasted is reduced and the incidence of food poisoning kept low. A wider range of foods is available including foods 'out of season' and foods from overseas that could not be transported and stocked in former times. The widespread use of preservatives, refrigerators, 'deep freeze' equipment and canned and dehydrated food has made it easy for the consumer or caterer to have available a wide range of wholesome food at all times of the year.

As well as suppressing the growth of microorganisms an effective method of food preservation must retain, as far as possible, the original characteristics of the food and impair its nutritive value as little as possible.

Chemical preservatives

Chemicals have been used in the preservation of foods for many centuries; sodium chloride, sodium and potassium nitrate, sugars, vinegar, alcohol, wood smoke and various spices have come to be regarded as traditional preservatives.

An example of the preservative action of concentrated sugar solutions has already been encountered in Chapter 7 in connection with jams and other sugar preserves. Condensed sweetened milk, which contains large amounts of sugar, is another excellent example of this principle. It can be kept for several weeks after opening the can without growth of microorganisms occurring. Microorganisms cannot tolerate high concentrations of alcohol and this is why fortified wines, such as sherry and port wine, keep better than unfortified wines. Similarly, vinegar discourages the growth of many microorganisms and it performs this function in 'pickled' foods.

PRESERVATION BY SALTING AND SMOKING   Meat and fish have been preserved by salting (or curing) since ancient times. The method is still used, often in combination with drying or smoking, in even the most primitive societies where salt is available. In medieval Britain weaker animals were killed off in the autumn because insufficient feeding stuff was available to keep them alive through the winter. Whole bullocks were salted and little or no fresh meat was available during the winter months.

Meat and fish are now preserved by refrigeration and the importance of curing as a method of preservation has diminished. Nevertheless, bacon and ham, salami-style sausage and corned beef all still feature in our diet. 'Lighter' cures are used today so that bacon, for example, is not as salty or as dry as in the past. In consequence it is more prone to bacterial spoilage and has to be treated almost as carefully as fresh pork.

Dry salting, in which meat or fish was buried in granular salt containing some sodium nitrate/nitrite, has almost died out. In wet curing processes a concentrated salt solution, or brine, is used. Sodium nitrate is traditionally added to the brine and some of it is reduced to sodium nitrite during the curing process. It is actually the nitrite which acts as the preservative and nowadays sodium nitrite itself is often used in place of sodium nitrate.

Cured meat (mainly pork) is usually made today by injecting the meat with a concentrated salt solution containing about five per cent sodium nitrate/nitrite and, for 'sweetcure' bacon, a little sugar. The meat is then immersed in a similar solution for a few days. Prepacked sliced bacon sometimes starts out as sliced raw pork and the curing solution is included in the pack. Curing takes place (or is supposed to take place) within the pack. Inevitably the product is somewhat wetter than traditionally cured bacon and it is not to everyone's taste.

Curing changes the colour of uncooked meat as a result of partial conversion of the protein *myoglobin* to the redder *nitrosomyoglobin* by nitrites present in the curing liquor. When bacon or ham is cooked (or 'corned beef' is further preserved by canning) the colour deepens owing to formation of a more complex nitroso-protein. The presence of nitrites in food may be harmful owing to the risk of nitrosamine formation (see p. 342).

Although nitrite ions (NO_2^-) are the main anti-microbial agents in cured meats the other salts present also help because they dissolve in the meat fluids to form a concentrated solution in which microbes cannot flourish. The dissolved salts 'capture' some of the water molecules so making them unavailable to microorganisms. The apparent water content (as far as the microorganism is concerned) is lower than the actual water content. The amount available is expressed as the water activity a_w of a sample of food. Water itself has an a_w value of 1.00 and a saturated salt solution an a_w value of 0.75.

The water activity of some foods is given in Table 16.1.

Bacteria flourish best on food with a high a_w value, provided of

Table 16.1 Water content and water activity of some foodstuffs

	Water content (%)	Typical a_w value
Uncooked meat	55–60	0.98
Cheese (Cheddar)	35–40	0.97
Bread	38–40	0.95
Jam	33–35	0.95
Cured meat	30–35	0.83
Honey	20–23	0.75
Dried fruit	18–20	0.76
Flour	14–16	0.75

course, that other conditions are also favourable. Many bacteria will not grow below an a_w value of 0.95 and an a_w value of 0.91 is the lowest water activity level tolerable by normal bacteria. Yeasts and moulds can tolerate much lower a_w values than bacteria. The minimum a_w figures tolerable by normal yeasts and moulds are 0.88 and 0.80 respectively.

Smoking is another ancient technique of chemical food preservation. Originally, smoke from an open fire was probably used but smoking was later carried out by hanging meat or fish (usually heavily salted) above smouldering wood chips in smoke houses. Traditionally smoked food has an outer layer consisting of condensed tars, phenols and aldehydes which have a powerful anti-microbial effect as well as a characteristic taste. The preservative effect is more-or-less limited to the surface of the food but spoilage of the interior is delayed because the outer layer acts as a bactericidal skin. Smoking is now used mainly to give flavour and colour to meat and fish and its preservative effect is of secondary importance.

Smoke contains many organic compounds and over 200 components have been identified. Among them are polycyclic hydrocarbons which are known to be carcinogenic. It is possible, therefore, that eating large amounts of traditionally smoked food over a long period could be harmful. As a precautionary measure smoke substitutes are now often used in place of real smoke. The smoke substitutes (known, somewhat improbably, as 'liquid smoke') are made by condensing the volatile substances present in smoke and separating the water-soluble components from the non-soluble carcinogenic polycyclic hydrocarbons.

Genuinely smoked foods have a brown smoky appearance as well as a smoky taste. 'Counterfeit' smoked food is often dyed to give the impression of thorough smoking. Kippers, for example, may be dyed with the permitted colour Brown FK – literally Brown For Kippers!

USE OF PERMITTED PRESERVATIVES Historically, many toxic substances have been used as food preservatives. Borates, fluorides and various phenols have all been used, but in the course of time it became apparent that their efficiency in killing microorganisms was coupled with considerable toxicity to man. This did not deter unscrupulous individuals from using them, however, often in injurious amounts. Preservatives were often used to mitigate the effect of unhygienic practices in the production and distribution of food. Milk, for example, remains fresh for comparatively long periods if first treated with formalin and this practice was once prevalent. Formalin is an aqueous solution of formaldehyde; it is extremely toxic and is widely used for preserving zoological specimens. The addition of any preservative to milk is now forbidden.

In Britain the use of chemicals to preserve food is strictly controlled by the Preservatives in Food Regulations (1979). In these regulations the word preservative means 'any substance which is capable of inhibiting, retarding or arresting the growth of microorganisms or any deterioration of food due to microorganisms or of masking the evidence of any such deterioration'.

The Regulations list the *permitted preservatives* and foods in which they may be used and specifies the maximum permissible amount of preservative which may be present. For the purposes of the regulations the following substances are not regarded as preservatives:

(a) any permitted antioxidant;
(b) any permitted artificial sweetener;
(c) any permitted bleaching agent;
(d) any permitted colouring matter;
(e) any permitted emulsifier;
(f) any permitted improving agent;
(g) any permitted miscellaneous additive;
(h) any permitted solvent;
(i) any permitted stabilizer;
(j) vinegar;
(k) any soluble carbohydrate sweetening matter, potable spirits or wines;
(l) herbs, spices, hop extract or essential oils when used for flavouring purposes;
(m) common salt;
(n) any substance added to food by the process of curing known as smoking.

It should be noted that most of the traditional preservatives are included in this list and hence, from a legal point of view, are not regarded as preservatives at all.

The permitted preservatives and *examples* of the foods in which they may be used are listed in Table 16.2. Each permitted preservative is identified by a serial number which may be used in place of the full

Table 16.2 Permitted preservatives

Permitted preservative	Examples of use
E200 sorbic acid	Soft drinks, fruit yoghurt, processed cheese
E201 sodium sorbate E202 potassium sorbate E203 calcium sorbate	Frozen pizza, flour, confectionery
E210 benzoic acid E211 sodium benzoate E212 potassium benzoate E213 calcium benzoate E214 ethyl 4-hydroxybenzoate E215 sodium ethyl 4-hydroxybenzoate E216 propyl 4-hydroxybenzoate E217 sodium propyl 4-hydroxybenzoate E218 methyl 4-hydroxybenzoate E219 sodium methyl 4-hydroxybenzoate	Beer, jam, salad cream, soft drinks, fruit pulp, fruit-based pie fillings, marinated herring and mackerel
E220 sulphur doxide E221 sodium sulphite E222 sodium hydrogen sulphite E223 sodium metabisulphite E224 potassium metabisulphite E226 calcium sulphite E227 calcium hydrogen sulphite	Dried fruit, dehydrated vegetables, fruit juices and syrups, sausages, fruit-based dairy products, cider, beer and wine, also used to prevent browning of raw peeled potatoes and to condition biscuit dough
E230 diphenyl E231 2-hydroxydiphenyl E232 sodium 2-phenylphenate	Surface treatment of citrus fruits
E233 thiabendazole	Surface treatment of bananas
E234 nisin	Cheese, clotted cream
E239 hexamine	Marinated herring and mackerel
E249 potassium nitrite E250 sodium nitrite E251 potassium nitrate E252 sodium nitrate	Bacon, ham, cured meats, corned beef and some cheeses
E280 propionic acid E281 sodium propionate E282 calcium propionate E293 potassium propionate	Bread and flour, confectionery, Christmas pudding

name on food labels. The serial numbers are in the 200s and if the preservative is approved for use throughout the European Economic Community the number is preceded by an E. Although there are 35 permitted preservatives many of them are actually alternative forms of a smaller number of 'parent' compounds.

Sorbic acid and its salts (E200–E203) are inhibitors of mould and yeast growth. The acid is used in soft drinks and yoghurt and the salts mainly in pastry-type products. Sorbic acid is a non-toxic unsaturated acid and it is probably dealt with by the body in the same way as naturally occurring unsaturated fatty acids.

$$CH_3-CH=CH-CH=CH-COOH$$

Sorbic acid

Benzoic acid, C_6H_5COOH, is the parent compound of another group of widely used permitted preservatives comprising the acid itself, its salts and esters and the salts and esters of 4-hydroxybenzoic acid (E210–E219).

Benzoic acid   Methyl 4-hydroxybenzoate

Benzoic acid is present naturally in some foods. In the body it combines with the amino acid glycine and is excreted as hippuric acid.

$$C_6H_5COOH + H_2NCH_2COOH \rightarrow C_6H_5CONHCH_2COOH$$

Benzoic acid   Glycine   Hippuric acid

The derivatives of benzoic acid are metabolized in the same way.

Sulphur dioxide and sulphites (E220–E227) are among the most widely used preservatives. Sulphur dioxide itself, SO_2, is a gas which has been used since antiquity to prevent the growth of unwanted organisms in wine. The free gas is now rarely used as a food preservative because its salts – sulphites, hydrogen sulphites and metabisulphites – are more convenient and equally effective. Unfortunately, sulphur dioxide and sulphites have a disagreeable taste and an after-taste which can be detected by some people at very low concentrations. When sulphites are used in foods which are to be boiled or cooked in some other way, however, most of the sulphur dioxide is driven off and so after-taste does not present a problem. Another disadvantage of sulphites is the fact that they rapidly destroy thiamine. Peeled potatoes kept in a sulphite solution (to prevent browning) lose a considerable part of their thiamine content.

Diphenyl and its derivatives (E230–E232) and *thiabendazole* (E233) are used to prevent moulds and other imperfections developing on the peel of citrus fruits and bananas.

Diphenyl   *o*-hydroxydiphenyl

Nisin (E234) is the only antibiotic which may be used as a food preservative. It is produced by certain strains of the organism *Streptococcus lactis* and it occurs naturally in milk and Cheshire and Cheddar cheeses. Its presence makes these cheeses relatively immune to spoilage from gas-forming bacteria. These bacteria, which are mainly *Clostridia*, cause blow-holes and sometimes cracks to appear in the cheese. Nisin is effective against a very limited range of organisms. It is not effective against Gram-negative organisms, moulds or yeasts but only against certain species of Gram-positive organisms. For this reason it is not suitable as a general-purpose food preservative but is attractive as a 'mopping-up' preservative for heat-processed foods such as canned foods, as heat-resistant spores are found only among the Gram-positive bacteria.

There are no medical uses for nisin and there is no danger that if bacteria develop a resistance to it they will also be resistant to other antibiotics. Nisin is a polypeptide and when eaten it is digested and absorbed without ill-effect in the same way as other polypeptides.

Nisin is also used as a preservative for cheese and clotted cream. As it is harmless to man no maximum permitted quantity is specified. This would, in any case, be difficult to establish because of the variable amounts which may be present through natural causes. The quantity of nisin required to prevent clostridial spoilage is about 2–3 ppm. Nisin prevents the development of bacterial spores; it does not kill them.

Hexamine (E239), or hexamethylenetetramine, $(CH_2)_6N_4$, is a bacteriostat derived from formaldehyde. It has limited applications; in weakly acid solution it is slowly converted to formaldehyde and this is probably the effective preservative.

The use of *sodium and potassium nitrite/nitrate* (E249–E252) for curing meat has already been discussed (see p. 336). They are also used in some cheeses.

Nitrites inhibit the growth of *Clostridium botulinum* bacteria which are responsible for the deadly form of food poisoning known as botulism. Without the use of nitrite many canned meats – especially large cans where heat processing on its own would be less effective in killing the extremely heat resistant *Clostridium botulinum* spores – would be less safe to eat.

Unfortunately, nitrites in food may be partly converted to nitrosamines through reaction with amino compounds. Nitrosamines have been shown to cause cancer when fed to animals but the amounts used in these feeding experiments were much greater than those likely to be consumed when cured meats are eaten. Nevertheless, as a precautionary measure the addition of nitrates or nitrites to food made for babies or young children is now prohibited.

The use of nitrites and nitrates in foods as preservatives is a subject of much debate and research. It highlights the dilemma of whether or not the use of such additives is justified. On balance – considering their contribution in reducing health hazards by destroying *Clostridium botulinum* – it would seem that their continued use is justified until any ill effects of the presence of minute amounts of nitrosamines in the body are known. In reviewing the use of nitrites and nitrates in food it has been recommended that the amount of nitrite permitted in a given food should be reduced to the *minimum* needed to prevent the growth of *Clostridium botulinum* in that food. There is no clear evidence that nitrates and nitrites in the average British diet are dangerous. But uncertainties still remain.

Even if the use of nitrates and nitrites as preservatives were no longer to be permitted we should still consume small quantities. Nitrates are widely used as agricultural fertilizers and find their way into our water supplies and vegetables. It has been estimated that the average Briton eats about 60 mg of nitrate per day and gets about another 10 mg from water. Vegetables are the main source in the diet, accounting for about 75 per cent of the total intake. They contain inhibitors to the formation of nitrosamines, however, and so their nitrate content may not be a problem. Beer brewed with water in which nitrate levels are high may be the major source of nitrate in the diet of 'heavy' beer drinkers.

Propionic acid, CH_3CH_2COOH, and its salts (E280–E283) are used in bread and other baked flour products as mould suppressants. Some bakers (perhaps bread manufacturers would be a more accurate term) use acetic acid, CH_3COOH, in the form of vinegar as an anti-mould agent rather than the more effective propionic acid. Vinegar is a *traditional* preservative – not a *permitted* preservative – and hence it is not associated on the label with one of the E numbers which consumers unjustly regard with such suspicion.

ANTIOXIDANTS The preservatives discussed above prevent or reduce attack by microorganisms, but they do not prevent deterioration of food through oxidation. Fatty foods and foods such as cakes and biscuits, which contain fat, are particularly prone to this type of spoilage and become rancid or 'tallowy' on keeping. The rancidity is caused by oxidation of the unsaturated fatty acid radicals in the triglycerides of which fat is composed. Fat-soluble vitamins are also destroyed by oxidation.

Antioxidants which occur naturally in fats tend to prevent the oxidative changes which produce rancidity. Chief among these is vitamin E, which is found widely distributed in vegetable oil-bearing tissues and to a smaller extent in animal tissues. These natural antioxidants, however, are usually not present in sufficient amounts to prevent completely oxidative changes which occur when food is stored and so antioxidants which are, in effect, preservatives with a special function, may be added.

Table 16.3 gives details of the antioxidants permitted in Britain and examples of the foods in which they may be used. The E numbers for antioxidants are in the 300s.

Table 16.3 Permitted antioxidants

Permitted antioxidant	Examples of use
E300 L-ascorbic acid	Fruit drinks; also used to improve flour and bread dough
E301 sodium L-ascorbate	
E302 calcium L-ascorbate	
E304 L-ascorbyl palmitate	Scotch eggs
E306 extracts of natural origin rich in tocopherols	Vegetable oils
E307 synthetic alpha-tocopherol	Cereal-based baby foods
E308 synthetic gamma-tocopherol	
E309 synthetic delta-tocopherol	
E310 propyl gallate	Vegetable oils, chewing gum
E311 octyl gallate	
E312 dodecyl gallate	
E320 butylated hydroxy anisole (BHA)	Beef stock cubes; cheese spread
E321 butylated hydroxy toluene (BHT)	Chewing gum
E322 lecithin	Low fat spreads; as an emulsifier in chocolate
diphenylamine	To prevent 'scald' (a discolouration on apples and pears)
ethoxyquin	To prevent 'scald'

Dehydration

Microorganisms require water in order to grow and reproduce; preservation by dehydration makes use of this fact. The water content of the food is reduced to below a certain critical value (which varies from food to food) and growth of microorganisms becomes impossible.

Dehydration is a time-honoured method of preserving food; sun-drying of fish and meat was practised as long ago as 2000 BC and dried vegetables have been sold for about a century and dried soups for much longer. A cake of 'portable soup', believed to have formed part of Captain Cook's provisions for his voyage round the world in 1772, is still in existence. It resembles a cake of glue and chemical analysis has shown that it has changed little in composition with the passage of years.

Dried fruits have been produced for many years by drying in the sun, but such unsophisticated techniques are not suitable for the dehydration of most other types of food. In modern practice many types of equipment are used for dehydrating food. Drying is usually accomplished by passing air of carefully regulated temperature and humidity over or through the food in tray driers, tunnel type driers or rotating drum driers. Heated vacuum driers are also used; the temperature necessary for dehydration under reduced pressure is much lower than that which would be required at ordinary pressures. In vacuum-drying the atmosphere above the food contains a much lower concentration of oxygen than in the normal methods of drying and this reduces the extent to which oxidative changes occur.

A modern development of vacuum-drying is *freeze-drying* in which *frozen* food is dried under high vacuum. It may seem surprising that frozen food can be dried at all but it is common knowledge that frozen puddles gradually 'dry out' in winter time and that washing will dry slowly on a clothes line even though it is frozen stiff. This is an example of sublimation – the ice becomes converted to water vapour without passing through the liquid phase. Drying by this method is very slow at normal pressures but it is speeded up tremendously in *accelerated freeze drying* (AFD) by reducing the pressure at which sublimation occurs and by supplying heat to provide the latent heat of sublimation of the ice. The rate of input of heat is carefully controlled so that the temperature of the food does not rise above the freezing point.

Freeze-drying is particularly attractive for drying heat-sensitive foods. Dehydration occurs without discolouration and sensitive nutrients such as vitamins remain unharmed. In most methods of dehydration the food has to be sliced or minced to present the maximum possible surface area to the hot air current which carries away the moisture. Large pieces of food, such as complete steaks, can be freeze-dried, however, and this is a great advantage. As ice at the surface of the food sublimes during freeze-drying the drying front recedes into the food until all the water has been abstracted. The highly porous product contains only a few per cent of water and it can be stored for long periods in moisture-proof packs at normal temperatures. Freeze-dried food can be rapidly rehydrated, by adding cold water, and the product closely resembles the starting material.

Freeze-drying is a relatively slow process and one which requires expensive equipment; freeze-dried products are therefore more expensive than foods dried by more conventional means.

Before vegetables are dehydrated, whether by freeze-drying or other methods, they are scalded or 'blanched' by immersion in boiling water or by treatment with steam. This inactivates oxidative enzymes such as *catalase*, *phenolase* and *ascorbic acid oxidase* and improves the stability of the dehydrated product. With coloured vegetables, blanching also improves the colour of the product. Some loss of water-soluble vitamins occurs during water-blanching but this can be minimized by allowing the concentration of water-soluble substances in the blanching water to build up. Sodium sulphite is usually added to the water used for blanching vegetables because this improves both the colour and ascorbic acid retention. In steam-blanching, vegetables may be treated with a sodium sulphite spray before steaming. Losses due to solution of water-soluble substances are much less with steam-blanching than with water-blanching. Blanching also destroys a large proportion of microorganisms present. For example, the microbial count is reduced by a factor of 2000 for peas and over 40 000 for potatoes.

It is not necessary to remove all the water from food in order to prevent the multiplication of microorganisms. Bacteria will not multiply in food with a water activity a_w (see p. 337) below 0.91. The minimum a_w level tolerable by most yeasts and moulds is 0.88 and 0.80 respectively. Most dehydrated foods contain less than 25 per cent water and have a water activity below 0.6. Freeze-dried foods contain practically no moisture.

Multiplication of microorganisms should not occur in properly processed dehydrated food but they are not immune to other types of food spoilage. Those containing fats are prone to develop rancidity after a period, particularly if the water content is reduced to too low a figure. This is true of potatoes but for non-fatty vegetables, such as cabbage, as much water as possible should be removed, because this helps to conserve ascorbic acid. The storage life of dehydrated food is much increased, and the loss of vitamin A and ascorbic acid much decreased, in the absence of oxygen. By completely filling the container with compressed dehydrated food the amount of oxygen can be reduced to a minimum. Replacement of the air in the container with nitrogen is far preferable: most dehydrated foods can be stored for two years or more in sealed tins in which the air has been replaced by nitrogen.

One of the great advantages of dehydrated foods is that they occupy very little space. Dehydrated potato in powder form, for example, has a volume only ten per cent that of ordinary potatoes.

Refrigeration and freezing

Microorganisms do not multiply nearly as rapidly at low temperatures as at normal temperatures. This is taken advantage of in the domestic refrigerator which is used for keeping foods for short periods. The temperature in such a refrigerator is usually about $5°C$, which is sufficient to chill the food and reduce the activity of microorganisms but insufficient to give a long storage life. This is because microorganisms are not killed and can still grow and reproduce but at a much slower rate. Moreover, enzyme action continues, although at a reduced rate, leading to chemical changes in the food and loss in quality.

Commercial refrigeration or chilling is applied to many foods including meat, eggs, fruit and vegetables. When meat is chilled the temperature is reduced to about $-1°C$ and it can remain in good condition for up to a month.

For large-scale use, chilling can be advantageously combined with gas storage, that is, storage in an atmosphere which has been enriched in carbon dioxide. Microbes produce carbon dioxide by their own respiration and addition of this gas to the atmosphere surrounding them retards their growth. Chilled beef, for example, will keep for ten weeks in an atmosphere containing 10–15 per cent carbon dioxide. Higher concentrations of carbon dioxide would be even more effective, but they are not used because they cause the meat to become brown, owing to the conversion of the haemoglobin to *methaemoglobin*.

Although chilling to about $5°C$ enables food to be stored for short periods it must be frozen and stored at a low temperature if long-term storage is required. Microorganisms, which are the main spoilage agents, become inactive at about $-10°C$ while enzymes, which cause chemical spoilage and consequent loss of quality, are largely inactivated below $-18°C$. Domestic freezers store food at about $-18°C$ but a temperature of $-29°C$ is employed commercially to ensure high quality and a long storage life.

Most fresh foods contain at least 60 per cent water, some of which – known as *bound water* – is tightly attached to the constituent cells the rest – known as *available* or *freezable water* – being mobile. On average, plant cells contain six per cent bound water and animal cells 12 per cent. Available water does not freeze at $0°C$ because of the solids dissolved in it which lower the freezing point. For example, at $-5°C$, 64 per cent of the water in peas is frozen, at $-15°C$, 86 per cent is frozen while at $-30°C$, 92 per cent (virtually all the available water) is frozen.

The *rate* at which foods are frozen is important. Good quality is only retained if freezing is quick, usually defined as meaning that the temperature at the thermal centre of the food pack should pass through the freezing zone $0-4°C$ within 30 minutes. It is within this temperature range that most of the available water is frozen and most heat (latent heat of freezing) must be removed.

The way that freezing rate affects quality can be appreciated from Fig. 16.1. Plant cells have relatively large vacuoles which contain most of the available water. During fast freezing tiny ice crystals are formed within the vacuoles and because they have little time to grow they do not distort the cellular structure. However, if freezing is slow, crystals start to form in the intercellular spaces outside the cell walls and as they grow they draw water from within the cells leaving the cells dehydrated and distorted. Some ice crystals may also be formed within the vacuoles. The process in animal cells – which have smaller vacuoles and which contain less water – is broadly similar.

If food is immersed in liquid nitrogen (boiling point $-196°C$), or if it is sprayed with liquid nitrogen, the liquid nitrogen boils as a result of rapid heat transfer from the food. Nearly instantaneous freezing occurs and the food retains its original shape and appearance. This technique, known as *cryogenic freezing*, or *immersion freezing*, is relatively expensive but it is useful for high cost products.

Loss of nutritional value on freezing and subsequent storage is

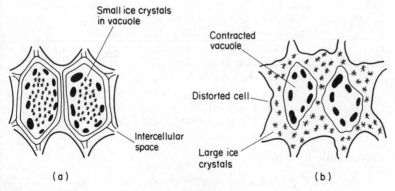

Figure 16.1 Plant cells: **(a)** after quick freezing; **(b)** after slow freezing.

small, but some losses do occur in the preliminary preparation of fruit and vegetables and during the storage of most frozen food.

In good commercial practice there is little delay between harvesting vegetables and freezing and consequently nutritional loss is insignificant. Vegetables, and some fruits such as apples, are blanched with boiling water or steam before freezing to destroy enzymes and some microorganisms. This causes some loss of water-soluble vitamins, mainly ascorbic acid and also, to a lesser extent, thiamine. The actual loss will depend on the way blanching is carried out but, overall, blanching conserves ascorbic acid by reducing the final cooking time required and by inactivating ascorbic acid oxidase, thus reducing loss of ascorbic acid on storage.

After freezing, foods are usually stored at $-18°C$ in home freezers or at $-29°C$ commercially. At these low temperatures there is a very slow and gradual loss of quality but little loss in nutritional value. Ascorbic acid is lost only very slowly on storage. If the temperature rises above $-18°C$ food starts to deteriorate more rapidly. For example, although strawberries can be stored successfully at $-20°C$ for over a year, at $-10°C$ they show some flavour deterioration after a few months.

When frozen foods are thawed there is often some loss of liquid – known as *drip* – which causes loss of soluble nutrients from the food. The extent to which drip occurs depends on the rate at which freezing is carried out, the duration and temperature of storage and the cellular nature of food. Plant material is more liable to drip than animal food because plant cells have larger vacuoles containing more available water (Fig. 16.1) and consequently suffer greater distortion on slow freezing. Fruit, particularly soft fruit such as strawberries, may suffer extensive drip and consequent loss of vitamin C on thawing. (Thawing of vegetables may also cause some loss of vitamin C and they are best cooked without thawing.) Soft fruits also suffer partial collapse of their cell structure on thawing, which makes them mushy. When frozen meat is thawed there may be considerable loss of soluble nutrients, including protein and B vitamins. However, loss of nutrients in drip may be avoided if the drip from meat is incorporated in gravy and the liquid (often syrup) from fruit is eaten.

In conclusion, it may be said that nutritional loss in food which has been properly frozen and stored is very small, and its nutritional value may well be superior to that of equivalent 'fresh' food which may have suffered a delay of several days between harvesting and consumption.

Preservation by heating

CANNING   Canning, which is the principal method by which foods are preserved by heat treatment, developed from bottling and, in essence, both processes are the same. The principle is delightfully simple – the food is sealed in a can which is then heated to such a temperature that all harmful microorganisms and spores capable of growth during storage of the can at normal temperatures are killed. As no microorganisms can gain access to food while the can remains sealed decomposition does not occur.

Almost any type of food may be canned and its nature largely determines what precanning operations are carried out. Food is first cleaned and inedible parts such as fruit stones, peel or bones are removed as far as possible. Fruit and vegetables may be subjected to a preliminary blanching before canning in order to soften them and enable a larger quantity to be pressed into the tin without damage. With vegetables, blanching also serves to displace air and causes a certain amount of shrinkage. The food is placed in the can which is then filled to within about half an inch of the top with liquor – usually sugar syrup in the case of fruits or brine in the case of vegetables. The lid is now placed loosely in position and the can and its contents are heated to about 95°C by hot water or steam. This process, known as 'exhausting', causes the air in the headspace of the can to expand and displaces any remaining air from the fruit or vegetable tissues. Exhausting also reduces strain on the can during subsequent heat treatment. It also substantially reduces the amount of oxygen in the headspace and so minimizes internal corrosion of the can and oxidation of nutrients, particularly ascorbic acid, after sealing. The can is sealed when exhausting is complete and it is then ready for heat sterilization or 'processing' as it is called.

Most canned food is processed in batch-type cookers which are large-scale, steam heated versions of the domestic pressure cooker. The temperature of processing is controlled by adjusting the pressure at which the equipment operates.

A great deal of work has been carried out to determine the optimum conditions for processing canned foods. Over-processing has an adverse effect on quality and it is desirable to reduce the time and temperature of processing as much as possible. Processing conditions must be severe enough, however, to ensure that all harmful microorganisms in the canned food are destroyed or inactivated. Bacterial spores are easily killed by heating in acid conditions, and the temperatures at which fruits are processed are not as high as those used for vegetables and meat.

Canned vegetables and meat are usually processed at 115°C, whereas fruits may be processed in boiling water. The size of the can and the physical nature of the food it contains are other factors which influence the amount of heat processing needed because they both affect the rate of heat penetration. If a liquid is present heat is distributed to all parts of the can by convection currents. With solid foods, on the other hand, the rate of heat penetration is slower and the time of heating must be correspondingly greater. Processing times for non-solid foods can be reduced by up to two-thirds by agitating the contents of the can, as this assists heat penetration.

HTST canning.   By substantially increasing the temperature at which it is carried out it is possible to reduce considerably the duration of heat processing: in theory, if very high temperatures could be used processing times could be very short indeed. In practice, however, the rate at which heat penetrates to the centre of the food in a can imposes a limitation on such high temperature short time (HTST) processes and so they can only be used for processing food *before* canning. Sterilization is carried out at about 120°C in special equipment designed to achieve a high rate of heat transfer. The food is then cooled somewhat before sealing into cans which have been previously sterilized with superheated steam. This procedure, known as *aseptic canning*, can be used at present only for liquid or semi-solid foods where a high rate of heat transfer to a thin film of the food is possible. The heating time varies from six seconds to about six minutes depending upon the type of food being canned.

An advantage of HTST processes is that as the food is cooked in thin layers there is less likelihood of some of it being *over*-processed to ensure that all of it is adequately processed. This, of course, is the situation with normal 'in-can' processing. Another advantage is that large cans, convenient for large-scale catering, can be used because there are no problems about heat penetration to the centre of the can.

Heat resistance of microorganisms.   Bacteria, moulds and yeasts are rapidly killed by the temperatures used in canning foods (see p. 330). Bacterial spores, however, can be very resistant to high temperatures. The death of bacterial spores in heat treated food follows a logarithmic course in which equal proportions of surviving cells die in each successive unit of time. Thus if 10 000 spores per unit volume were initially present and 9000 were killed by exposure to a particular temperature for one minute, 900 would be killed in the second minute, 90 in the third minute, 9 in the fourth minute and so on. One thousand times as many spores would be killed during the first minute's exposure as during the fourth.

The heat resistance of a particular microorganism can be expressed in terms of its *thermal death time* (TDT) at a particular temperature. The TDT is the time required to achieve sterility in a culture or preparation of bacterial spores containing a known number of organisms in a specific medium. At a particular temperature the TDT of a given spore preparation depends upon the number of spores present. The relationship between TDT and temperature is logarithmic and when TDTs are plotted on a logarithmic scale against temperature on a linear scale a straight line results, as shown in Fig. 16.2. A thermal death time curve is characterized by the TDT at some particular temperature, usually 121°C, and the slope. The symbol *F* is used to designate the TDT at 121°C and *z* to designate the slope. The slope of the curve is defined as the temperature difference, in Fahrenheit degrees, required to produce a tenfold increase or decrease in TDT, i.e. for the TDT to traverse one log cycle. In Fig. 16.2, *F* is 2.78 minutes and *z* is 18.

Thermal death time curves can also be drawn for vegetative bacteria (as opposed to bacterial spores). The TDT of vegetative cells is almost zero at 121°C and so 65.5°C is used as the reference temperature for these organisms. To avoid confusion the TDT of vegetative cells is given the symbol F^1.

Thermal death times may be used to compare the heat resistance of organisms, but it is also convenient to have some method of comparing the effectiveness of various heat processing *procedures*. This is done by adapting the concept of the TDT and using the symbol *F* to denote the *sterilizing value* or *lethality* of a process. The *F* value of a process is defined as the number of minutes at 121°C (assuming instantaneous heating and cooling at the beginning and end of the exposure) which will have a sterilizing effect equivalent to that of the process. If the sterilizing value is calculated on the basis of $z = 18$ the symbol F_0 is used. The symbol F_1 is used for the lethality of pasteurizing processes, the reference temperature in this case being 65.6°C.

Conditions drastic enough to kill the spores of the organism *Clostridium botulinum*, which are heat-resistant above pH 4.5, and particularly dangerous (see p. 365), will also kill all other harmful organisms. The heat-resistance of these spores forms a standard of

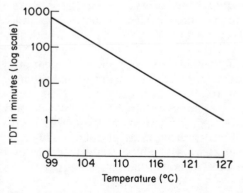

Figure 16.2 Thermal death time curve for *Clostridium botulinum* spores.

comparison against which the efficiency of a heat treatment process can be judged. Figure 16.2 shows that a preparation of *Clostridium botulinum* containing 6×10^{10} spores was killed by exposure to $121°C$ for 2.78 minutes. This figure has been approximated to three and used as the minimum sterilizing value acceptable for a safe process, i.e. an $F_0 = 3$ process. Such a process is referred to as a minimum safe process or, less formally, as a 'botulinum cook'. The spore preparation mentioned in Fig. 16.2 passed through 12 decimal reductions (i.e. reduced successively to one-tenth 12 times) before no spores were detected and hence a process with an F_0 value of 3 is sometimes referred to as a 12D process. The possibility of a spore of *Clostridium botulinum* surviving such process is less than one in 10^{12}. From this it is apparent that the risk of an outbreak of botulism occurring as a result of eating properly canned food is very small.

A few viable microorganisms may remain in canned food but they are unobjectionable and, in normal circumstances, will not cause spoilage. Spores are unable to develop in acid foods and the processing given to fruits is designed primarily to kill moulds, yeasts and non-sporing bacteria, the presence of bacterial spores being quite acceptable. In 'non-acid' foods only highly resistant thermophiles will survive the high processing temperatures and under normal storage conditions they will be unable to develop.

Canned food cannot really be said to be sterile but it is as sterile as it need be – a condition euphemistically described as 'commercial sterility'.

Nutritive value of canned foods.   Some nutrient loss occurs during heat processing and more thiamine may be lost from meat during processing than would be lost during normal cooking. Reduction in ascorbic acid content also occurs during processing but much more disappears during the first few weeks of storage as a result of oxidation by the small amount of oxygen remaining in the headspace of the can. Further destruction of thiamine may occur during storage but in normal conditions this should not exceed 10–15 per cent during two years' storage.

Apart from the losses of thiamine and ascorbic acid mentioned above, canned foods are quite as good, from a nutritional point of view, as corresponding fresh foods. Indeed, canned fruits and vegetables may be better because they are often canned within a few hours of being picked and this reduces precanning losses of ascorbic acid to a minimum. The total loss of ascorbic acid in canned fruit and vegetables may be much less than in 'fresh' vegetables bought in a semi-fresh condition and cooked at home.

Spoilage of canned foods.   Properly canned food remains edible for very long periods if the cans are not corroded. In 1958 a number of cans which had been sealed for many years were examined. A tin of plum-pudding prepared in 1900 was opened and the contents were found to be in excellent condition. The meat in two cans sealed in 1823 and 1849 was found to be free from bacterial spoilage but the fat was partially hydrolysed into glycerol and fatty acids. The contents of a number of cans taken to the Antarctic by Shackleton in 1908 and Scott in 1910, and brought back to this country in 1958, were found, with some exceptions, to be in first-class condition.

When spoilage of canned foods occurs it is commonly caused by a defect in the can. Spoilage may also arise from inadequate heat treatment which is insufficient to kill all the microorganisms present in the food. Certain heat-resistant bacterial spores produce acids when they germinate in foods and if this happens a *flat sour* results. No gas is produced and the spoilage is not evident until the can is opened and the unwholesome smell of the contents becomes evident. The organism particularly responsible for flat sours is *Bacillus stearothermophilus*, the spores of which are able to survive exposure to 120°C for 20 minutes. Non-acid foods such as peas are most likely to be affected. The organism finds its way to the food via infected equipment or ingredients such as sugar or flour, and spoilage of this type may be an indication of low standards of hygiene at the canning plant.

Another type of spoilage to which canned foods are prone is the *hydrogen swell* or *hard swell*. This is caused by heat-resistant bacteria such as *Clostridium thermosaccharolyticum* which produce hydrogen gas as they grow in the canned food. The ends of the can may bulge, as a result of increased pressure, to produce what is known as a *blown can*.

Improperly canned food sometimes smells offensively of bad eggs and may be very dark in colour. This is an example of sulphide spoilage (or, in more descriptive American parlance, a sulphide stinker!) and is caused by the presence in the can of *Clostridium nigrificans*. This organism produces hydrogen sulphide gas which is responsible for the foul smell. Not enough gas is produced to cause distortion of the can. Spoilage of this type is not common in Britain.

The three types of spoilage organism mentioned above are not harmful but they make the canned food unfit to eat. The fact that these organisms have survived the heat treatment process, however, indicates that the food has been inadequately heat treated and there is the possibility that more harmful organisms such as *Clostridium botulinum* may also be present. Spoiled canned food should never be eaten.

Canned foods should be stored in dry, fairly cool conditions because storage at higher temperature will encourage the growth of any thermophiles which have survived heat processing. Cans stored in damp conditions may become rusty and, in time, penetration of the can and spoilage of the contents may occur.

HEAT TREATMENT OF MILK   Milk is such a rich source of nutrients that it is an ideal medium for the growth of microorganisms. Although milk should be practically free of bacteria at the time it is obtained from a clean and healthy cow, it is almost impossible to maintain it in this condition. Bacteria from the milk container, from the milker or milking machine and from the air pass into the milk, where they find congenial conditions in which to flourish. In addition to this, unhealthy cows contribute disease-bearing bacteria to the milk, the most dangerous of which is the *tubercle bacillus*. In the past this microorganism has caused thousands of deaths annually in both cattle and humans. In Britain in 1931 2000 people died from tuberculosis contracted from milk. Other organisms such as *Brucella abortus*, which causes the disease brucellosis or undulant fever, and *Streptococcus pyogenes*, which causes sore throats and scarlet fever, may also pass from infected cows to man via untreated milk.

Most milk in Britain is now heat treated to ensure that harmful organisms are destroyed before it is consumed. As well as preventing the spread of disease, heat treatment of milk also considerably improves its keeping properties since lactic bacilli which cause milk to become sour are also killed.

Pasteurization of milk.   Milk is pasteurized by heating it to at least 72°C for not less than 15 seconds, after which it is rapidly cooled to less than 10°C. Over 99 per cent of the bacteria present are killed and although the product is not completely sterile all harmful organisms are destroyed. The organisms which are not killed, together with any heat-resistant bacterial spores, are inactivated by the rapid cooling which follows pasteurization.

The temperatures used in pasteurization are not high enough to cause any significant physical or chemical changes in the milk, so that there is no noticeable change in palatability due to pasteurization. In some cases pasteurized milk has a flavour different from that of fresh milk but this is due to faulty pasteurization; either the milk has been heated to a temperature higher than normal or the process has been carried out in unsuitable equipment and the milk has thereby become tainted.

Pasteurization causes some slight decrease in nutritional value but, unless the recommended temperature and time of pasteurization are exceeded, only ascorbic acid and thiamine are appreciably affected, some 10–20 per cent of each being lost. Milk is not an important source of these vitamins, however, and even if this loss did not occur, it would only supply a small proportion of the body's needs. In addition, ascorbic acid is destroyed by storage in direct light, especially sunlight, so that even non-pasteurized milk is a doubtful source of this vitamin.

Sterilized milk. Milk that has been homogenized, filtered and subjected to heat treatment so it will remain in good condition in an unopened bottle for at least a week and usually for several weeks, is sold as sterilized milk.

The object of homogenization is to break up the oil globules so that they will remain uniformly distributed through the milk and not form a layer at the surface. This is achieved by heating the milk to 65°C and forcing it through a small aperture under high pressure. This breaks up the oil droplets and the fine emulsion so formed is stabilized by protein which is adsorbed at the surface of the oil droplets. Although such milk is said to be homogenized it is not homogeneous in the scientific sense, for by definition an emulsion must contain two separate phases. After homogenization the milk is filtered, sealed into narrow-necked bottles and heated to at least 100°C and maintained at this temperature for up to an hour. In practice, higher temperatures and shorter heating times are used, a typical example being 112°C for 15 minutes.

Ultra-high temperature (UHT) sterilization of milk is carried out before the milk is bottled. The milk is first homogenized and it is then heated to not less than 132°C for 1–3 seconds by flowing over a heated surface. A completely sterile product is thus obtained which after cooling is packed in sterile containers.

Sterilization by traditional methods causes a change in the flavour and physical composition of the milk and also causes a slight decrease in nutritional value owing to loss of vitamins. Lactalbumin and lactoglobulin are coagulated, some calcium phosphate is precipitated and about 30 per cent thiamine and 50 per cent ascorbic acid are destroyed. Milk sterilized by the UHT process resembles pasteurized milk much more closely, both in flavour and vitamin retention. The main virtues of sterilized milk are that its cream content is uniformly distributed, it is safe and it can be kept for considerable periods.

Preserved forms of milk. Milk is a valuable food but, as we have already seen, it is perishable and even after pasteurizing needs rapid distribution and careful handling. It is also bulky and expensive to transport. For these reasons, and also because seasonal surpluses of milk occur, effective methods of preserving milk in concentrated form are most desirable. Milk may be concentrated by removing a proportion of its water and it may be rendered safe by suitable heat treatment. The water is removed by evaporation, this being carried out in closed vacuum pans under reduced pressure. The temperature is kept below 70°C so that the proteins are not coagulated and the production of a cooked flavour is avoided. Evaporated milk, also called unsweetened condensed milk in Britain, is pasteurized milk which after evaporation is homogenized and then sterilized in sealed cans. Condensed sweetened milk is made in a similar way, except that sugar is added and homogenization omitted. After evaporation further heat treatment is not necessary, because of the preservative action of the sugar.

Evaporated milk still contains about 68 per cent water and a much greater proportion of the water in fresh milk may be removed by drying it, usually by spray-drying. In this process the milk is first concentrated under vacuum at a low temperature and it is then dried by spraying it in the form of minute droplets into a current of hot air. The product is almost 100 per cent soluble in water and is nutritionally only slightly inferior to pasteurized milk. Dried full cream milk can be stored for fairly long periods but develops a tallowy taste after being stored for 9–24 months, owing to oxidative changes. Such changes can be prevented by packing the dried milk in sealed containers in which the air has been replaced by nitrogen. Properly stored spray-dried milk does not deteriorate as a result of the growth of micro-organisms because its water activity is too low. Very few micro-organisms survive the spray-drying process and those that do gradually die.

Although spray-dried milk is completely soluble in water it is not easily wetted by water and this causes difficulty in reconstituting it in such a way that no lumps form. 'Instant' milk powders which can be reconstituted with ease are made from spray-dried skim milk by rewetting the particles in warm moist air and allowing them to clump together into porous spongy aggregates which are subsequently redried.

Food preservation by irradiation

Certain radioactive isotopes emit electromagnetic radiation called γ-rays (gamma-rays) which are extremely effective in killing micro-organisms and they can be used for preserving food. At present, however, the preservation of food by irradiation is not permitted in Britain (see below).

Cobalt-60 and caesium-137 are two radioactive isotopes available as byproducts of the nuclear power industry. Both of them emit γ-rays with sufficient energy to kill all the microorganisms found in food, but not of such high energy that the irradiated food is itself made radioactive.

The unit of radiation dose is the gray (Gy). One gray is the dose of radiation received by one kilogram of matter when it absorbs one joule of radiation energy.

γ-rays are able to penetrate food (or any other substance, for that matter) to a considerable depth. Consequently food can be processed in bulk or in packages made of any material and of any size without fear that the innermost parts will not be properly processed. The food is carried past the radio-isotope on a conveyor and after a very brief exposure the process is complete.

Three levels of radiation treatment, known in order of severity as radurization, radicidation and radappertization are used for extending the storage life of food.

Radurization (low doses – below 1 kGy)
Inhibits sprouting of vegetables such as onions and potatoes.
Retards ripening of fruit.
Kills insects in grain, rice and spices.

Radicidation (moderate doses – 1–10 kGy)
Kills most microorganisms and so extends storage life and reduces risk of food poisoning.
Kills parasites in meat (e.g., larvae of *Trichina spiralis* which cause Trichinosis).

Radappertization (high doses – above 10 kGy)
Completely sterilizes food by killing all microorganisms. Irradiation equivalent of canning: for a 12D process (see p. 352) a dose of about 50 kGy is required.

As well as inhibiting sprouting and ripening, and killing pests and microorganisms, irradiation causes minor chemical changes. Some large molecules, such as those of carbohydrates or proteins, may be split and some destruction of vitamins may also occur. New compounds – called *radiolytic products* – may be formed, but they will be present to only a very minor extent – perhaps one or two parts per million. Nevertheless, it is possible that they might be carcinogenic or be harmful in other ways. Even such minor changes in the composition of a food can affect its flavour and texture. Irradiated meats, for example, have been said by some to have a 'goaty' flavour or, more vividly, a 'wet dog' taste. Vegetables may become soft and spongy after drastic irradiation owing to partial breakdown of their cellulose cell walls.

An official committee – The Advisory Committee on Irradiated and Novel Foods (ACINF) – has exhaustively examined all the information currently available about the safety of irradiated food. They came to the conclusion that there are no health hazards associated with eating food which has been irradiated up to an average overall dose of 10 kilograys by γ-rays of energy below 5 million electronvolts.

Nevertheless, there is still a great deal of suspicion about irradiated food. Because of fears that it may be harmful its sale is not at present permitted in Britain. Irradiated food is sold in several other countries and no harmful effects have been reported.

SUGGESTIONS FOR FURTHER READING

COOK, D.J. AND BINSTEAD, R. (1975). *Food Processing Hygiene*. Food Trade Press, Orpington.

DESROSIER, N.W. (Ed.) (1985). *Microbiology of Frozen Foods*. Elsevier, Amsterdam.

HALLIGAN, A.C. (1987). *Food Spoilage – The Role of Micro-organisms*. Leatherhead Food Research Association. (A short elementary guide for non-microbiologists.)

HERSOM, A.C. AND HOLLAND, E.D. (1980). *Canned Foods: Thermal Processing and Microbiology*, 7th edition. Churchill Livingstone, Edinburgh.

INSTITUTE OF FOOD SCIENCE AND TECHNOLOGY. (1988). *Preservatives in Food*. London.

MINISTRY OF AGRICULTURE, FISHERIES AND FOOD (1979). *Refrigerated Storage of Fruit and Vegetables*. HMSO, London.

REED, G. (Ed.) (1975). *Enzymes in Food Processing* , 2nd edition. Academic Press, London. (A comprehensive text dealing with both properties and applications.)

ROBINSON, E.K. (Ed.) (1985). *Microbiology of Frozen Foods*. Elsevier, Amsterdam.

THORNE, S. (Ed.) (1981). *Developments in Food Preservation*, Vol. I. Applied Science, Barking.

THORNE, S. (1986). *The History of Food Preservation*. Parthenon Publishing Group, Carnforth, Lancs.

TILBURY, R.H. (1980). *Developments in Food Preservatives*, Vol. I. Applied Science, Barking.

UNILEVER (1988). *Food Preservation*, Unilever Educational Booklet. Unilever, London.

WEBB, T. AND LANG, T. (1987). *Food Irradiation: The Facts*. Thorsons, Wellingborough.

WHO/FAO/IAEA (1981). *Wholesomeness of Irradiated Food*. HMSO, London.

Food poisoning and food hygiene

FOOD POISONING

Food poisoning may occur if food containing poisons of chemical or biological origin is eaten. A surprising number of foods in their natural state contain toxic substances in small amounts. The presence of solanine in green potatoes, oxalates in rhubarb and spinach and goitrogens in cabbage, cauliflower and Brussels sprouts have already been referred to. There are numerous other examples and Table 17.1 shows some of the toxins present in common foods.

The toxic substances listed in Table 17.1 are present in food in such small amounts that in most cases they do no harm. Coffee 'addicts' for example, are able to drink numerous cups of coffee each containing about 100 mg caffeine without apparent harm. Yet a single dose of 10 g might have fatal consequences. (Those who do not drink coffee need not feel smug. A cup of tea contains 50–80 mg caffeine and a can of cola drink about 35 mg.) Similarly, many people regularly consume ethyl alcohol, which is undoubtedly toxic, without great harm.

Just as the body is able to deal with regular but small amounts of caffeine or ethyl alcohol, so can it deal with small amounts of the other toxins listed in Table 17.1. When one food is eaten to excess, however, even substances normally regarded as nutrients can be toxic and at least one death has occurred through retinol (vitamin A) poisoning. Admittedly this is a rather extreme example as the person concerned, a health-food enthusiast, had drunk a gallon of carrot juice daily for ten days. It is recorded that retinol poisoning has also been caused by eating large amounts of polar bear liver though this is hardly likely to be a problem in Britain!

The toxic substances listed in Table 17.1 cause so little trouble that the term *food poisoning* is normally used to refer to illness caused by bacteria or, less commonly, by viruses or moulds.

Bacterial food poisoning

This is the most common type of food poisoning and it is caused by

Table 17.1 Toxins present in some foods

Almonds, lima beans, kirsch, fruit stones and seeds	Cyanogens which produce cyanides	Inhibition of respiratory system with possible fatal consequences
Nutmeg, mace, black pepper, parsley, celery seed	Myristicin	Headaches, cramps, nausea, hallucination
Green or sprouting potatoes	Solanine and chaconine	Stomach upsets, nervous effects
Rhubarb (especially leaves)	Oxalic acid	Interference with calcium absorption
Alcoholic drinks	Ethanol (ethyl alcohol)	Personality changes, vomiting, unconsciousness, subsequent 'hangover'
Cabbage and other brassicas	Goitrogens	Interference with iodine absorption by thyriod gland
Most *raw* beans (especially soya)	Protease inhibitors	Interference with protein digestion and absorption
Bread and other cereal products	Phytic acid	Complexes with iron and calcium and may interfere with their absorption
Raw red beans	Haemagluttenins	Cause red blood cells to cling together
Mushroom of *Amanita* variety	Amanitin	Inactivates metabolic enzymes causing severe illness with possible fatal results
Some shellfish	Alkaloid related to strychnine	Severe and sometimes fatal illness
Fish especially puffer fish and scombroid fish (e.g., tuna, mackerel and bonito) if not fresh	Various toxins derived from food eaten by fish	Severe digestive upset
Mustard	Sanguinarine	Fluid retention (dropsy)
Parsnips, celery, parsley	Psoralens	Genetic mutations
Tea, coffee, cola drinks	Caffeine	Diuretic and stimulant
Some cheeses, yeast extract, red wines	Tyramine	Increased blood pressure, migraine. Interferes with some antidepressant drugs

the presence in food of harmful bacteria or poisonous substances produced by them. The general term 'food poisoning', which is used in this chapter refers, more specifically, to bacterial food poisoning. The characteristic symptoms of food poisoning – abdominal pain and diarrhoea, usually accompanied by vomiting, which follow from one to 36 hours after eating such food, will be familiar to most readers from personal experience.

An outbreak of food poisoning may be caused by food which appears to be quite wholesome in spite of the fact that it is heavily infected by bacteria. In fact, food which has obviously 'gone off' or 'gone bad' is unlikely to be eaten and hence will not give rise to food poisoning.

Most bacterial food poisoning incidents occur as a result of unhygienic practices and this means that they are preventable. Harmful bacteria, or *pathogens* as they are called, find their way into food in a number of ways. Meat and meat products may be infected at source, i.e. they may come from animals which are themselves hosts to food poisoning bacteria. The human body is another potent source of food poisoning organisms, which are transferred easily from the mouth, nose and bowel to food. Pathogens can be 'carried' and passed on to others by individuals who are themselves not ill. Such carriers may have recently suffered an attack of food poisoning and still be harbouring the organisms in their body. In some instances carriers of food poisoning organisms act as 'hosts' over a period of many years having themselves acquired an immunity to the organism concerned. More often than not they are unaware of their role as reservoirs of infection. Animals may also harbour food poisoning organisms and pass them on to human beings via food with which they come into contact. Rats, mice, cockroaches and domestic pets can all be instrumental in transmitting food poisoning in this way.

High-risk foods. Some foods are categorized as high-risk foods because they are particularly likely to become infected with pathogens and are intended to be eaten without further cooking. The more important high-risk foods are:

1. cooked meat and poultry;
2. cooked meat products (e.g., pies, gravy, soups and stock);
3. milk, cream, custards and dairy produce, artificial cream;
4. cooked rice;
5. shellfish;
6. cooked eggs and egg products (e.g., custards, mayonnaise).

TOXIC FOOD POISONING   Some bacteria produce toxins or poisons outside their cells when they are growing and multiplying in food;

these toxins are known as *exotoxins*. Exotoxins are not living cells; they are poisonous chemicals. The incubation period – that is the period of time between the entry of the poison into the body and the appearance of the first symptoms – is normally short with toxic food poisoning. The toxins produce irritation of the stomach and vomiting occurs, often within two hours of eating the food. Abdominal pain and diarrhoea normally follow.

Exotoxins are less easily destroyed by heating than the bacteria from which they come. Thus if food is only heated sufficiently to kill the bacteria the exotoxins may survive and still cause food poisoning when the food is eaten. Food poisoning bacteria are killed in 1–2 minutes in boiling water, whereas it may take up to 30 minutes to destroy exotoxins.

INFECTIVE FOOD POISONING   This type of food poisoning (or, more correctly, food infection) is caused by eating food containing live bacteria in sufficient numbers to cause illness. Infective food poisoning bacteria produce toxins within their own cells. Such toxins – called *endotoxins* – are not as heat-resistant as the exotoxins referred to above and if infected food is heated to a temperature high enough to destroy the bacteria the endotoxins are also destroyed. Food poisoning may occur if infected food is *not* heated to a high enough temperature during cooking or if infected food (e.g., cooked meat) is eaten without further cooking. When this happens the live organisms containing their endotoxins enter the gut and as they die the endotoxins are released and cause illness. The incubation period for infective food poisoning is normally longer than that for toxic food poisoning because it takes time for the endotoxins to be released. The symptoms of fever, headache, diarrhoea and vomiting do not usually appear for about 12 hours. The eight main types of bacteria which cause food poisoning are listed in Table 17.2.

Salmonella.   Food poisoning caused by the *Salmonella* group of bacteria is called salmonellosis. There are many different strains of *Salmonella*, some of which take their names from the places where they were first observed. Examples are *Salmonella typhimurium*, *Salmonella enteritidis*, *Salmonella newport*, *Salmonella dublin* and *Salmonella eastbourne*. Bacteria can survive outside the body for long periods and on warm moist food multiply rapidly. Food must be grossly infected with a large number of live bacteria (normally over 100 000 per gram) before illness occurs.

Multiplication of salmonellae can be prevented by keeping food below 5°C. Meat which has been cooked and is not to be eaten at once should be cooled quickly so that the temperature zone in which salmonellae multiply rapidly is passed through as quickly as possible.

Table 17.2 Bacterial food poisoning

Bacteria responsible	Source and foods commonly affected	Illness	
		Incubation period	Duration
Infective food poisoning			
Salmonellae, especially *Salmonella typhimurium* and *Salmonella enteritidis*	Raw or inadequately cooked meat, milk, eggs, poultry. 'Carried' by pets and rodents	6–72 hours but usually 12–30 hours	1–8 days
Listeria monocytogenes	Pre-cooked chilled foods. Untreated dairy products	See text	
Escherichia coli	Excreta and polluted water. Raw or inadequately cooked meat and poultry	10–72 hours but usually 12–24 hours	1–5 days
Campylobacter jejuni	Raw or inadequately cooked foods of animal origin, raw or inadequately heat-treated milk	3–5 days	2–3 days
Toxic food poisoning			
Staphylococcus aureus	Human nose, mouth, skin. Boils and cuts. Raw milk and cheeses made from raw milk	2–6 hours	6–24 hours
Bacillus cereus	Rice, cornflour, vegetables, dairy products	1–6 hours (vomiting type) 8–16 hours (diarrhoea type)	24 hours
Clostridium perfringens	Animal and human excreta. Soil, dust. Raw or inadequately cooked meat and poultry. Gravy, stews, large joints of meat	8–22 hours but usually 12–18 hours	12–48 hours
Clostridium botulinum	Soil, meat, fish and vegetables. Inadequately processed canned food	2 hours–8 days but usually 12–36 hours	Death within 7 days or slow recovery

The foods most often infected are meat (particularly processed meats such as pies and brawn), eggs and egg-products, custard cakes, trifles and artificial cream. Poultry and other animals often act as carriers of salmonellae and eggs are a common source of infection. They may be infected with *Salmonella enteritidis* in the chicken's oviduct before they are laid or they may be infected after laying by contamination by poultry excreta. Foods containing raw eggs (e.g., mayonnaise or cake icing) or inadequately cooked eggs (e.g., lightly boiled or scrambled eggs, meringues and omelettes) may cause salmonellosis. For this reason no unpasteurized liquid or frozen whole egg may be used in the preparation of food in Britain. Dried egg may also be infected with salmonellae (spray-drying does not kill it) and so it should be used as soon as it is reconstituted and must be well cooked.

Rats and mice may be carriers of *Salmonella typhimurium* and are a common source of infection especially where it is possible for their excreta to come into contact with food. Domestic animals may also excrete salmonellae without exhibiting symptoms of food poisoning. This is one good reason for excluding dogs from food shops.

Raw meat is often infected by salmonellae and meat products, particularly if they are made from meat 'trimmings' and scraps from the outside of meat carcasses, may be heavily contaminated. Fish and poultry are also often contaminated with salmonellae and there is a particular danger of food poisoning when frozen poultry are only partly thawed before cooking, especially if they are undercooked. Salmonellae are easily killed and provided that cooking is thorough so that the temperature at the centre of the food is high enough (i.e. at least 65°C) all such bacteria will be destroyed.

Clostridium perfringens.   This bacterium, which causes toxic food poisoning, is the second most common cause of food poisoning in Britain. It is found in soil, in human and animal intestines and excreta and on raw meat and poultry. The spores are very heat-resistant and remain active after boiling or slow roasting. They are able to grow readily in cooked meat which is cooled slowly or kept in a warm place. *Clostridium perfringens* can reproduce every ten minutes at its optimum temperatures of 43–47°C. It can continue to grow at temperatures of up to 50°C, and as it is an anaerobe it grows in the absence of oxygen. These are just the conditions likely to be found at the centre of a large joint of meat cooling slowly. When infected food is eaten the enterotoxin is released in the gut and causes food poisoning.

Food poisoning from *Clostridium perfringens* may occur in large-scale catering establishments where meat is cooked some time before it is required, allowed to cool, and then reheated before serving. Large pieces of meat used in catering establishments are often tightly tied into rolls before cooking. As a result contamination at the outer surfaces may be transferred to the interior of the piece of meat where the temperature during cooking is insufficient to kill the spores.

Clostridium perfringens does not grow below 10°C and if cooked meat cannot be eaten at once it should be quickly cooled and held below this temperature. Similarly, meat and meat dishes which have to be reheated before consumption should be *heated* and not warmed. The Food Hygiene (General) Regulations, 1970, which lays down minimum standards of hygiene for food premises in Britain, requires that high-risk foods in catering premises be kept below 10°C or above 63°C other than in the course of preparation or when exposed for sale.

Staphylococcus aureus. Food poisoning produced by this organism is caused by toxins produced by the bacteria growing in food before it is eaten. The bacteria are found in the human nose and throat. They are also present in abundance in whitlows, infected burns and wounds and in nasal secretions following colds. They find their way into foods mainly via the hands of infected persons. The bacteria can be killed fairly easily by heating food but the toxin is more heat-resistant and is only completely destroyed by boiling for at least 30 minutes. Staphylococcal food poisoning is usually caused by eating cream-filled cakes, custard cakes or by cooked meats which have been contaminated by a food handler. Staphylococci are able to grow in higher concentrations of salt than other food poisoning bacteria with the result that they are often responsible for food poisoning involving salty foods (especially meat), such as ham and bacon.

Listeria monocytogenes. This organism may cause the disease *listeriosis* if food contaminated by it is eaten. It is found in a wide range of foodstuffs and raw chickens are commonly infected with it: it may also be present in untreated milk (and dairy products made from raw milk), vegetables and seafood.

Listeria bacteria can multiply at temperatures below those found in many domestic refrigerators and commercial chilled food cabinets. They may stay dormant for several days at these temperatures and then multiply rapidly. Some pre-cooked chilled meals and chilled prepacked salads have been found to be dangerously contaminated by the organism. Cook-chill foods should not be stored for more than five days at 4°C and they should be eaten within 12 hours if their temperature reaches 5°C. Listeria bacteria are also fairly heat-resistant and may sometimes even survive in pasteurized milk.

Listeriosis is a serious food-borne illness. A toxic enzyme produced by the bacteria may cause serious illness if it enters the blood system. Healthy adults are generally resistant to listeriosis but pregnant women may miscarry. Newborn babies may develop septicaemia or meningitis and the disease may be fatal to frail people or those whose immune system is already suppressed.

Fortunately, listeriosis occurs relatively infrequently and only 249 cases were reported in England and Wales in 1987. Nevertheless, because of the danger to vulnerable groups and its relatively high fatality rate, it must be regarded as a serious food-borne illness.

Clostridium botulinum. Food poisoning from *Clostridium botulinum* – known as botulism – is extremely serious. The bacteria produce a toxin which is the most virulent poison known, one gram of which would be sufficient to kill 100 000 people. The mortality rate from botulism is about 65 per cent, but fortunately poisoning of this type is rare in Britain. *Clostridium botulinum* occurs in the soil and on vegetables which have been in contact with contaminated soil. It is also found in fish and the intestines of pigs and certain other animals. Like *Clostridium perfringens* it is a spore-forming anaerobe.

The spores of *Clostridium botulinum* are extremely heat-resistant and may survive in canned or bottled foods which have been inadequately heat treated. As the organism is anaerobic it may develop within the can if the pH is above 4.5 and the majority of cases of botulism are caused by improperly canned contaminated meats, vegetables or fish. Commercial canners are well aware of the dangers of botulism and ensure that canned foods with a pH above 4.5 are heat processed sufficiently to destroy all *Clostridium botulinum* spores. If non-acid foods are bottled or canned at home, however, insufficient heat processing may inadvertently be given and some spores may survive. For this reason bottling or canning of non-acid foods by amateurs is not advisable.

Bacillus cereus. This organism causes toxic food poisoning through production of exotoxins when it grows on food. Like the Clostridia bacteria it can form spores where conditions for growth are unfavourable but unlike them it is an aerobe and so requires air for growth. The spores of *Bacillus cereus* are often found in cereals, especially in rice, cornflour and spices. They survive normal cooking and rapid growth and exotoxin production will occur if cooked food is not cooled rapidly. One of the exotoxins is not destroyed by exposure to $126°C$ for 90 minutes. Clearly, once this toxin is produced in food reheating is unlikely to destroy it.

In order to prevent food poisoning from *Bacillus cereus* cooked food should be cooled rapidly and stored in a refrigerator. If the food is reheated, this should be done rapidly and thoroughly and the food eaten without delay.

Food poisoning caused by *Bacillus cereus* usually has a very short incubation period which is followed by vomiting and expulsion of the toxin and bacteria. Alternatively, there may be no vomiting in which case a longer incubation period is followed by diarrhoea.

Campylobacter jejuni. This organism is a very common cause of infective food poisoning in Britain. It is present in a wide range of animals and in untreated water. Most cases of *Campylobacter* food poisoning have been caused by poultry and raw or inadequately heat treated milk.

Escherichia coli. This organism is a common and normally harmless inhabitant of our intestines. There are many subtypes or strains, however, and exposure to an unfamiliar one of these – through foreign travel for instance – may cause illness. The resulting intestinal disorder known, amongst other colourful names, as traveller's diarrhoea, is caused by the presence of living bacteria on food eaten. Raw meat and poultry are the main sources.

Other types of food poisoning

Food poisoning can be caused by moulds and viruses as well as by bacteria.

Poisonous mycotoxins are produced by some moulds and serious illness may be caused if affected foods are eaten (see p. 332). Viruses are extremely tiny particles (about 100–10 000 times smaller than the average bacterium) which span the boundary between animate and inanimate things. They have none of the usual characteristics of living things and in some respects may be regarded as large organic molecules similar to protein molecules. They can, however, reproduce themselves inside a living cell and if they are taken into the body in or on food, however, they may invade the body's cells and multiply inside them.

Viruses may get onto food through faecal contamination by food handlers and so attention to personal hygiene is important. Most attacks of viral food poisoning are caused, however, through eating shellfish such as oysters, mussels and cockles which have grown in sewage-polluted coastal waters.

FOOD HYGIENE

The subject of food hygiene is essentially concerned with how treatment of food affects the health of the consumer. High standards of hygiene minimize food spoilage and help to ensure that when food is eaten it is as wholesome and free from pathogenic bacteria as possible.

Many factors may affect the quality and wholesomeness of food. Among them are:

(a) the way in which it is grown or, in the case of animals, reared and fed;
(b) the design and cleanliness of farm buildings, slaughterhouses and factories in which it is processed;
(c) the premises, equipment and conditions in which it is stored;
(d) the care taken by food handlers to avoid contamination from other foods;
(e) the personal hygiene of food handlers.

It is not possible to deal properly (or indeed at all) with all these matters here. The last two points listed, however, are of concern to all food handlers and, indeed, to all those who have an interest in food.

The basic principles of good food hygiene practice can be summarized as follows:

1. Avoid contamination of food by bacteria.
2. Prevent the multiplication of any bacteria which nevertheless gain access to food.
3. Destroy any remaining bacteria, where possible, by thorough cooking.

These three foundations of good hygienic practice are amplified below.

Avoid contamination

- Keep raw and cooked foods separate. The same working surfaces and equipment should not be used for both raw and ready-to-eat foods. Surfaces and equipment used for raw foods (including vegetables) should be thoroughly cleaned afterwards.
- Keep animals out of food areas and make sure that birds and insects do not have access.
- Dispose of waste food promptly.
- Do not use washbasins for food preparation or food-preparation sinks for washing hands.
- Maintain high standards of personal hygiene. Thorough hand-washing is essential especially after handling raw meat or using a toilet. *Clean* towels or disposable towels or hot-air driers should be used for hand-drying to prevent reinfection of clean hands by dirty towels.
- Clean protective clothing (including head covering) should be worn in food areas and should be removed when leaving those areas.
- Hands should be kept away from bacteria-rich areas of the body

including the mouth and nose. It is for this reason that smoking by food handlers is prohibited by law. Soiled handkerchiefs and uncovered wounds are other potent sources of infection.

Prevent multiplication of microorganisms

● Keep foods cold or hot. Avoid the temperature zone 5–65°C in which bacteria flourish. Food should be in this temperature danger-zone for as little time as possible.

Cook food thoroughly

● Cooking times and temperatures must be sufficient to ensure that all bacteria and their toxins are destroyed. If bacteria present in food (or their spores) are not killed when the food is cooked they may multiply when it is cooling through the 'danger-zone', or if it is kept in this temperature zone before being served.
● Meat and poultry should be completely defrosted before cooking. If this is not done some bacteria or bacterial spores may survive the cooking process.

The incidence of food poisoning

Food poisoning bacteria were described as long ago as 1888 and during the subsequent 60 years the main types of bacteria causing food poisoning were identified. Since the Second World War there has been a notable increase in recorded incidents of food poisoning, and this has occurred in spite of an increased understanding and awareness of the causes of food poisoning and considerably improved standards of living and of hygiene.

Between four and five thousand cases of food poisoning were notified yearly in England and Wales during the 1960s, although the actual numbers of cases were certainly considerably higher as milder cases are often not reported. During the 1970s there was a gradual rise in reported cases of food poisoning and this continued into the 1980s. In 1981, for example, some 10 000 cases were reported and in 1988 the figure was about 30 000 – an alarming increase in seven years. It is estimated that only about ten per cent of all cases are reported so the actual number of cases probably exceeds 300 000 per annum. Food poisoning may be regarded as a preventable disease which is not being prevented.

There are many reasons for the increase in food poisoning in Britain over recent years, but some of the more important are as follows:

1. *A great increase in communal feeding.* Large scale catering, whether in hospitals, schools, canteens, or restaurants, means that a single infected food can produce many cases of food poisoning.

2. *Varied menus and rapid food service.* In order to have a wide menu and be able to produce dishes quickly food may be precooked and kept warm until it is required or it may be reheated rapidly and perhaps inadequately in a microwave oven or under an infra-red grill when it is ordered.

3. *Increased use of convenience foods.* Although factory processes are carefully controlled and carried out under hygienic conditions, one source of infection can lead to the contamination of thousands of prepacked items. In addition, the use of convenience foods, especially meat products, which are eaten cold or only warmed through increases the risk of food poisoning.

4. *Increase in factory farming.* The intensive rearing of poultry and animals increases the possibility of large scale infection of such food supplies, especially by Salmonella.

5. *Rapid increase in consumption of take-away meals.* Such food may be kept warm for long periods or briefly reheated in the home thus allowing rapid bacterial growth. For example, 'take-away' rice may be contaminated with *Bacillus cereus*.

6. *Changing patterns of shopping and food storage in the home.* A weekly, rather than a daily, shopping routine means that food has to be stored for greater periods of time. Incorrect storage conditions may encourage the growth of bacteria. Increasing use of freezers means that meat, especially poultry, needs to be thawed before it is cooked. Incomplete thawing followed by normal cooking may not kill all bacteria, especially in the centre of the food.

7. *Increased use of packed meals.* The incidence of food poisoning is much higher in summer than in winter largely because of inadequate refrigeration. The increasing tendency to have a packed meal, often including cooked meat in sandwiches, in the middle of the day increases the risk of food poisoning in summer.

8. *Use of staff untrained in hygiene in catering establishments.* Large-scale handling of food by staff not trained or conscious of hygiene requirements is a major source of infection. In such circumstances *cross-contamination*, that is the transfer of bacteria from a contaminated source to an uncontaminated source, can easily occur and spread the risk of infection.

In Britain salmonellae are responsible for most food poisoning, and account for 70–80 per cent of all cases. *Clostridium perfringens* is next in importance, accounting for 15–25 per cent. Outbreaks of food poisoning caused by this organism often involve food which has been prepared in bulk and hence tend to affect a large number of people.

The foods most commonly causing food poisoning in Britain are meat and poultry which together account for at least 80 per cent of all cases. *Clostridium perfringens* poisoning mainly arises from infected meat and poultry which is reheated while Salmonella outbreaks mainly involve cold cooked meat or poultry. Staphylococci, on the other hand, may cause infection in salty meats such as ham or bacon.

SUGGESTIONS FOR FURTHER READING

ALCOCK, P.A. (1986). *Food Hygiene: A Study Guide.* Stanley Thornes, Cheltenham.

ALCOCK, P.A. (1981). *Food Hygiene Manual.* H.K. Lewis, London.

BRITISH ASSOCIATION (1978). *Salmonella, the Food Poisoner* (A review). British Association for the Advancement of Science, London.

CHRISTIE, A.B. AND CHRISTIE, M.C. (1977). *Food Hygiene and Food Hazards*, 2nd edition. Faber & Faber, London.

DAVENPORT, J.K. (1982). *Food Hygiene in the Catering and Retail Trades.* H.K. Lewis, London.

HOBBS, B.C. AND ROBERTS, D. (1987). *Food Poisoning and Food Hygiene*, 5th edition. Edward Arnold, London.

SPRENGER, R.A. (1985). *Hygiene for Management*, 2nd edition. Highfield Publications, Rotherham.

TRICKETT, J. (1986). *The Prevention of Food Poisoning.* Stanley Thornes, Cheltenham.

CHAPTER 18

Food contaminants and additives

Many people are concerned about the presence of 'chemicals' in their food and worry about the possible ill effects of eating them. 'E numbers' (which signify the presence of an additive in a food) have become notorious and the proverbial benefits of an apple a day have been replaced by the widely held belief that an E number a day is more than enough! Yet, as anyone who has read the preceding chapters will be aware, food itself is composed of chemicals and so, for that matter, are our bodies. What is usually meant by 'chemicals' in connection with food are those substances which do not normally form a part of the food in its natural or traditional state. Their presence may arise as a result of accidental contamination, or they may be deliberately added to improve processing or keeping qualities or to supplement the nutrients already present. Deliberate adulteration practised with intent to deceive is fortunately not as common today as in former years. The addition of alum to flour and water to milk, and other crude methods of adulteration, though once commonplace no longer happens today. Many people, however, take the view that the adulterants of former times have been replaced by twentieth century additives. What is the true position? Let us first consider what 'chemicals' may be present in food and why they are there. We will then be better able to assess whether their presence is justified or not.

FOOD CONTAMINANTS

Food can become contaminated in a multitude of ways but there are some sources of contamination which are of outstanding importance today, and it is with these that we shall be concerned here.

AGRICULTURAL CONTAMINATION Many crops are treated with insecticides to prevent infestation by insects, fungicides to prevent growth of fungi, and weedkillers or growth regulators to kill weeds selectively. The treatment of growing crops with chemicals is by no means new. Insecticides containing sulphur have been used for well over a hundred years and the arsenical sprays *Paris Green* and

Bordeaux Mixture, which contain copper, were first used over a hundred years ago.

If chemicals were not used by farmers the amount of food available would fall considerably and its cost increase dramatically. The nineteenth century Irish potato famine, which was caused by potato blight, is a striking example of the misery which can be caused by crop failure. Potato blight is now completely controlled by fungicidal spraying and the Colorado beetle, a modern menace which could wreak equal havoc on the potato crop, by the use of insecticides. Pesticides and fungicides are invaluable aids to efficient food production, not only when crops are growing, but also after harvesting when they are being stored.

Modern agricultural chemicals are for the most part complex organic compounds and they are becoming increasingly complex. Many of them are toxic to animals and human beings, but they are usually applied to the plants before the part which is eaten has appeared, or at least a sufficient length of time before harvesting to permit their removal by rain. The amounts remaining on food by the time it reaches the table are exceedingly small – never more than a few parts per million and usually much less than this. Toxicologists have been able to determine an Acceptable Daily Intake (ADI) for the agricultural chemicals used. The ADI represents an amount which could be taken daily for an entire lifetime without appreciable risk. In practice the ADI is set at a very low level with a safety factor of at least 100, i.e. the ADI is set at least 100 times less than the 'no effect' level. Even so, the actual level of consumption has been found to be only a few per cent of the ADI.

The number of agricultural chemicals in use has increased dramatically in recent years. Whereas in 1926 there were a mere 12 chemicals in common use there are now over 600. In Britain the *Pesticides Safety Precautions Scheme* is a non-statutory scheme whereby new pesticides are evaluated first for safety then for efficiency so that no such substances come into use before they have been carefully assessed.

Antibiotics. As well as treating soil and crops with chemicals, farmers treat animals with chemicals in the form of antibiotics. Antibiotics are used to cure animal diseases such as mastitis, enteritis, pneumonia and infected wounds and feet. By using antibiotics many animal infections which formerly were a source of great trouble to farmers can now be easily controlled. Farmers are also tempted to dose pigs and poultry with antibiotics when there is nothing apparently wrong with them. It is supposed that by doing so the animals will be given blanket protection against disease or will be better able to resist stress and hence will grow faster or give higher yields of eggs.

Antibiotics are also used to stimulate growth: only three antibiotics may be used in this way and none of them are used for treating humans.

The use of antibiotics for treating infected animals meets with general approval. Indeed, the use of antibiotics in veterinary practice has been a great boon for farmers and, even more so, to the animals themselves. The use of antibiotics as animal food supplements is a different matter, however, and this practice is widely opposed.

Indiscriminate use of antibiotics by farmers is not approved by most veterinary workers and it is generally accepted that antibiotics should only be used to treat specific infections. The reason for this is that microorganisms which are continually exposed to low concentrations of an antibiotic may become antibiotic-resistant; when this occurs the antibiotic is valueless when it is required to treat an illness caused by the organism concerned. An organism which has acquired resistance to an antibiotic can, in some circumstances, transmit this resistance, merely by contact, to a previously sensitive organism. This is known as *infective drug resistance*.

When organisms become resistant to antibiotics it is more difficult for veterinary surgeons to deal with infected animals. Also, because there is an inevitable interchange of organisms between animals and man, similar difficulties may arise in treating infected humans. It would be most unfortunate if, as a result of veterinary use, the typhoid organism *Salmonella typhi* became resistant to the antibiotic chloramphenicol (Chloromycetin) as this is by far the most effective drug available for treating typhoid fever in man. There are, however, no indications at present that this is likely to occur. Chloramphenicol is extensively used for treating animals and there have been no cases of typhoid which did not respond to treatment with it.

A further objection to the widespread use of antibiotics by farmers is the danger that an individual who is allergic to a particular antibiotic may be made ill if he or she drinks milk in which the antibiotic is present. To prevent this happening farmers are required to withhold milk from treated cows until residues of antibiotic have reached insignificant levels. Usually no traces can be detected after 48 hours but in some cases up to 14 days may elapse before the milk is acceptable. Less than one per cent of samples of bulk milk tested are found to contain antibiotics but, even so, the danger of an allergic reaction still exists.

Lead.   Fruit and vegetables grown near large towns or busy roads may be seriously contaminated with lead deposited from the exhaust fumes of motor vehicles. Most petrol is 'doped' with tetraethyl lead to improve its 'anti-knock' properties, and on combustion the lead is converted to lead oxide and bromide which enter the atmosphere with the exhaust gases. Fortunately, fruit and vegetables are usually washed before eating and so surface contamination is removed in this way. Lead which has been absorbed from contaminated soil by the growing plant, however, cannot be removed and as it is a cumulative poison long-term consumption could be harmful.

There are no legal limits for the amount of lead in food (except for canned food and baby food).

Colouring matter. The addition of colouring matter to food is strictly controlled (see p. 378) but the restrictions do not apply to food for fish or animals. The food colour canthaxanthin is used in fish-food pellets to impart a colour to salmon and pink trout. It has been recommended that only specified amounts should be permitted and that if fish have been so treated it should be indicated on a label. Similar restrictions have been recommended for the use of colours in poultry feed to impart colour to egg-yolks.

RADIOACTIVE CONTAMINATION Contamination of food with *radioisotopes*, which enter the atmosphere as a result of accidents in nuclear power stations or nuclear fuel reprocessing plants, is another serious modern problem. Radioisotopes are unstable forms of atoms which change into more stable forms by emitting radiation and atomic particles. Although radioisotopes are unstable they can remain active and emit radiation for a very long period. The radiation and atomic particles emitted are extremely harmful to living cells. Radioisotopes which enter the atmosphere become widely distributed and may finally reach the ground many thousands of miles away. When radioisotopes fall on soil and vegetation they may be absorbed by plants. If these are eaten by man or animals and the nutrients absorbed the radioisotopes may become incorporated into the body tissues with harmful consequences.

The treatment of food with ionizing radiation for preservation purposes (see p. 356) does not produce radioisotopes if it is properly carried out and it does not make food radioactive.

CONTAMINATION FROM PACKAGING MATERIALS There has been some concern in recent years about the possible contamination of foodstuffs by migration of chemicals from the materials in which they are packed. Plastics are increasingly used as packaging materials and while the polymers themselves are non-toxic, compounds which may have been added to them to improve their properties may not be equally innocuous. Catalysts such as organic peroxides or complex metal salts may have been used to initiate polymerization and these remain in the polymerized product. Plasticizers are incorporated in many plastic materials to increase their flexibility. They are usually viscous organic liquids such as the esters of phthalic, phosphoric or ricinoleic acid. Plastics may be adversely affected by atmospheric oxidation, especially when they are in the form of thin sheets which present a large surface to the air, and antioxidants are used to mitigate this. Unfortunately, however, the antioxidants used are not those approved for use in foods. Stabilizers, which may be organo-tin salts or calcium salts of fatty acid, may be used in some plastic materials and pigments, anti-static agents, bactericides and fungicides may also be present. When plastics are used as packaging materials for food any of these substances may find their way into the food. Only very small amounts will be present in the food, but even these minimal amounts may prove to be toxic if ingested over a period.

Paper-based packaging materials are widely used for foodstuffs and even these may be a source of contamination if, as often happens, the paper or board has been treated to increase its strength when wet. Such wet-strength paper is made by impregnating paper or pulp with urea-formaldehyde or melamine-formaldehyde resin, and it is possible for formaldehyde to migrate from the treated paper to the foodstuff. The presence in foods of up to 5 ppm of formaldehyde arising from any resin used in the manufacture of wet-strength papers or of plastic food containers or utensils is permitted.

FOOD ADDITIVES

Additives are natural or synthetic substances which are added to foods to serve particular purposes. The legal definition of an additive (Food Labelling Regulations, 1984) is

> any substance not commonly regarded or used as a food, which is added to or used in or on food at any stage to affect its keeping qualities, texture, consistency, appearance, taste, odour, alkalinity or acidity or to serve any other technological function in relation to food.

Most processed and manufactured foods contain additives and in Britain, the use of additives in the following categories is controlled by law.

1. Preservatives
2. Antioxidants
3. Emulsifiers and stabilizers
4. Colours
5. Sweeteners
6. Miscellaneous additives

In each category only substances known as *permitted additives* may be used in food. The first four categories correspond with the European Economic Community classification. Category 6 – miscellaneous additives – includes solvents, mineral oils, flavour modifiers, anti-caking agents and raising agents. Any substance which does not fall into one of the controlled categories may be added to food subject to the general limitation imposed by the Food Act that it must not 'render the food injurious to health'.

Most permitted additives have been given a serial number. Where usage is also controlled by the EEC the serial number is prefixed by an E.

Some permitted additives are naturally occurring substances. Pectin (E440a), for example, which is used as a gelling agent, is obtained from apples. Other permitted additives are manufactured 'copies' of naturally occurring substances. An example is L-ascorbic acid (E300) which is manufactured on a large scale for use as a food additive, and is identical in every respect with naturally occurring vitamin C. In addition to the natural or nature-identical compounds there are a substantial number of permitted additives which do not occur naturally and are purely synthetic in character. This does not mean that they are harmful, but it is mainly to this type of additive that people object. Azodicarbonamide (E927), which is used as a flour improver (see p. 136) is an example of this type of additive.

Preservatives

Preservatives prevent microbial spoilage of food, as described in Chapter 16. There are 35 permitted preservatives and all except one of them (nisin) may also be used throughout the EEC. A list of permitted preservatives is given in Table 16.2 (see p. 339). Their serial numbers are in the 200s.

Antioxidants

Antioxidants prevent fats and oils from going rancid and their use is described in Chapter 16. There are 16 permitted antioxidants. All but two of them (diphenylamine and ethoxyquin) are approved for use throughout the EEC. A list of permitted antioxidants is given in Table 16.3 (see p. 343). Their serial numbers are in the 300s.

Emulsifiers and stabilizers

Emulsifiers are used to make stable emulsions or creamy suspension from oils and fats and water (see p. 61). They are also used in baked food to slow down the rate of staling. Stabilizers are used to improve the stability of emulsions and prevent the separation of their components. There are 54 permitted emulsifiers and stabilizers and all but four of them may be used throughout the EEC. They include: edible gums, such as locust bean gum (carob gum), tragacanth and acacia (gum arabic); alginic acid and alginates; cellulose derivatives; mono- and diglycerides of fatty acids such as glyceryl monostearate; pectins; various sorbitan derivatives and polyoxyethylene esters. The large number of permitted emulsifiers and stabilizers reflects their wide- spread use in a large variety of different types of manufactured foods where 'creaminess' or 'spreadability' are required. Salad cream, ice-cream, instant desserts, cheese, fish and meat spreads, margarine and low-fat spreads all contain emulsifiers or stabilizers, or both.

The serial numbers of permitted emulsifiers and stabilizers are in the 400s.

Colours

Dyes and pigments used in food are known collectively as food colours or, more tersely, simply as colours. They are added to food to make it more attractive to the purchaser or consumer or to replace natural colour lost during food processing. Canned strawberries, for example, would be greyish-brown if colours were not added and canned peas a brownish-green.

The addition of colours to foods is controlled by law in most countries and in Britain only 50 *permitted colours* may be used. Their serial numbers range from 100 to 180: about two-thirds of them may be used throughout the EEC.

Many permitted colours are of natural origin and they include saffron, cochineal, carotenes (and the closely related xanthophylls) and anthocyanins, which colour ripe fruit. Other permitted colours are inorganic substances such as the white pigment titanium dioxide and finely divided aluminium, silver and gold which are used in cake decoration.

The list of permitted colours includes some synthetic dyes. Thou- sands of such substances are known but most of them are too toxic to be used in foods and some are known to be carcinogenic (i.e. able to cause cancer). For this reason only 15 have been approved for use in food. Most of them are approved for use throughout the EEC (i.e. they have E numbers), but four of them are not and they can be used only in Britain. Colouring matter must not be added to meat, fish, poultry, fruit, cream, milk, honey, vegetables, wine, coffee, tea and condensed or dried milk. Also, by agreement with the manufacturers, colours are not used in foods made for babies and infants.

Sweetening agents

Sweetening agents, or sweeteners as they are generally known, fall into two categories. *Intense sweeteners* are many times sweeter than sucrose and they are therefore used in very low concentrations. *Bulk sweeteners* are about as sweet as sucrose and hence are used in roughly equal amounts. The relative sweetness of naturally occurring sugars and artificial sweeteners is considered in Chapter 6 (see p. 99).

There are four permitted intense sweeteners:

1. *acesulfame potassium*, used in canned foods, soft drinks and table-top sweeteners;
2. *aspartame*, used in soft drinks, yoghurts, dessert and drink mixes and sweetening tablets;
3. *saccharin* (and its sodium and calcium salts), used in soft drinks, cider, sweetening tablets;
4. *thaumatin* used in sweetening tablets and yoghurt.

Aspartame (which is sold under the trade name Nutrasweet) is a dipeptide made from the amino acids L-aspartic acid and L-phenylalanine. It is broken down in the body to its constituent amino acids. People who suffer from the genetic disease phenylketonuria (PKU) are unable to metabolize phenylalanine and some doubts have been expressed about whether aspartame can safely be consumed by them. The Government committee responsible for overseeing the use of chemicals in food has concluded, however, that aspartame can safely be used by those suffering from PKU.

None of the permitted intense sweeteners has been given a serial number and they are not approved for use throughout the EEC.

The bulk sweeteners (sometimes known as nutritive sweeteners) are mostly hydrogenated sugars: *hydrogenated glucose syrup*, *isomalt* and *mannitol* (E421) are used in sugar-free confectionery; *sorbitol* (E420) is used in sugar-free confectionery and jams for diabetics; and *xylitol* is used in sugar-free chewing gum.

Bulk sweeteners do not require insulin to be metabolized and hence they can be used by diabetics.

Miscellaneous food additives

There are many other uses for additives outside the five main categories discussed above and over 120 miscellaneous additives have been approved for use in foods in Britain. They are principally used as processing aids and the main types are listed in Table 18.1. Only substances which have been approved may be used for the purposes listed in this table.

FLAVOUR MODIFIERS   A flavour modifier is a substance which is capable of enhancing, reducing or otherwise modifying the taste or odour, or both, of a food. Flavour modifiers, or enhancers, are

Table 18.1   Additives used as processing aids

Acids and bases	Flavour modifiers	Mineral oils
Anti-caking agents	Flour bleaching agents	Packaging gases
Anti-foaming agents	Flour improvers	Propellants
Buffers	Glazing agents	Release agents
Bulking agents	Humectants	Sequestrants
Firming agents	Liquid freezants	Solvents

themselves practically tasteless but they intensify the flavour of soups, meat and other savoury foods.

The most widely used flavour enhancer is monosodium glutamate (E621), or MSG as it is commonly known. MSG is the sodium salt of glutamic acid, an amino acid which is a common component of proteins (see p. 160). It occurs in soy sauce, which is made by fermenting or hydrolysing soya beans and it has long been used in this form as a flavour enhancer in Chinese food. MSG itself is made on a large scale by allowing a specific bacterium to grow in a solution containing ammonium ions.

Most dehydrated soups and stock cubes contain MSG and it is also present in many other 'meaty' prepared foods. Some people react unfavourably to MSG and suffer from what has come to be known as the 'Chinese restaurant syndrome' if they eat food containing it. This manifests itself in a variety of ways including palpitations, chest or neck pain and dizziness. The cause is not known and the ill effects soon disappear. MSG is not thought to be harmful but, as a precautionary measure, it is not now added to food manufactured for babies and infants.

The flavour-enhancing capacity of MSG is much intensified if it is used in conjunction with *inosine monophosphate* (E631) or *guanosine monophosphate* (E627) which are known as IMP and GMP respectively.

Sodium 5'-ribonucleotide (E685) is another powerful permitted flavour enhancer. Ribonucleotides are compounds formed from the sugar ribose, phosphoric acid and an organic base such as guanine. They occur in all animal tissues and are present in yeast extracts and contribute substantially to their characteristic meaty taste. As soon as an animal is killed, however, enzymes called *phosphatases* begin to break down the ribonucleotides and by the time the food reaches the table a great deal of the flavour-enhancing properties of the ribonuc-

Table 18.2   Permitted solvents

Ethyl alcohol	Glycerol	Glyceryl triacetate
Ethyl acetate	Glyceryl monoacetate	Isopropyl alcohol
Diethyl ether	Glyceryl diacetate	Propylene glycol

leotides may be lost. By adding ribonucleotides to foods of animal origin such loss of flavour may be made good.

Ribonucleotide flavour enhancers are used in soups, meat and fish pastes, all types of canned meat products, sausages, meat pies and other processed food products of which the main ingredient is meat or fish. As their flavour enhancing power is very great – ten times that of MSG – only very small quantities (as little as 20 ppm) are required.

SOLVENTS Flavourings and colours are often added to foods in the form of concentrated solutions in organic solvents. Solvents may also be used to facilitate the incorporation of other ingredients into food and only the nine solvents listed in Table 18.2 may be used. Water, acetic acid, lactic acid, any propellant or any permitted food additive (e.g., a preservative or emulsifying agent) are not regarded as solvents. None of the permitted solvents have been given serial numbers.

MINERAL OILS Mineral oils and waxes are hydrocarbons of fairly high molecular weight. They are familiar in everyday life as lubricating oils, medicinal paraffin, petroleum jelly (e.g., Vaseline) and wax (e.g., candles).

Hydrocarbons are not metabolized and hence they have no nutritive value. Their presence in food is not desirable because they may interfere with the absorption of the fat-soluble vitamins A and D and, if present in a highly emulsified form, they may become deposited in certain organs.

Mineral oils may become incorporated into foods, in small amounts, as a result of food processing operations. Oils used for lubricating machinery may find their way into food and so moving parts which actually come into contact with food should wherever possible be lubricated with vegetable oil. Unfortunately, although vegetable oils are effective as a lubricant when first applied, they tend to become 'gummy' and to have the reverse effect after a time.

Dried fruits such as prunes, currants, raisins and sultanas may be given a surface coating of permitted mineral oil (E905) to prevent them sticking together in a tight mass during storage as a result of exudation of sugar syrup through surface cracks. The coating of mineral oil also improves the appearance of the fruit and deters insect infestation. Up to 0.5 per cent may be present. Citrus fruits may be given a coating of mineral oil to replace the natural protective oils which are lost during cleaning operations for removal of dirt and moulds. Some cheeses are given a coating of mineral oil to minimize drying out and loss of weight. It also prevents mould growth and improves the appearance of the cheese. A thin coating of mineral oil may be used on eggs which are to be preserved by chilling.

The addition of mineral oil to foods as described above is no longer permitted in Britain.

POLYPHOSPHATES   A range of phosphoric acids and their sodium, potassium, and calcium salts, are used as food additives. Twenty-six such substances, from the simple acid orthophosphoric acid, H_3PO_4, to complex polyphosphates are permitted for use in food.

Phosphates, particularly polyphosphates, are added to flesh foods such as meat, fish and poultry in order to permit increased retention of water and greater solubility of proteins and, it is claimed, to improve texture.

Chickens and other poultry can be made to absorb water with the aid of polyphosphates. If this is done it must be declared on a label and if the amount of added water exceeds five per cent that also must be stated. Water added to poultry without using polyphosphates, however, need not be declared even though some frozen poultry may contain as much as eight per cent.

Flavouring agents

At present flavouring agents, or flavourings as they are referred to by the food industry, may be used in foods without restriction. It is intended that they will, in time, be subject to control in the same way as the permitted additives already discussed. The problems to be overcome in doing so, however, are very great because of the large number of flavourings available and the minute concentrations in which they are often used. The blending and use of flavourings is more akin to an art than a science and hence it is more difficult to control in a scientific way. Flavour is one of the most important attributes of food and is detected by the senses of taste and smell. Taste itself is made up of the four primary tastes – sweet, sour, salt and bitter – which are detected by taste buds situated in the mouth mainly on the tongue, palate and cheeks. Smell is detected by extremely sensitive cells situated at the top of the nasal cavity.

Flavouring agents have been used from earliest times to increase the attractiveness of food. Originally flavourings were the dried, and sometimes powdered, forms of spices, herbs, berries, roots and stems of plants. For example, spices, such as pepper, cloves and ginger, were prized for their ability to add interest and palatability to a monotonous diet. Condiments and spices were invaluable not only for their flavour but also because they were able to disguise the tainted flavour of meat which was past its best.

As the demand for flavouring agents increased methods for extracting the active principles were devised. The most important types of natural flavouring agent are *essential oils* which are extracted from plant tissues. These oils (which are chemically quite different from the oils and fats discussed in Chapter 4) are volatile (easily vaporized) and have a flavouring power many times that of the raw material from which they come. Synthetic flavours which are usually copies of the essential oils are cheaper and far more convenient to use than the corresponding natural flavours. They are usually dissolved in ethyl alcohol or another permitted solvent but powdered versions, made by spray-drying the flavouring material in a gum arabic solution, are also available. Such powders are widely used in powdered convenience foods, such as instant dessert mixes.

Many thousands of flavourings are available for use by the food industry and a classification of natural and synthetic flavourings is given in Table 18.3.

Many of the simpler artificial fruit flavours are *esters* which are formed when carboxylic acids react with alcohols. Ethyl acetate, for example, which is a fruity-smelling liquid, is produced when acetic acid and ethyl alcohol are heated together:

$$CH_3COOH + C_2H_5OH \rightleftharpoons CH_3COOC_2H_5 + H_2O$$

Ethyl acetate

The reaction also occurs at normal temperatures, but at a much slower rate, and esters, particularly ethyl acetate, are formed in this way during the maturing of wine and contribute to the much prized bouquet. Other esters can be easily and cheaply made in the same way by heating together the appropriate acid and alcohol, usually in the presence of a catalyst.

At one time synthetic fruit flavours were usually made up of a single ester and consequently they were rather poor imitations. As the composition of natural flavours has become known it has become possible to blend together synthetic materials so as to imitate the natural flavour more closely. Esters are the most widely used type of flavouring and Table 18.4 shows some of those commonly used.

It is possible to analyse natural flavours by vapour-phase chromatography, which is a very sensitive technique for splitting up mixtures of volatile compounds into their components. The results of such investigations show that natural flavouring agents are usually complex

Table 18.3 Classification of flavouring agents

Natural	Synthetic		
Herbs, spices	Acids	Ethers	Esters
Essential oils,	Acetals	Ketals	Others
extracts, distillates	Alcohols	Ketones	
Foods (e.g., cocoa)	Aldehydes	Lactones	

Table 18.4 Esters used as flavouring agents

Name	Formula	Use
Ethyl formate	$HCOOC_2H_5$	Rum, raspberry and peach essences
Ethyl acetate	$CH_3COOC_2H_5$	Apple, pear, strawberry and peach essences
Pentyl acetate	$CH_3COOC_5H_{11}$	Pear, pineapple and raspberry essences
Pentyl butyrate	$C_4H_9COOC_5H_{11}$	Banana, pineapple and peach essences
Allyl caproate	$C_5H_{11}COOC_3H_5$	Pineapple essence

mixtures of many different substances, some of which are present in extremely small amounts. Occasionally, however, the flavour of a natural flavouring agent depends very largely on the presence of a single substance which may be the main ingredient, such as *eugenol* in oil of cloves which constitutes 85 per cent of the oil, or a minor ingredient, such as *citral* in oil of lemon which constitutes only five per cent of the oil.

In the rare cases where a natural flavour is composed wholly or predominantly of one chemical, a synthetic substitute can be made fairly easily. Usually, however, a large number of substances must be blended together to obtain a good imitation of the natural flavour.

Added nutrients

Nutrients are added to some processed foods to improve their nutritional quality. In some cases the added nutrients replace those lost during processing. In the manufacture of dehydrated potato, for example, a great deal of the ascorbic acid (vitamin C) originally present is lost and synthetic ascorbic acid is added to replace it and to act as an antioxidant. The addition of iron, thiamine and niacin to all flour (except wholemeal) to replace losses due to milling is another example (see p. 134).

Nutrient additives may be used not only to maintain nutritional value but also to improve it. This is known as food *fortification* or *enrichment*. Such addition of nutrients may either be a legal requirement to safeguard public health or it may be done voluntarily by the manufacturer. The enrichment of margarine with vitamins A and D so as to make its vitamin content equivalent to that of summer butter is obligatory, as is the addition of nutrients to flour. Fortification of breakfast cereals with a range of vitamins and minerals or the addition of vitamin C to soft drinks and fruit juices, on the other hand, is not legally required.

The definition of additives in the Food Labelling Regulations, 1984, specifically excludes nutrients and so, illogical though it may be, added nutrients are not legally regarded as additives.

SAFETY OF ADDITIVES

There is widespread and understandable concern, amounting almost to hysteria in some cases, about the presence of 'chemicals' in food. Additives, with their accompanying 'E numbers', have acquired an undeservedly poor reputation with the man in the street. Readers of this book will, it is hoped, be better informed and will appreciate that one of the main reasons for the presence of additives in food is to ensure that it is safe to eat and that perishable foods are available in good condition throughout the year.

The object of the regulations which control the use of additives in food is to prevent abuses which might otherwise occur and the consequent dangers to health. Nevertheless, the general public views the presence of additives in food and the regulations which control their use as aspects of a vast conspiracy practised by an unscrupulous food industry without regard to possible harm. To what extent, if any, can these suspicions be justified? Is it *possible* that the use of food additives could make our food less safe? Or is the alternative possibility, that additives make our food safer, cheaper, more varied and more attractive likely to be nearer the truth?

Many people feel that the addition to food of chemicals of any sort, for whatever purpose, is a practice which deserves nothing but condemnation. Such people are apt to accuse food scientists of 'tampering' with food and hence in some way making it unwholesome. They often do not realize that many of the traditional foods we regard as 'natural foods' may be far removed from their original condition as a result of treatment with chemicals which, because of long-standing usage, are not now regarded as additives. Traditional techniques for processing foods, such as cooking, smoking, pickling and curing bring about chemical and physical changes and the idea that modern food processing techniques are in a different, and possibly more harmful, category from these time-honoured methods is quite without foundation.

It is undoubtedly true that were it not for the food scientist there would be even less food available for the world's ever growing population than there is today. The use of agricultural chemicals and food additives has been largely responsible for the notable improvement in nutritional standards in the Western world which has taken place in the last 50 years or so.

It is generally agreed that the use of additives in food is justified when they serve one or more of the following functions.

1. Maintenance of nutritional quality.
2. Improvement of keeping quality or stability with a reduction of wastage.

3. Enhancement of attractiveness in a manner which does not lead to deception.
4. Provision of essential aids to processing.

Situations in which the use of food additives would not be in the interests of the consumer, and should not be permitted, include the following.

1. When faulty processing and handling techniques are disguised.
2. When the consumer is deceived.
3. When the result is a substantial reduction in the nutritive value of a food.

It is important that the risks of possible ill effects from the presence of chemicals in foods should be negligible compared with the benefits which ensue, and the food additives approved for use in Britain perform useful functions without harming the vast majority of consumers. It has to be accepted, however, that some sensitive people may react adversely to some food additives. The European Community Scientific Committee for Food has estimated that 0.03–0.10 per cent of the population may be sensitive to a food additive. This is unfortunate, but it has to be borne in mind that the number of people who have adverse reactions to other foods – to strawberries or shellfish for example – may exceed these figures. It has been estimated that from 0.3 to 2.0 per cent of children in Britain suffer from dietary intolerance. These figures exceed by a factor of 10–20 the numbers thought to be sensitive to food additives.

Clearly, additives should neither be condemmed nor condoned as a class, but each should be considered and judged upon its merits. Some chemicals are so objectionable that their presence in food is indefensible. Formaldehyde, fluorides and similar toxic compounds which were formerly used as preservatives fall into this category; so do dyes known to be carcinogenic to man.

Only slightly less objectionable than the above are those substances which may *conceivably* cause harm when eaten. Synthetic colouring materials are somewhat suspect because, as already pointed out, so many of them are toxic. Many thousands of synthetic dyes are known and it would appear to be easy to draw up a list of a few dozen which would be adequate for colouring food and which could be guaranteed to be safe. It is not possible, however, in spite of the most rigorous tests, to be sure that a compound which is apparently harmless is in fact so. This is why the lists of permitted colours vary from country to country. Norway and Sweden prohibit the presence of all synthetic colours in food and several British permitted colours are not approved for use in the other countries of the EEC. The number of colours approved for use in Britain is steadily decreasing as new evidence of their toxicity becomes available. Pressure of public opinion (well informed or not) and a 'bad press' has caused several major supermarket chains to impose bans on the presence of certain additives in their 'own brand' products.

Tartrazine (E102), a bright yellow permitted colour used extensively in soft drinks and confectionery has acquired a reputation for provoking food intolerance and causing hyperactivity in children. The Food Advisory Committee (which gives expert scientific advice to the Ministry of Agriculture, Fisheries and Food), however, does not consider that tartrazine poses more problems than any other colour or food ingredient.

Food colours are used in low concentrations and it has been estimated that in Britain the average daily intake is between 10 and 50 mg. The consumption of individual colours is estimated to be never greater than ten per cent (and in most cases less than one per cent) of the Acceptable Daily Intake (ADI) figure worked out by toxicologists (see p. 373).

Additives are sometimes used to give manufactured foods properties associated with the presence of traditional ingredients. Emulsifying agents used to decrease the amount of fat needed in the manufacture of cakes and bread come into this category. Although compounds of this type reduce the amount of fat required to produce familiar physical properties in cake, they do not, of course, fulfil its nutritional functions. The amount of fat lost in this way, when considered in terms of an individual, is not large, but for someone living on an inadequate diet it may not be insignificant. Most people, however, eat too much fat and a reduction in intake would not be undesirable.

Synthetic cream and meringues are today often made from cellulose derivatives which have absolutely no nutritional value. The use of colouring matter in cakes to give an impression of richness is also open to criticism. There are numerous other examples of the use of additives which contribute nothing to the nourishment of the consumer. The more luxurious classes of foodstuffs lend themselves particularly well to sophistication and it may be argued that these are not, in any case, eaten primarily for their nutritive value. This is a specious argument, however, and does not wholly remove the impression that the use of such substances is tantamount to a confidence trick.

One might suppose that the addition of nutrients to food would be above suspicion but this is not always so, and even here there is need for caution. The quantities of nutrients employed, however innocent they seem, should bear some relationship to the body's needs. The bad effect of an excessive intake of vitamin D, mentioned in Chapter 12, shows that it *is* possible to have too much of a good thing!

THE RISK/BENEFIT BALANCE   It is generally accepted that no food additive can be proved to be absolutely safe. It is impossible to prove that an additive causes *no* harm and in this situation it is helpful to balance possible risks against benefits conferred.

The assessment of a risk/benefit balance is well exemplified by the use of nitrites and nitrates to preserve meat products discussed in Chapter 16. The benefit of using these additives is well established; they destroy *Clostridium botulinum* bacteria and so prevent food poisoning. There is the risk, however, that nitrites in food are partly converted into nitrosamines which are known to produce cancer in animals. At the low levels permitted in specified foods nitrites are not known to have caused any harm to any human being. Nevertheless, the risk remains and it is a matter of judgement whether the risk is considered to be justified in the light of the known benefits.

With some types of food additive it is more difficult to assess the risks and benefits. Food colours provide a suitable example. The risk of using food colours is clear from the knowledge that *some* synthetic dyes are carcinogens. The benefits arising from the use of food colours, on the other hand, are harder to assess, though it is true that they improve the attractiveness and appearance of food. The risk in using a known carcinogen would so obviously outweigh the benefits that the use of such a substance would not be considered. Difficulties arise with substances which are not *known* to be carcinogenic but cannot be said *with certainty* not to be. The decision to permit the use of such a substance as a food additive is at best a compromise based on the risk/benefit considerations outlined above. Assessment of risk can only be made on the scientific evidence available, or obtainable, at the time. Nevertheless, evidence accumulated later during its use, and possibly by using improved scientific techniques, may show that continued use of the additive is not advisable. Great harm may have been done to consumers in the meantime and this is why such extreme caution is necessary before the use of a food additive is permitted.

The use of synthetic colours is particularly difficult to justify. The risks, though small, are thought by many people to outweigh substantially the somewhat dubious cosmetic benefits of the additives. The number of permitted synthetic colours has decreased from 32 in 1957 (when the first list of permitted colours appeared) to the current number of 15. Perhaps, in the fullness of time, the use of synthetic dyestuffs in food will be prohibited in Britain and such a move would, it is felt, meet with general approval.

Pessimistic readers may feel inclined to forgo all foods containing additives, but that would prove to be rather difficult. Fortunately, the risks, though real, are (for most people) exceedingly small and they are insignificant when measured against other hazards. Smokers, in particular, have less cause to worry about the dangers of food additives than the threat to health from smoking, which is probably many thousands of times greater. Consumption of excessive amounts of alcohol and, possibly, coffee or cola drinks is likewise not free from risk. Of more general concern, (and far outweighing the dietary risks already mentioned) are the life-threatening risks associated with obesity, hypertension and coronary heart disease. All of these conditions may be diet-related (see Chapter 15) and they undoubtedly pose a far greater threat to corporate health than the presence in food of minuscule amounts of additives.

Food laws and the control of food additives in Britain

The adulteration of food has its origins in antiquity, and in England the earliest action to prevent adulteration was taken by the guilds. For example, in the reigns of Henry II and Henry III pepper was widely used to preserve meat and other foods but because it was expensive adulteration was common. To remedy this the guild of Pepperers was granted powers to sift spices so as to control quality. In the sixteenth and seventeenth centuries adulteration of food was prevalent and practices such as the dilution of mustard with flour, the addition of leaves (e.g., scorched oak leaves) to tea, the mixing of sand, ashes and sawdust to bread doughs and the addition of water to milk became commonplace. Some forms of adulteration, such as the addition of sulphuric acid to vinegar, were even more serious and made the food poisonous.

Adulteration of food continued unabated in the eighteenth and nineteenth centuries and following increasing expressions of public concern and the publishing of the findings of a Select Parliamentary Commission the Adulteration of Food and Drink Act was effected in 1860. This was extensively amended and strengthened by the 1875 Sale of Food and Drugs Act which made the appointment of Public Analysts obligatory and gave Inspectors the right to take samples for analysis. Over 100 years have elapsed since the passage of the 1875 Act and in this period a vast amount of legislation controlling the composition, production, distribution and sale of food has been enacted.

The statute currently in force is the Food Act, 1984, which consolidates previous legislation and takes into account Britain's membership of the EEC. Membership imposes obligations on Britain to conform with Community Law. *Regulations* of the Council and Commission of the European Community are binding and apply in their entirety to member states. A *directive* is also binding as to its results but the means of achieving the result is left to each member state.

The fundamental intention of the 1984 Food Act is to ensure that food shall be in as wholesome a condition as possible when it is eaten. The Act and associated legislation prescribes legally enforceable standards of composition and treatment and makes infringement a criminal offence. It prohibits the addition to food of any substance which would make it 'injurious to health'. Ministers are also required to 'have regard to the desirability of restricting, so far as practicable, the use of substances of no nutritional value as foods or ingredients of foods'. The Food Act empowers Food and Health Ministers to make regulations concerning foods and these, after approval by Parliament, are published as legally binding Statutory Instruments (SI). The more important regulations which control the use of food additives in England and Wales are listed in Table 18.5. Similar legislation exists in Scotland and Northern Ireland.

Table 18.5   Food additive regulations in England and Wales

The Antioxidants in Food Regulations 1978 (SI 1978 No. 105)
The Colouring Matter in Food Regulations 1973 (SI 1973 No. 1340)
The Emulsifiers and Stabilizers in Food Regulations 1980 (SI 1980 No. 1833)
The Mineral Hydocarbons in Food Regulations 1966 (SI 1966 No. 1073)
The Miscellaneous Additives in Food Regulations 1980 (SI 1980 No. 1834)
The Preservatives in Food Regulations 1979 (SI 1979 No. 752)
The Solvents in Food Regulations 1967 (SI 1967 No. 1582)
The Sweeteners in Food Regulations 1983 (SI 1983 No. 1211)

SUGGESTIONS FOR FURTHER READING

ACCUM, F. (1966). *Treatise on Adulterations of Food, 1820.* Reissued by Mallinckodt, USA.
BLAXTER, K. (Ed.) (1980). *Food Chains and Human Nutrition.* Applied Science Publishers, (Includes a discussion of pesticides and radioactivity.)
BRITISH NUTRITION FOUNDATION (1987). *Food Processing – a Nutritional Perspective.* BNF Briefing Paper 11. BNF, London.
BRITISH NUTRITION FEDERATION (1975). *Why Additives? – The Safety of Food.* Forbes Publications, London.
CONSUMERS' ASSOCIATION (1980). *Pesticide Residues and Food.* Consumers' Association, London.
COUNSELL, J.N. (Ed) (1981). *Natural Colours for Food and Other Uses.* Applied Science, Barking.
FARRER, K.T.H. (1987). *A Guide to Food Additives and Contaminants.* Parthenon Publishing Group, Carnforth, Lancs.
FOOD ADDITIVES AND CONTAMINANTS COMMITTEE (1973). *Interim Report on the Review of Colouring Matter in Food Regulations.* HMSO, London.
FOOD ADDITIVES AND CONTAMINANTS COMMITTEE (1982). *Report on the Sweeteners in Food.* HMSO, London.
GRENBY, T.H. (Ed.) (1983). *Developments in Sweeteners – 2.* Applied Science, Barking.
HANSSEN, M. (1984). *E for Additives.* Thorsons, Wellingborough.

HEATH, H.B. (1978). *Flavour Technology*. Avi, USA.

JUKES, D.J. (1978). *Food Legislation in the UK*, 2nd edition. Butterworth & Co., Sevenoaks.

LUCAS, J. (1975). *Our Polluted Food; a survey of the risks*. Charles Knight, Croydon. (A survey of agricultural chemicals, radioactivity and mineral elements.)

MERORY, J. (1973). *Food Flavourings: Composition, Manufacture and Use* 2nd edition, vol. 1. Avi, USA.

MERORY, J. (1980). *Food Flavourings: Composition, Manufacture and Use*, 2nd edition, vol. 2. Avi, USA.

MILLER, M. (1985). *Danger! Additives at Work*. London Food Commission, London.

MILLSTONE, E. (1986). *Food Additives*. Penguin Books, Harmondsworth.

MILLSTONE, E. AND ABRAHAMS, J. (1988). *Additives: A Guide for Everyone*. Penguin Books, Harmondsworth.

MINISTRY OF AGRICULTURE, FISHERIES AND FOOD (1976). *Food Quality and Safety: A Century of Progress* (Symposium Report). HMSO, London.

MINISTRY OF AGRICULTURE, FISHERIES AND FOOD (1984). *Look at the Label*. HMSO, London.

MINISTRY OF AGRICULTURE, FISHERIES AND FOOD (1987). *Food Additives, The Balanced Approach*. HMSO, London.

MINISTRY OF AGRICULTURE, FISHERIES AND FOOD (1987). *Survey of Consumer Attitudes to Food Additives*. HMSO, London.

NATIONAL ACADEMY OF SCIENCES (USA) (1983). *Toxicants Naturally Occurring in Foods*.

TAYLOR, R.J. (1980). *Food Additives*. John Wiley & Sons, Chichester.

WALFORD, J. (1980). *Developments in Food Colours* – I. Applied Science, Barking.

WHEELOCK, V. (1980). *Food Additives in Perspective*. University of Bradford Briefing Paper, Bradford.

General reading list

BANWART, G.J. (1979). *Basic Food Microbiology*. Avi, USA.

BARASI, M.E. AND MOTTRAM, R.F. (1987). *Human Nutrition*, 4th edition. Edward Arnold, London.

BENDER, A.E. (1982). *Dictionary of Nutrition and Food Technology*, 5th edition. Newnes-Butterworth, Sevenoaks.

BENDER, A.E. (1978). *Food Processing and Nutrition*. Academic Press, London.

BERK, Z. (1976). *Braverman's Introduction to the Biochemistry of Foods*, 2nd edition. Elsevier, Amsterdam. (Contains some fairly advanced chemistry.)

BINGHAM, S. (1987). *The Everyman Companion to Food and Nutrition*. Dent & Sons, London.

BIRCH, G.G. AND PARKER, K.J. (1980). *Food and Health: Science and Technology*. Applied Science, Barking.

BIRCH, G.G., SPENCER, M. AND CAMERON, A.G. (1986), *Food Science*, 3rd edition, Pergamon Press, Oxford.

BOARD, R.G. (1983). *Modern Introduction to Food Microbiology*. Blackwell, Oxford.

BRITISH NUTRITION FOUNDATION (1984). *Eating in the Early 1980s*. BNF, London.

CLARKE, D. AND HERBERT, E. (1986). *Food Facts. A Study of Food and Nutrition*. Macmillan Educational, London.

CLYDESDALE, F. (Ed.) (1979). *Food Science and Nutrition; Current Issues and Answers*. Prentice-Hall International, Hemel Hempstead.

COTTRELL, J. (1987). *Food and Health: Now and the future*. Parthenon Publishing Group, Carnforth, Lancs.

COTTRELL, J. (1987). *Nutrition in Catering*. Parthenon Publishing Group, Carnforth, Lancs.

COULTATE, T.P. (1989). *Food: The Chemistry of its Components*, 2nd edition, Royal Society of Chemistry, London.

FENNEMA, O.R. (Ed.) (1985). *Food Chemistry*, 2nd edition. Marcel Dekker Inc., USA.

FRAZIER, W.C. AND WESTHOFF, A. (1978). *Food Microbiology*, 3rd edition. McGraw-Hill, USA.

GIBNEY, M.J. (1986). *Nutrition, Diet and Health*. Cambridge University Press, Cambridge.

GRISWOLD, R.M. *et al.* (1980). *The Experimental Study of Foods*, 2nd edition. Constable & Co., London.

HAMILTON, E.M.N. AND WHITNEY, E.N. (1982). *Concepts and Controversies in Nutrition*, 2nd edition. West Publishing Co., USA.

HARTLEY, D. (1956). *Food in England*. Macdonald, London. (An interesting historical survey.)

HEINMANN, W. (1980). *Fundamentals of Food Chemistry*. John Wiley & Sons, USA.

HOFF, J.E. (Compiler) (1973). *Food: Readings from Scientific American*. W.H. Freeman, Oxford.

JAY, J.M. (1986). *Modern Food Microbiology*, 3rd edition. Van Nostrand Reinhold, Wokingham.

KARE, M.R. AND BRAND, J.G. (Eds) (1987). *Interaction of the Chemical Senses with Nutrition*. Academic Press, London.

LANKFORD, R.R. AND JACOBS-STEWARD, P.M. (1986). *Foundations of Normal and Therapeutic Nutrition*. John Wiley & Sons, Chichester.

MINISTRY OF AGRICULTURE, FISHERIES AND FOOD (1985). *Manual of Nutrition*, 9th edition. HMSO, London.

MULLER, H.G. AND TOBIN, G. (1980). *Nutrition and Food Processing*. Croom Helm, Beckenham.

PARRY, T.H. AND PAWSEY, R.K. (1984). *Principles of Microbiology for Students of Food Technology*, 2nd edition. Hutchinson Educational, London.

PASSMORE, R. AND EASTWOOD, M.A. (1986). *Human Nutrition and Dietetics*, 8th edition. Churchill Livingstone, Edinburgh. (A comprehensive standard text.)

PAUL, A.A. AND SOUTHGATE, D.A.T. (1978). *McCance and Widdowson's 'The Composition of Foods'*, 4th edition. HMSO, London.

PYKE, M. (1981). *Food Science, and Technology*. 4th edition. John Murray, London.

RANKIN, M.D. (Ed.) (1984). *Food Industries Manual*, 21st edition. Leonard Hill, Glasgow.

RITSON, R. (Ed.) (1986). *The Food Consumer*. John Wiley & Sons, Chichester.

SOYER, A. (1977). *The Pantropheon or A History of Food and its Preparation in Ancient Times*. (Published 1853). Paddington Press, New York.

STEWART, G.F. AND AMERINE, M.A. (1982). *Introduction to Food Science and Technology*, 2nd edition. Academic Press, London.

TANNAHILL, R. (1988). *Food in History*. Penguin Books, Harmondsworth. (An interesting summary that includes a final section on 'The scientific revolution'.)

TANNENBAUM, S.R. (Ed.) (1979). *Nutritional and Safety Aspects of Food Processing*. Marcel Dekker, USA.

TRUSWELL, A.S. (1986). *ABC of Nutrition*. British Medical Association, London. (A collection of articles from the British Medical Journal.)

TUDGE, C. (1985). *The Food Connection*. BBC, London.

WEININGER, J. AND BRIGGS, G.M. (Eds) (1983). *Nutrition Update*, vol. 1. John Wiley & Sons, Chichester.

WEININGER, J. AND BRIGGS, G.M. (Eds) (1985). *Nutrition Update*, vol. 2. John Wiley & Sons, Chichester.

WINICK, M. (Ed.) (1985). *Nutrition in the Twentieth Century*. John Wiley & Sons, Chichester.

YUDKIN, J. (1985). *The Penguin Encyclopaedia of Nutrition*. Penguin Books, Harmondsworth.

Appendices

Appendix I Recommended daily amounts of food energy and some nutrients for population groups in the UK (1979). By courtesy of H.M. Stationery Office

Age range years	Occupation category	Energy MJ	Energy kcal	Protein (g)	Calcium (mg)	Iron (mg)
Boys						
under 1		(a)	(a)	(a)	600	6
1		5.0	1200	30	600	7
2		5.75	1400	35	600	7
3–4		6.5	1560	39	600	8
5–6		7.25	1740	43	600	10
7–8		8.25	1980	49	600	10
9–11		9.5	2280	57	700	12
12–14		11.0	2640	66	700	12
15–17		12.0	2880	72	600	12
Girls						
under 1		(a)	(a)	(a)	600	6
1		4.5	1100	27	600	7
2		5.5	1300	32	600	7
3–4		6.25	1500	37	600	8
5–6		7.0	1680	42	600	10
7–8		8.0	1900	47	600	10
9–11		8.5	2050	51	700	12
12–14		9.0	2150	53	700	12
15–17		9.0	2150	53	600	12
Men						
18–34	Sedentary	10.5	2510	63	500	10
	Moderately active	12.0	2900	72	500	10
	Very active	14.0	3350	84	500	10
35–64	Sedentary	10.0	2400	60	500	10
	Moderately active	11.5	2750	69	500	10
	Very active	14.0	3350	84	500	10
65–74	Assuming a	10.0	2400	60	500	10
75 +	sedentary life	9.0	2150	54	500	10
Women						
18–54	Most occupations	9.0	2150	54	500	12(b)
	Very active	10.5	2500	62	500	12(b)
55–74	Assuming a	8.0	1900	47	500	10
75 +	sedentary life	7.0	1680	42	500	10
Pregnancy		10.0	2400	60	1200	13
Lactation		11.5	2750	69	1200	15

(a) See Table 15.4
(b) These recommendations may not cover heavy menstrual losses.
(c) Children and adolescents in winter and housbound adults may need a supplement of 10 μg daily.

Appendix 1 *Continued*

Vitamin A retinol equivalents (μg)	Vitamin D (μg)	Thiamine (mg)	Riboflavin (mg)	Niacin equivalents (mg)	Vitamin C (mg)
450	7.5	0.3	0.4	5	20
300	10.0	0.5	0.6	7	20
300	10.0	0.6	0.7	8	20
300	10.0	0.6	0.8	9	20
300	(c)	0.7	0.9	10	20
400	(c)	0.8	1.0	11	20
575	(c)	0.9	1.2	14	25
725	(c)	1.1	1.4	16	25
750	(c)	1.2	1.7	19	30
450	7.5	0.3	0.4	5	20
300	10.0	0.4	0.6	7	20
300	10.0	0.5	0.7	8	20
300	10.0	0.6	0.8	9	20
300	(c)	0.7	0.9	10	20
400	(c)	0.8	1.0	11	20
575	(c)	0.8	1.2	14	25
725	(c)	0.9	1.4	16	25
750	(c)	0.9	1.7	19	30
750	(c)	1.0	1.6	18	30
750	(c)	1.2	1.6	18	30
750	(c)	1.3	1.6	18	30
750	(c)	1.0	1.6	18	30
750	(c)	1.1	1.6	18	30
750	(c)	1.3	1.6	18	30
750	(c)	1.0	1.6	18	30
750	(c)	0.9	1.6	18	30
750	(c)	0.9	1.3	15	30
750	(c)	1.0	1.3	15	30
750	(c)	0.8	1.3	15	30
750	(c)	0.7	1.3	15	30
750	(c)	1.0	1.6	18	60
1200	(c)	1.1	1.8	21	60

Appendix II Metric and imperial units

	Non-metric	Metric equivalent
Energy	1 kilocalorie (kcal)	4200 joules (J) 4.2 kilojoules (kJ)
Temperature	32° Fahrenheit (F) 212° Fahrenheit (F) To convert °F into °C: − 32° and then ×$\frac{5}{9}$	0° Celsius (C) 100° Celsius (C)
Volume	1.8 pints 1 pint 1 gallon	1 litre (l) 1000 millilitres (ml) 568 millilitres (ml) 4.5 litres (l)
Weight	1 ounce (oz) 1 pound (lb) 2.2 pounds (lb)	28.4 grams (g) 454 grams (g) 1 kilogram (kg)
Length	1 inch (in) 1 foot (ft) 39.4 inches (in)	2.5 centimetres (cm) 30.5 centimetres (cm) 100 centimetres (cm) 1 metre (m)

Index

Numbers in **bold** refer to tables